Contributions from Science Education Research

Volume 15

Contributions from Science Education Research is the international, multidisciplinary book series of the European Science Education Research Association (ESERA). The aim of the series is to synthesize, for the benefit of the scholarly community, the findings of high quality, theoretically-framed research in the domain of science education as well as comprehensive explorations of specific methodological strands in science education research. The series aims to publish books that are innovative in attempting to forge new ways of representing emergent knowledge in the field. The series includes edited collections of chapters, monographs and handbooks that are evaluated on the basis of originality, scientific rigor and significance for science education research. The book series is intended to focus mainly on work carried out in Europe. However, contributions from researchers affiliated with non-European institutions and non-members of the European Science Education Research Association are welcomed. The series is designed to appeal to a wide audience of researchers and post-graduate students in science education. Book proposals for this series may be submitted to the Publishing Editor: Claudia Acuna E-mail: Claudia.Acuna@springer.com

Gultekin Cakmakci • Mehmet Fatih Tasar
Editors

Connecting Science Education with Cultural Heritage

Selected Papers from the ESERA 2023 Conference

Springer

Editors
Gultekin Cakmakci
Hacettepe University
Çankaya, Ankara, Türkiye

Mehmet Fatih Tasar
Georgia State University
Atlanta, GA, USA

ISSN 2213-3623 ISSN 2213-3631 (electronic)
Contributions from Science Education Research
ISBN 978-3-031-90943-6 ISBN 978-3-031-90944-3 (eBook)
https://doi.org/10.1007/978-3-031-90944-3

This Springer imprint is published by the registered company Springer Nature Switzerland AG
The registered company address is: Gewerbestrasse 11, 6330 Cham, Switzerland

List of Reviewers

Albert Zeyer
Ana Sofia Afonso
Anastasios Zoupidis
Ayelet Baram-Tsabari
Bahadir Aktan
Betul Sen Gumus
Bodil Sundberg
Carla Morais
Christina Siry
Claudio Fazio
David Treagust
Eilish McLoughlin
Federico Corni
Filiz Demirci
Gultekin Cakmakci
Hakki Ilker Kostur
Heidi Cian
Ibrahim Delen
Ilídio André Costa
Ingo Eilks
İsmail Dönmez
Italo Testa
Jens Dolin
Jesper Bruun
Kalypso Iordanou
Katerina Pia Günter
Kayahan Ince
Laura Branchetti
Leticia Garcia-Romano
Lorenz Kampschulte
Lucy Avraamidou

Lukas Rokos
Maria Evagorou
María Napal
María Ruth Jimenez Liso
Mariona Espinet Blanch
Marta Romero Ariza
Massimiliano Malgieri
May Lee
Mehmet Fatih Tasar
Mehmet Sogut
Metin Sardag
Mustafa Sozbilir
Nadja Belova
Oya Aglarci Ozdemir
Radu Bogdan Toma
Sibel Erduran
Suzanne Kapellari
Tugba Ecevit

Contents

Chapter 1
Introduction: Connecting Science Education with Cultural Heritage

Gultekin Cakmakci and Mehmet Fatih Tasar

The European Science Education Research Association (ESERA) was established during the European Conference on Research in Science Education, which took place in Leeds, England, in April 1995. Since then, ESERA has organized biennial conferences. The 15th Biennial Conference of the European Science Education Research Association (ESERA 2023) was held in Cappadocia, Türkiye, from August 28 to September 1, 2023. Hacettepe University, Gazi University, and Nevşehir Hacı Bektaş Veli University hosted the ESERA 2023 Conference.

Spanning across both Europe and Asia, Türkiye has served as a meeting place for many peoples and cultures throughout centuries. Having colleagues from over 65 countries at this conference in Cappadocia, an ancient historical heritage site in Central Anatolia, allowed us to demonstrate our commitment to mutual understanding, respect, and cooperation, which is especially needed in today's world. Therefore, the ESERA 2023 conference's theme was "*Connecting Science Education with Cultural Heritage*."

Approximately 1300 submissions were received. After review, 695 papers, 54 symposia, two panels, 20 workshops, and 150 posters featuring three plenary speakers and two ESERA community plenary speakers were accepted for presentation at the conference. There were about 1100 participants.

As always, ESERA supports early-career researchers. Everything must be done to maintain their enthusiasm for cultivating a science education culture. In line with this commitment, we held five workshops for early-career researchers and

G. Cakmakci (✉)
Hacettepe University, Ankara, Turkey
e-mail: cakmakci@hacettepe.edu.tr

M. F. Tasar
Georgia State University, Atlanta, GA, USA
e-mail: mtasar@gsu.edu

G. Cakmakci, M. F. Tasar (eds.), *Connecting Science Education with Cultural Heritage*, Contributions from Science Education Research 15,
https://doi.org/10.1007/978-3-031-90944-3_1

practitioners as part of the ESERA 2023 conference program. This initiative brings fresh energy to the new generation of researchers and practitioners in STEM education worldwide. One outcome of this endeavor is the opportunity for early-career researchers to serve on the editorial board for the development of five ESERA 2023 Conference eProceedings (Curaoglu et al., 2024, Kaya et al., 2024a, 2024b; Sardag et al., 2024a, 2024b).

After the conference, strand chairs were asked to rank and select the best three conference papers in their respective strands as the most suited for inclusion in this book. Afterward, we invited the author(s) of the first ranked nominee paper from that strand to contribute their work to this book. In the case of unavailability, authors of the subsequent papers in the rank were invited from the same group of three. We also invited plenary speakers to submit chapters for consideration for inclusion in the book.

This book is part of the series "*Contributions from Science Education Research*," published by Springer in collaboration with ESERA. It collects a selection of papers presented at the 15th ESERA Conference, which focuses on connecting science education with cultural heritage. Considering the strands and plenary talks, the book now contains 23 chapters, categorized into four sections:

- Foundations and Emerging Trends in Science Education Research
- Teaching, Learning, and Student Engagement in Science Education
- Teacher Agency, Professional Development, and Leadership in Science Education
- Societal and Global Challenges in Science Education

1.1 Foundations and Emerging Trends in Science Education Research

In Chap. 2, **Pimentel and Osborne** address misinformation in science by examining how science education improves young people's ability to evaluate false information. Students must understand scientific practices and digital media literacy to navigate unreliable information effectively. Science education should emphasize content, credibility evaluation, media analysis, and understanding scientific consensus. Policy changes are essential to prioritize these skills. While science education alone cannot eradicate misinformation, it is crucial for nurturing informed, critical thinkers in today's complex world.

Ginzburg and colleagues discuss the need to reconceptualize the 'Nature of Engineering' in Chap. 3. This chapter establishes a theoretical foundation for a nuanced understanding of the Nature of Engineering and integrates elements of the Nature of Science through the Family Resemblance Approach. They review and analyze current studies on engineering education in schools and the incorporation of engineering into the science curriculum.

Science education research (SER) has evolved over the years, partly influenced by curriculum reforms, policies, and advancements in science and teaching methodologies. In Chap. 4, **Sozbilir and Teke** utilize bibliometric analysis to examine the trends in SER literature from 2000 to 2022. Their findings seek to guide future research and policy decisions in science education, ensuring alignment with technological and societal advancements.

In Chap. 5, **MacLeod and Avraamidou** reflect on the development of the research area of science identity over the past two decades and its role in science education research. They also raise important research questions and avenues for future studies to explore further.

In Chap. 6, **Caramaschi and colleagues** analyze 223 essays within the context of a future-oriented science education project, comparing Thematic Analysis (a qualitative analytical method) with Latent Semantic Analysis (a computational, data-driven method). They emphasize the significance of both bottom-up and top-down analysis in qualitative and computational data-intensive methods.

1.2 Teaching, Learning, and Student Engagement in Science Education

Educational videos provide valuable resources, but their content and structure can result in different educational outcomes. In Chap. 7, **Mutschler and colleagues** investigate whether the sequence of content—example-rule versus rule-example—impacts learning success, especially concerning learners' prior knowledge.

Kekulé and Krejčí employ eye-tracking to observe students' strategies when reading a physics textbook supplemented with pictures and questions. Chapter 8 discusses students' interaction with pictures that can convey varying amounts of information, ranging from purely decorative pictures (no information) to those crucial for understanding the text.

In Chap. 9, **Frågåt and colleagues** discuss how computational thinking (CT) can be integrated into inquiry-based science teaching to enhance subject-specific learning and identify which CT categories are applied when using such an approach. They present this through a case study where students use programmable micro:bits to explore the concept of an electrical circuit.

Students' multimodal meaning-making of nature of science (NOS) involves different modes of representation. In Chap. 10, **Cheung and colleagues** present how students connect individual scientific practices into a coherent picture of scientific methods using various modes of representation to reflect on their understanding of NOS.

Petersen and colleagues conducted a study combining surveys and observations to explore science in Denmark's preschool and early childhood education and care (ECEC). In Chap. 11, they discuss these identified categories of activities, which they called S'ence of science activities: i) Show science, ii) Slow science, iii) Spontaneous science, iv) Silent science, and v) Skewed science. They also discuss them against theoretical frameworks.

In Chap. 12, **Ruiz-González and colleagues** discuss design-based research by presenting an evaluation of a teaching-learning sequence on sound. They explicitly highlight the iterative methodology employed and the empirically based decisions made to refine and further implement this teaching-learning sequence.

The theory of evolution is regarded as one of the most challenging topics in schools. In Chap. 13, **Dagan and colleagues** discuss the perspectives of secondary school students from various religious and ethnic backgrounds on this topic.

In Chap. 14, **Christiansen** presents a three-year Assessment of Science Competence project and discusses part of it, which uses a design-based research approach to develop and test formative assessment tools for assessing science competencies in Danish primary and lower secondary education.

1.3 Teacher Agency, Professional Development, and Leadership in Science Education

In Chap. 15, **Guler** presents a longitudinal study conducted over two years based on the concepts of teacher agency to investigate how (if at all) science teachers develop their agency and agentic practice in the pedagogy of scientific argumentation over time.

In Chap. 16, **Vendrasco and Marzabal** examine mediators' perceptions of how their professional knowledge has evolved over time in Chile's science centers, museums, aquariums, botanical gardens, and zoos. They identify three stages of professional development for mediators: insertion, stabilization, and diversification.

The Next Generation Science Standards in the United States is one of the pioneering programs focusing on science and engineering practices. That demands teachers to have a stronger understanding of science and engineering practices. In Chap. 17, **Hill and colleagues** present their research on teachers' perceptions of authentic science and engineering practices and practices-based curricula. Their findings suggest that the research experiences for teachers in academic labs and science education-focused professional development programs can successfully support teachers in building a curriculum that engages students in the practices; however, these are limited to practices directly associated with their research.

Inquiry-based learning has been a significant focus in various curricula, but implementing this approach in the classroom presents challenges. In Chap. 18, **Zur and Tall** utilize the Family Resemblance Approach and thematic analysis to identify the aspects of the nature of science that teachers emphasize and contextualize. Their findings indicate that teachers prioritize the cognitive-epistemic aspects of NOS over its social-institutional dimensions, such as the cultural and social norms of the scientific community.

In Chap. 19, **Brinks and Balslev** examine the future of science education leadership by challenging traditional practices and offering a new perspective. They demonstrate this through a case study of the Danish Academy of Science Education

(NAFA), which seeks to improve science education in Denmark. They argue that adaptive, dynamic learning models link formal and informal education and promote collaborative network literacy, vital for educational partnerships marked by professional diversity and high complexity.

In Chap. 20, **Upadhyay and colleagues** present insights from a workshop with forty teachers at a public school. The workshop examined how the local cultural heritage of Gaijatra can be integrated into science teaching and learning. The findings indicate that for many teachers, Gaijatra holds both personal and pedagogical significance; however, it faces challenges from mandated external exams despite its recognized positive impact on student engagement and learning.

In Chap. 21, **Ferrari and colleagues** discuss equitable quality education (EQE) perspectives among a group of faculty and novice science teachers from the Harvard Teacher Fellows (HTF) program and how these perspectives relate to the formative cohesion of a program that identifies the achievement of EQE as its mission. While aligned with HTF's mission, the study suggests integrating broader social and political dimensions into science teacher education for a more comprehensive approach.

1.4 Societal and Global Challenges in Science Education

In Chap. 22, **Tasquier and colleagues** present a thematic analysis of micro-narratives from Italian and Norwegian students collected through a digital questionnaire focused on sustainability stories. The study indicates that students recognise the importance of narratives in complex socio-scientific issues; however, it also uncovers significant challenges in cultivating a sense of agency among young people.

In the last chapter, **Goldman and colleagues** explore how policymakers, professional development leaders, and teachers involved in climate change education (CCE) perceive CCE in relation to its primary learning goals, appropriate approaches for implementation, and the factors that enable and inhibit the implementation of climate change curricula. This is a cross-country comparison of the perspectives of key players in CCE from five countries: Australia, England, Finland, Indonesia, and Israel. The findings suggest a consensus on the significance of CC but reveal varying perceptions of the nature of CC knowledge and its role in the curriculum. A diverse array of constructivist, student-centred pedagogies is utilized in teaching CCE.

1.5 Concluding Remarks

This edited volume presents most recent innovative research in science education, exploring the intersection of science and cultural heritage. It features twenty-three chapters of high-quality studies derived from presentations at the 15th Biennial

ESERA Conference in Cappadocia, Türkiye, hosted by Hacettepe University, Gazi University, and Nevşehir Hacı Bektaş Veli University, Türkiye.

The 2023 ESERA Conference emphasized the theme "the connection between science education and cultural heritage." The chapters in this text provide fresh insights into developing science education literature from various perspectives, as showcased at the conference. This book includes contributions from every continent, offering a diverse range of international studies that feature original, innovative, and rigorous methodologies. For those interested in science education research, this compilation is an invaluable resource. This book presents a profound exploration that will resonate with scholars and researchers worldwide, making it an essential addition to any academic library.

References

Curaoglu, O., Kaya, G., Sardag, M., & Altintas, B. (Eds.). (2024). *Proceedings book series-IV of the ESERA 2023 conference. Connecting science education with cultural heritage*. Nobel Bilimsel Eserler.

Kaya, G., Sardag, M., & Donmez, I. (Eds.). (2024a). *Proceedings book series-V of the ESERA 2023 conference. Connecting science education with cultural heritage*. Nobel Bilimsel Eserler.

Kaya, G., Sardag, M., & Gul, K. (Eds.). (2024b). *Proceedings book series-III of the ESERA 2023 conference. Connecting science education with cultural heritage*. Nobel Bilimsel Eserler.

Sardag, M., Kaya, G., & Evecit, T. (Eds.). (2024a). *Proceedings book series-II of the ESERA 2023 conference. Connecting science education with cultural heritage*. Nobel Bilimsel Eserler.

Sardag, M., Kaya, G., & Sogut, M. (Eds.). (2024b). *Proceedings book series-I of the ESERA 2023 conference. Connecting science education with cultural heritage*. Nobel Bilimsel Eserler.

Gultekin Cakmakci is a Professor of Science Education and the Deputy Chair of the Research & Innovation Coordination Board at Hacettepe University. His research interests focus on engineering education, developing scientific literacy among students and the public, as well as designing, implementing, and evaluating STEM instruction. He is the founding editor of the Journal of Research in STEM Education.

Mehmet Fatih Tasar is a Professor of Science Education at Georgia State University in the USA. He was the chief editor of The International Handbook of Physics Education Research, published by the American Institute of Physics Publishing. Additionally, he is a co-editor for the International Journal of Physics and Chemistry Education, Action Research and Innovation in Science Education, and the Hellenic Journal of STEM Education.

Part I
Foundations and Emerging Trends in Science Education Research

Chapter 2
Misinformation in Science: The Response of Science Education

Daniel Pimentel and Jonathan Osborne

2.1 Introduction

The ability to evaluate sources of scientific claims has become ever more important in an age where misinformation is widespread and easy to access. Underlying growing concern about misinformation is the use of social media and the internet to present and circulate false information (Höttecke & Allchin, 2020; Kozyreva et al., 2020; Vosoughi et al., 2018). One study found that on social media, "it took the truth about six times as long as falsehood to reach 1500 people and 20 times as long as falsehood to reach the same audience" (Vosoughi et al., 2018). The spread and scale of misinformation are alarming because, unlike most other forms of mass media, information shared online is often *not* curated by professional gatekeepers (Höttecke & Allchin, 2020). The internet allows users to bypass experts and certified authorities who often function to sort truths from falsehoods and deceptive half-truths. While conspiracy theories and the use of deceptive tactics are not new, the internet and social media have provided individuals with a means to avoid traditional gatekeepers and a much louder megaphone to spread false and, sometimes dangerous, ideas.

Young people are more likely to get their information from YouTube and TikTok than from conventional media (National Science Board, 2020; Pew Research Center, 2022). While some of the channels on these platforms are credible sources of information, many of them are not. Given the absence of expert gatekeepers on such

D. Pimentel (✉)
The University of Alabama, Tuscaloosa, AL, USA
e-mail: drpimentel@ua.edu

J. Osborne
Stanford University, Stanford, CA, USA
e-mail: osbornej@stanford.edu

G. Cakmakci, M. F. Tasar (eds.), *Connecting Science Education with Cultural Heritage*, Contributions from Science Education Research 15,
https://doi.org/10.1007/978-3-031-90944-3_2

platforms, the onus is then put on students to make judgments about the credibility of information and its source. Contrary to the common conception that young people are fluent with digital technologies, the research evidence shows that students struggle to identify deceptive information online (Palfrey & Gasser, 2011; Wineburg & McGrew, 2019).

Disinformation is false or misleading information created and shared to deceive audiences (Wardle & Derakhshan, 2017). Evaluating the credibility of such online information is a difficult task. In addition, a lack of education about how to approach the evaluation of online sources has left students more vulnerable to bad actors who exploit online communication for their own ends (Chinn et al., 2020; Jewett, 2020; Lynch, 2016; Nichols, 2017; Sinatra & Hofer, 2021). Such exploitation is a concern for science educators because one way that bad actors accomplish their deception is by questioning reliable, consensually agreed-upon scientific knowledge. Ironically, much of this questioning takes advantage of individuals' trust in science by cloaking deceptive claims in the language and style of science communication (Oreskes, 2019; Oreskes & Conway, 2010). Allchin has coined the term "science deceit" to describe the use of persuasive tactics (e.g., mimicking style or use of false experts) to mislead audiences about scientific information (Allchin, under review). Often, purveyors of disinformation encourage their audience to evaluate the evidence for themselves, encouraging them to "do your own research (DYOR)." This tactic is used to circumvent the legitimate conclusions of those with relevant expertise (e.g., Chinn & Hasell, 2023) and is susceptible to confirmation bias and data voids (i.e., when searches generate few results or results dominated by false/misleading content) (Vetter, 2023).

In contrast to disinformation, misinformation is false or misleading information shared by an individual or group that does not necessarily have an intent to deceive others (Wardle & Derakhshan, 2017). There are several reasons why an individual might believe and share misinformation, including confirmation bias (e.g., Zhou & Shen, 2021), motivated reasoning (e.g., Kahne & Bowyer, 2016), and identity protective thinking (e.g., Kahan, 2017). In addition to these factors, misinformation about science may be particularly challenging for students to evaluate due to our current approach to scientific literacy in schools. Students are often not provided with opportunities to engage in critique in science classrooms (Osborne, 2010) and when they are, it often occurs with sources of information that are assumed to have been vetted and curated by the teacher. In the science classroom, students are rarely tasked with evaluating diverse and authentic sources of information from epistemically "unfriendly" environments, where one cannot take the accuracy or credibility of sources as a given (Chinn et al., 2020).

Providing students with opportunities to develop the knowledge and competencies required to distinguish the good, from the ill-informed to the misinformed is essential for avoiding deception. A focus on information evaluation aligns with the stated goals of many science education policy documents. For example, in the US, *A Framework for K-12 Science Education* articulates that a key goal of K-12 science education is to ensure that students, "are careful consumers of scientific and technological information related to their everyday lives" (National Research Council,

2012, p. 1). Similarly, the 2025 version of the PISA assessment framework has added a focus on students' ability to, "research, evaluate and use scientific information for decision making and action" (OECD, 2022). However, as yet, what it means to teach and develop these competencies has not been articulated. To date, teaching about information evaluation has taken a backseat to a focus on disciplinary content and it is often absent from the science curriculum altogether. For science education, the increasing use of digital media requires that we prepare students not only to evaluate science claims but to evaluate claimants—that is, does a particular source have legitimate expertise to speak with the authority on the topic? This, we argue, requires clarifying our conception of scientific literacy and how we prepare students.

2.2 Scientific Literacy and the Competent Outsider

From its beginnings, science education has been perceived as a preparatory step for the next generation of scientists (Dainton, 1968; National Research Council, 2012; The Royal Society, 2014). However, this rationale does not justify teaching science to all, as most students will not go on to become professional scientists or engineers. The concept of scientific literacy was developed to link science education with "literacy" (commonly associated with basic skills in reading and writing), portraying science education as an essential component of education for the average citizen (Rudolph, 2023). Developing students' scientific literacy has been used as the primary justification for science education *for all* public school students.

Scientific literacy has many meanings, however, which makes it a rather elusive goal for science education (e.g., Osborne, 2023; Rudolph, 2023). In his review of the term, DeBoer (2000) identified nine different meanings of the concept, ranging from goals as varied as teaching about science as a cultural achievement to preparing students for science-related work. In 2007, Roberts categorized the varying goals of science education into two "visions"—Vision I, which positions students as novice scientists to develop a foundational understanding *of* science ideas, and Vision II, which is largely focused on developing citizens who have understandings *about* science and how to use it in their everyday lives (Roberts, 2007).

An assumption underlying Vision I is that engaging students as novice scientists will benefit all students. The tacit belief underlying this vision is that it enables students to evaluate the science they encounter in their daily lives. The result is that students sit through a "long parade of concepts and theories" that are sufficient to make them "marginal insiders" (Feinstein, 2011). However, the idea that science education can provide students with learning experiences that develop sufficient understanding to become "critical consumers" of complex scientific arguments about contemporary topics such as climate change, vaccination, environmental choices and sustainability, and personal health is misguided. As Ryder (2001) has shown, many contemporary science-related issues require knowledge that school science does not, and could not, provide. The body of scientific knowledge has continued to grow exponentially (Bornmann et al., 2021; Zhang et al., 2015), however time in the science curriculum and individual human capacities to process information are limited. Understanding and knowledge are limited both by our finite cognitive capabilities and by the complexity of the environment. We cannot be experts in everything. Even if more information could be included in the curriculum and learned sufficiently, studies suggest that increased scientific knowledge alone does not improve decision-making about everyday scientific issues (e.g., Dalyot & Baram-Tsabari, 2023; Jho et al., 2014).

Vision I goals remain prominent throughout science policy documents, however. In the US, the Next Generation Science Standards (NGSS) emphasize the internal practices and core disciplinary ideas of science. These standards largely focus on engaging students as novice scientists and promote the idea that the average citizen can acquire sufficient scientific knowledge to critically examine scientific arguments on a given scientific issue. For example, the NGSS practice of argumentation from evidence sustains the impression that students will attain adequate disciplinary knowledge from their K-12 science education to evaluate scientific evidence themselves (Allchin & Zemplén, 2020). Such an impression is problematic in light of evidence that intermediate levels of scientific knowledge can lead to overconfidence in one's understanding (Lackner et al., 2023). Overconfidence in one's abilities might, in turn, leave students vulnerable to deceit, exploitative tactics, and appeals to "doing their own research."

By focusing on the practices of individual scientists, what often gets overlooked in Vision I is any treatment of the social practices of the scientific community that are essential to the production of reliable knowledge (Longino, 1990). Most contemporary science curricula focus on the experimental processes required to collect and analyze data. How those findings are translated into a scientific publication is rarely if ever discussed. Neither are the social practices and the checks and balances that lead to the emergence of a scientific consensus, even in undergraduate courses. Nor are the credentials that might permit anybody to claim scientific expertise in any domain. Yet, it is these indicators that are central to making judgments about the credibility of any proponent of a scientific claim, not the evidence, which is rarely accessible or understandable for those without expertise in that domain (Dagher & Erduran, 2016; Höttecke & Allchin, 2020).

Unlike Vision I, the primary challenge with Vision II is that science educators have struggled a) to agree on what it is and b) to articulate any of the competencies needed to achieve it (Roberts, 2007). Vision I uses the practicing scientist as an "expert model" to inform curricular goals, however the expert model for Vision II is less obvious. Polman et al. (2014) argue that a promising candidate is the science journalist. In many ways, science journalists reflect Noah Feinstein's notion of the "competent outsider" (2011), people who have "learned to recognize the moments when science has some bearing on their needs and interests and to interact with sources of scientific expertise in ways that help them achieve their own goals" (Feinstein, 2011, p. 180). Many of the practices used by science journalists can be useful for everyday citizens, such as the ability to "search effectively for and recognize relevant, credible information sources," "accurately represent up-to-date science," and attribute scientific information to "multiple, credible, attributed sources"

(Polman et al., 2014, p. 11). Such practices might be incorporated into Vision II lessons that engage students with scientific information and misinformation found in everyday places, like the media. However, science media education remains largely absent from the science curriculum (Reid & Norris, 2016).

The NGSS includes practices, such as "obtaining, evaluating and communicating information" that might appear to address the issue of misinformation, however, the field needs a much more detailed elaboration of the capabilities and knowledge required from a Vision II perspective of scientific literacy. The NGSS standards state, for instance, that students should "gather, read and evaluate scientific information from multiple authoritative sources," but they do not specify the knowledge or principles by which a student should judge a source to be "authoritative" (Achieve, 2012). Appendix F of the NGSS standards documents also states that students should, "[assess] the evidence… of each source" without acknowledging that when evaluating socially relevant scientific claims, scientific evidence is often quite complex and requires knowledge that goes beyond what students learn in K-12 science education.

Missing from the NGSS practices is an acknowledgment of the concept of epistemic dependence. Non-experts and experts alike are epistemically dependent on others with relevant expertise in a specialized domain (Hardwig, 1985; Norris, 1997). Given this reality, the public is confronted with the issue of whom to trust when it comes to scientific claims. Navigating such an information environment requires understanding how to decide which sources have the authority "to speak" for particular scientific claims (Allchin, 2022). Such competencies align with Feinstein's argument (2011) for putting the needs of the "competent outsider" to the fore. Today the crisis surrounding the ocean of misinformation has led us to ask again, what does the competent outsider need to know and be able to do? Feinstein did not spell out what those capabilities might be (Feinstein, 2011). Scholars such as Allchin and Höttecke have attempted to undertake this project by articulating frameworks and competencies needed for science media literacy and to put some flesh on the concept of the competent outsider (Allchin, 2022; Höttecke & Allchin, 2020).

In addition to such efforts, a third vision of scientific literacy has emerged over the past 20 years (Bencze et al., 2020; Sjöström & Eilks, 2018). Proponents of Vision III point to the fact that the production of scientific knowledge is a cultural act and findings have implications that necessarily *inform or demand action*. For example, the evidence for Anthropogenic climate change has implications for individual and policy choices, as do the scientific findings for mitigating COVID-19. Sjostrom and Elks, for instance, argue that "any kind of education should focus on making the young generation capable of a self-determined life in society, for being able to participate and solve problems in it" (2018, p. 81). However, solving problems related to climate change, for example, is a much bigger challenge when climate denialists use disinformation and deception to delay climate action by fostering doubt and uncertainty in the public (Oreskes & Conway, 2010). The ability to decipher credible sources of scientific information from less credible sources is only the first step in taking informed action, but it is a *necessary* first step. Developing such capabilities requires not just a focus on science as an academic or industrial endeavor but understanding science as an institution with its own set of social and institutional practices.

To this end, we argue that the science-related issues that confront today's citizens require knowledge of at least two domains: 1) *social practices and institutional mechanisms* that enable the production of reliable knowledge in science, and 2) *digital fact-checking strategies*. Both are domains of knowledge that are rarely addressed by K-12 science education.

2.3 The Social Practices and Institutional Mechanisms of Science

When evaluating reports about science-related issues, three key criteria are essential—(i) an assessment of the *credibility* of the source, (ii) an evaluation of whether they have legitimate *expertise*, and (iii) whether a *scientific consensus* exists (Osborne & Pimentel, 2023).

Conflicts of Interest Credibility is an a priori matter because there is little point in evaluating information if the source is not trustworthy. Establishing credibility within science is dependent on knowing key features about the social and institutional nature of the scientific endeavor. As one example, students should develop at least some understanding that scientific information can, and often is, shaped by funding (Allchin, 2020; Erduran & Dagher, 2014). Developing scientific knowledge requires sources of funding to pay for research personnel, equipment, and other associated costs. Conducting rigorous, high-quality scientific research is costly, leading many scientists to spend much of their time competing for funding from public and private sources. Relying on grant funding is a typical feature of science, however, it can create concerns about conflicting interests when wealthy funders attempt to influence findings and conclusions to advance their objectives (Allchin, 2020). Currently, K-12 science education places offers no explanation of the role that finances play in science. A scientifically literate citizen should be aware that funders with conflicts of interest may leverage their finances to distort or shape scientific findings to align with their interests (Allchin, 2020).

Relevant Expertise If there are no self-evident conflicts of interest, the next question that must be asked is, does the source have relevant expertise? Answering this question requires an understanding of the criteria for *expertise* and what to look for as evidence of relevant knowledge in the domain. Becoming a practicing scientist requires years of training—generally, a minimum of an undergraduate degree and, for most, a PhD. Even a PhD only marks specialist knowledge of one very small field, which is not readily transferable. A nuclear physicist, for instance, does not have the expertise in virology to override claims made by virologists (Oreskes & Conway, 2010). Expertise can also be acquired from professional training in science or from experience undertaking skilled work in certain contexts, such as nurses and midwives working with patients; farmers' knowledge of the environment; or fishermen's understanding of sustainable practices (Wynne, 1991). Claims made by those who lack any form of expertise or relevant experience should be weighed significantly less than those made by experts.

Even well-educated individuals can struggle when assessing sources for their expertise (Brashier & Marsh, 2020). To engage in such evaluations, students will need to learn the basic criteria for assessing expertise in science, such as judgments by other relevant experts, a past track record of work in the field, and an awareness of potential biases and interests (Goldman, 2001; Nichols, 2017). To apply such criteria, students might learn to ask questions such as, what credentials does a source of information have that warrants their claim to expertise? Does their area of expertise lie in the domain on which they are commenting? Students may also learn to ask questions such as, does the source work for a recognized scientific institution or hold an academic post in a research-based institution? Are they a representative of a leading scientific body such as the National Academy, the CDC, or the IPCC? By asking such questions and looking for answers, students can find evidence for or against a particular source's credibility.

Scientific Consensus The third criterion is the *importance of consensus* in the scientific community. In science, confidence in a particular theory emerges when it is tested repeatedly, found to be robust in the face of criticism, and sufficient to explain the relevant findings. Consensus is attained when the answer to an empirical question is so universally agreed upon that the field moves on to answer other unresolved questions. Most of the ideas discussed in K-12 science classrooms involve knowledge of this nature — knowledge that is largely settled within the scientific community. Science-in-the-making, in contrast, refers to science ideas that are equivocal and questionable, advanced by experts who may legitimately disagree—sometimes in public. Resolution takes time, further investigation, and multiple studies before a consensus emerges. To date, science education has placed little emphasis on engaging students with the differences between settled science and the processes of disagreement and argumentation among experts that are essential components of contemporary, cutting-edge science. As a result, students may leave the classroom assuming that science is only concerned with settled information, which is problematic given that the most pressing science-related social issues involve science-in-the-making.

When citizens are confronted with the need for scientific information in their everyday lives, the science may be uncertain and emergent (e.g., the prevalence of side effects in a new drug treatment). In these instances, an essential question is, "Is there a consensus or emergent consensus among the experts on this issue?" In many cases, like the existence of climate change or evolution, the answer is "yes." In the case of threats posed by new virus variants or the long-term effects of novel medical treatments, the answer is less certain and more equivocal. But, if the evidence is unequivocal, and the consensus among those with legitimate claims to expertise is overwhelming, any such questioning should be regarded with suspicion (Russell, 1928).

When it comes to evaluating science-related claims in the media, students should learn to ask whether a scientific consensus has emerged on the topic and what the degree of certainty is regarding that consensus. The existence of critics or doubters is normal and not necessarily a justification to call into question a scientific consensus (Oreskes, 2019). Even when legitimate concerns about a consensus have been raised by those with relevant expertise, it is often up to those within the scientific community to resolve those concerns, as they are the ones best positioned to engage the criticisms. Multiple studies suggest that communicating the scientific consensus about topics like climate change has a significant and enduring effect on individuals' beliefs, particularly among the doubtful or dismissive (Chinn et al., 2020; Goldberg et al., 2022). In contrast, if a search shows that the claimed science is controversial and lacks any consensus—as is the case with homeopathy or astrology—any claims should be approached with circumspection.

The importance of scientific consensus can be seen in efforts to confound it when spreading disinformation in the public media. One method for sowing "seeds of doubt" in a consensus is to generate the impression that a competing consensus exists (Oreskes & Conway, 2010). In this vein, false consensus documents against anthropogenic climate change, such as the Leipzig Declaration and the Oregon Protocol, were assembled and distributed to mislead the public into believing that there is an "alternate" consensus of experts. However, when investigated further, these lists consist overwhelmingly of non-experts and those with financial conflicts of interest. Such attempts demonstrate the epistemic authority of science by acknowledging the significance of consensus in science since naysayers are ready to promote and organize a false consensus to justify their claims. Students will need to learn that it is only a consensus of experts with the appropriate disciplinary knowledge that matters. Moreover, the greater the diversity of views included within that scientific community, the more likely that consensus can be trusted (Longino, 1990).

One paper alone, even if peer-reviewed, is not sufficient evidence to throw into question a consensus of a large, diverse group of experts. The process of peer review, in which new claims and findings are vetted by experts in the same domain of science, provides one important check against enduring error. The peer review process, however, is not designed to catch every logical or methodological error in a scientific study, let alone to detect deliberate fraud. Peer reviewers do not replicate the experiments or even the statistical analyses described. Rather, the authors' work is taken as having been conducted in good faith, by individuals of suitable skill who carry out the procedures correctly. As such, peer review does not guarantee correctness or replicability. Rather, it is each study that contributes to a larger body of evidence that is then reviewed by those with relevant expertise to develop consensus perspectives.

Much as a single study is not sufficient to throw a consensus into question, neither is a lone (or very small number) of challengers. In determining whether or not a scientific consensus has been reached, students will need to understand that there can be a range of expert views (Caulfield & Wineberg, 2023). A consensus indicates that the evidence is sufficiently compelling to settle a question. However, there are cases when a minority of respected experts support a perspective that runs counter to the majority perspective. Importantly, in these cases, many of those with relevant expertise consider these minority perspectives worthy of consideration. In contrast, fringe perspectives are often those advocated by individuals outside of an expert community and they are rarely considered worthy of consideration by those with relevant expertise. The idea that climate change is caused by fluctuations in solar output is one example of an idea that has moved to the fringe after it was discarded by climate scientists as an explanation for climate change (Union of Concerned Scientists, 2009). Teaching students to understand the range of expert opinions, or "reading the expert room", involves helping them understand that there are often varying degrees of certainty and uncertainty in an expert community (Caulfield & Wineberg, 2023, p. 106). However, a consensus is often established by extensive, careful, meticulous, empirical work that has been examined critically at all stages. One lone voice (or even the voices of a few on the fringe) does not have the same weight as the overwhelming majority of experts.

When sources of scientific information meet our aforementioned criteria (e.g., no egregious conflicts of interest, relevant expertise, and alignment with scientific consensus), then one can justify the credibility of their claims.

2.4 Fact-Checking Scientific Information on the Web

To evaluate the credibility of scientific information online, students will also need to be armed with the basics of digital media literacy and how to apply these to investigate criteria for the credibility of scientific information (described above). Interventions that teach students the following fact-checking strategies have been shown to significantly improve their ability to recognize misinformation (e.g., Wineburg & McGrew, 2019).

Lateral Reading Expert fact-checkers typically use a strategy known as "lateral reading," which involves leaving an unfamiliar website and opening new tabs to search for information about the page or post's source (Breakstone et al., 2021). Fact-checkers use this strategy to avoid relying on the web page itself when determining if it is a reliable source of information. Lateral reading allows the individual to take advantage of the interconnectedness of the web to search for information across multiple sources. A key component of the strategy involves postponing deep investment in the content of the source before one has deemed it to be credible. This strategy also actively works against the use of weaker heuristics (e.g., the look of the website), by directing the user away from the website in question to look for information about the source.

Students can learn to use sites such as Wikipedia, Sourcewatch.org, and Snopes.com to evaluate whether a source is credible before engaging with the information therein. In the case of scientific information, there are well-established scientific organizations whose credibility is dependent on providing reliable scientific information, such as the National Academies, the Royal Society, and NASA. While lateral reading is not specific to any one subject area, it is an essential ability when teaching students to evaluate scientific claims made on the internet.

Click Restraint When using lateral reading, fact-checkers hold a critical stance and practice patience when clicking on new links on a search results page (e.g., Wineburg et al., 2022). *Click restraint* involves scanning a results page and using the snippets of information found under each result before deciding which might be worth pursuing. Commonly, professional fact-checkers do not default to the top results but pause to consider which sources seem most likely to provide useful information (Wineburg et al., 2022). By holding back from clicking on the very first link in a search result, the user avoids the "weak" heuristic of the "top link bias," which leads people to believe that results at the top of a page are more reliable (Pan et al., 2007). This assumption is flawed because the order of results can be biased. Search algorithms can be manipulated by a careful choice of terminology (i.e., search engine optimization) or by artificially inflated volume statistics. In sum, the top results are not always the most relevant or informative. Instead, students should learn to survey the "information neighborhood" by "mining" the snippets of information, the URLs, and other information provided by the search browser (Wineburg et al., 2022, p. 897). This strategy can help students find better information and improve search efficiency by minimizing the amount of time required to answer a query.

Using Wikipedia A third fact-checking strategy, used in conjunction with lateral reading, involves knowing how to use Wikipedia wisely. Unfortunately, students are often taught not to use Wikipedia for academic work (e.g., McGrew & Byrne, 2020). While this advice may have been sound during the nascent stages of the website, Wikipedia has grown into a useful resource for establishing one's bearing in the new informational terrain. Professional fact-checkers regularly use Wikipedia as a starting point in their research to learn more about a source or topic (Wineburg & McGrew, 2019). Students can learn to use a range of features on the site to assist their fact-checking, like mining the references used to substantiate claims on a particular page, to help them navigate the information found there. Taking climate change as an example, students can use Wikipedia to learn about the substantive body of evidence supporting the scientific consensus on climate change, investigate whether or not a particular climate denialist has accepted fossil fuel monies, and learn more about a particular source's credentials. When taught to use Wikipedia in the classroom, students demonstrate an increased use of the resource as a starting point in their research (McGrew & Byrne, 2020).

2.5 Conclusion

In this chapter, we have advanced an argument for two domains that are essential for teaching students to become competent outsiders in science education: knowledge of the social practices of science and the fundamentals of digital media literacy. These capabilities are essential for equipping students with the tools they will need to navigate the epistemically "unfriendly" information environment they will encounter when they leave the science classroom. At the core of our argument is the idea that science education must transcend its traditional focus on content knowledge and disciplinary practices to encompass a broader purpose: preparing students to critically evaluate scientific information in real-world contexts. By equipping students with the tools necessary to discern credible sources from unreliable ones, science education can contribute to a society that is more informed about fundamental issues like personal health, environmental sustainability, and technological advancement.

To realize such a vision, science classrooms must engage students with authentic, often contentious, science-related issues. In the classroom, students should grapple with real-world sources of information, learn to evaluate them, and appreciate the processes by which scientific consensus is formed. For example, activities might task students with analyzing media coverage of emerging scientific topics or investigate the credentials of individuals challenging vaccination campaigns, the harm caused by tobacco smoking, intensive farming, eating less meat, the value of electric cars and more. Through such experiences, students not only develop essential skills but also come to understand the social and institutional dimensions of science that make it a uniquely reliable enterprise.

This vision may require shifts in the goals of science education policy and practice. Science standards will need to place greater emphasis on competencies such as evaluating the credibility of sources, understanding epistemic dependence in an interconnected society, and navigating uncertainty when reasoning about socioscientific issues (Sadler et al., 2007). Rather than placing so much emphasis on knowing the right answer, the approach will be more one which helps students establish why the wrong answer is wrong. Professional development will also need to prepare science educators to teach these skills.

To be clear, science education alone cannot solve the challenges posed by misinformation and disinformation. However, it holds a crucial role in equipping the next generation with the knowledge, skills, and dispositions needed to thrive in a world where the ability to discern credible information has clear implications for personal and societal well-being. By teaching students about criteria for evaluating the credibility of scientific information and strategies needed to apply these criteria on the web, we can give them a fighting chance to meet the challenges of a rapidly changing information landscape.

References

Achieve. (2012). *Next generation science standards*. https://www.nextgenerationscience.org

Allchin, D. (2020). From nature of science to social justice: The political power of epistemic lessons. In H. A. Yacoubian & L. Hansson (Eds.), *Nature of science for social justice* (pp. 23–39). Springer. https://doi.org/10.1007/978-3-030-47260-3

Allchin, D. (2022). Who speaks for science? *Science & Education, 31*(6), 1475–1492.

Allchin, D. (under review). *Science Denial, or Science Deceit?*

Allchin, D., & Zemplén, G. Á. (2020). Finding the place of argumentation in science education: Epistemics and whole science. *Science Education, 104*(5), 907–933.

Bencze, L., Pouliot, C., Pedretti, E., Simonneaux, L., Simonneaux, J., & Zeidler, D. (2020). SAQ, SSI and STSE education: Defending and extending "science-in-context". *Cultural Studies of Science Education, 15*, 825–851. https://doi.org/10.1007/s11422-019-09962-7

Bornmann, L., Haunschild, R., & Mutz, R. (2021, October 7). Growth rates of modern science: A latent piecewise growth curve approach to model publication numbers from established and new literature databases. *Humanities and Social Sciences Communications, 8*(1), 224. https://doi.org/10.1057/s41599-021-00903-w

Brashier, N. M., & Marsh, E. J. (2020). Judging truth. *Annual Review of Psychology, 71*, 499–515.

Breakstone, J., Smith, M., Connors, P., Ortega, T., Kerr, D., & Wineburg, S. (2021). Lateral reading: College students learn to critically evaluate internet sources in an online course. *Harvard Kennedy School (HKS) Misinformation Review, 2*, 20210201.

Caulfield, M., & Wineberg, S. (2023). *Verified: How to think straight, get duped less, and make better decisions about what to believe online*. University of Chicago Press.

Chinn, C. A., Barzilai, S., & Duncan, R. G. (2020). Education for a "post-truth" world: New directions for research and practice. *Educational Researcher, 50*(1). https://doi.org/10.3102/0013189X20940683

Chinn, S., & Hasell, A. (2023). Support for "doing your own research" is associated with COVID-19 misperceptions and scientific mistrust. *Harvard Kennedy School Misinformation Review*. https://doi.org/10.37016/mr-2020-117

Dagher, Z. R., & Erduran, S. (2016). Reconceptualizing the nature of science for science education: Why does it matter? *Science & Education, 25*, 147–164.

Dainton, F. S. (1968). *The Dainton report: An inquiry into the flow of candidates into science and technology*. HMSO.

Dalyot, K., & Baram-Tsabari, A. (2023). Justifying decision making in socio-scientific issues: The roles of reasoning and knowledge. *International Journal of Science Education, Part B, 15*(1), 1–21.

DeBoer, G. E. (2000). Scientific literacy: Another look at its historical and contemporary meanings and its relationship to science education reform. *Journal of Research in Science Teaching, 37*(6), 582–601.

Erduran, S., & Dagher, Z. R. (2014). *Reconceptualizing nature of science for science education*. Springer. https://doi.org/10.1007/978-94-017-9057-4

Feinstein, N. (2011). Salvaging science literacy. *Science Education, 95*(1), 168–185. https://doi.org/10.1002/sce.20414

Goldberg, M. H., Gustafson, A., van der Linden, S., Rosenthal, S. A., & Leiserowitz, A. (2022). Communicating the scientific consensus on climate change: Diverse audiences and effects over time. *Environment and Behavior, 54*, 1133–1165. https://doi.org/10.1177/00139165221129539

Goldman, A. I. (2001). Experts: Which ones should you trust? *Philosophy and Phenomenological Research, 63*(1), 85–110.

Hardwig, J. (1985). Epistemic dependence. *The Journal of Philosophy, 82*, 335–349.

Höttecke, D., & Allchin, D. (2020). Reconceptualizing nature-of-science education in the age of social media. *Science Education, 104*(4), 641–666. https://doi.org/10.1002/sce.21575

Jewett, A. (2020). *Science under fire*. Harvard University Press.

Jho, H., Yoon, H. G., & Kim, M. (2014). The relationship of science knowledge, attitude and decision making on socio-scientific issues: The case study of students' debates on a nuclear power plant in Korea. *Science & Education, 23*, 1131–1151.

Kahan, D. M. (2017). *Misconceptions, misinformation, and the logic of identity-protective cognition*. (Cultural cognition project. Working paper). Yale Law & Economics Research.

Kahne, J., & Bowyer, B. (2016). Educating for democracy in a partisan age: Confronting the challenges of motivated reasoning and misinformation. *American Educational Research Journal, 54*(1), 3–34. https://doi.org/10.3102/0002831216679817

Kozyreva, A., Lewandowsky, S., & Hertwig, R. (2020). Citizens versus the internet: Confronting digital challenges with cognitive tools. *Psychological Science in the Public Interest, 21*(3), 103–156. https://doi.org/10.1177/1529100620946707

Lackner, S., Francisco, F., Mendonça, C., Mata, A., & Gonçalves-Sá, J. (2023). Intermediate levels of scientific knowledge are associated with overconfidence and negative attitudes towards science. *Nature Human Behaviour, 7*(9), 1490–1501.

Longino, H. E. (1990). *Science as social knowledge*. Princeton University Press.

Lynch, M. P. (2016). *The internet of us: Knowing more and understanding less in the age of big data*. WW Norton & Company.

McGrew, S., & Byrne, V. (2020). Teaching Wikipedia: Supporting students in lifelong learning. In *The Interdisciplinarity of the learning sciences, 14th international conference of the learning sciences (ICLS) 2020* (Vol. 3, pp. 1565–1568). International Society of the Learning Sciences.

McIntyre, L. (2018). *Post-truth*. MIT Press.

National Research Council. (2012). *A framework for K-12 science education: Practices, crosscutting concepts, and core ideas*. National Academies Press.

National Science Board. (2020). *Science and engineering indicators 2020: The state of U.S. science and engineering (NSB-2020-1)*. National Science Foundation. https://ncses.nsf.gov/pubs/nsb20201/

Nichols, T. (2017). *The death of expertise: The campaign against established knowledge and why it matters*. Oxford University Press.

Norris, S. P. (1997). Intellectual Independence for nonscientists and other content-transcendent goals of science education. *Science Education, 81*(2), 239–258.

OECD. (2022). *The PISA 2025 Assessment Framework: Key competencies in reading, mathematics and science*. https://pisa-framework.oecd.org/science-2025/

Oreskes, N. (2019). *Why trust science?* Princeton University Press.

Oreskes, N., & Conway, E. M. (2010). *Merchants of doubt*. Bloomsbury Press.

Osborne, J. (2010). Arguing to learn in science: The role of collaborative, critical discourse. *Science, 328*(5977), 463–466.

Osborne, J. (2023). Science, scientific literacy, and science education. In *Handbook of research on science education* (pp. 785–816). Routledge.

Osborne, J., & Pimentel, D. (2023). Science education in an age of misinformation. *Science Education, 107*(3), 553–571.

Palfrey, J., & Gasser, U. (2011). *Born digital: Understanding the first generation of digital natives*. https://www.readhowyouwant.com

Pan, B., Hembrooke, H., Joachims, T., Lorigo, L., Gay, G., & Granka, L. (2007). In Google we trust: Users' decisions on rank, position, and relevance. *Journal of Computer-Mediated Communication, 12*(3), 801–823. https://doi.org/10.1111/j.1083-6101.2007.00351.x

Pew Research Center. (2022). *Teens, social media and technology*. https://www.pewresearch.org/internet/2022/08/10/teens-social-media-and-technology-2022/

Polman, J. L., Newman, A., Saul, E. W., & Farrar, C. (2014). Adapting practices of science journalism to foster science literacy. *Science Education, 98*(5), 766–791.

Reid, G., & Norris, S. P. (2016., March 1). Scientific media education in the classroom and beyond: A research agenda for the next decade. *Cultural Studies of Science Education, 11*(1), 147–166. https://doi.org/10.1007/s11422-015-9709-1

Roberts, D. A. (2007). Scientific literacy/science literacy. In S. Abell & N. G. Lederman (Eds.), *Handbook of research on science education* (pp. 729–780). Lawrence Erlbaum.
Rudolph, J. L. (2023). Scientific literacy: Its real origin story and functional role in American education. *Journal of Research in Science Teaching, 61*(3), 519–532.
Russell, B. (1928). *Sceptical essays*. George Allen & Unwin Ltd..
Ryder, J. (2001). Identifying science understanding for functional scientific literacy. *Studies in Science Education, 36*, 1–44.
Sadler, T. D., Barab, S. A., & Scott, B. (2007). What do students gain by engaging in socioscientific inquiry? *Research in Science Education, 37*, 371–391.
Sinatra, G. M., & Hofer, B. K. (2021). *Science denial: Why it happens and what to do about it*. Oxford University Press.
Sjöström, J., & Eilks, I. (2018). Reconsidering different visions of scientific literacy and science education based on the concept of Bildung. In Y. J. Dori, Z. R. Mevarech, & D. R. Baker (Eds.), *Cognition, metacognition, and culture in STEM education* (pp. 65–88). Springer.
The Royal Society. (2014). *Vision for science and mathematics education*.
Union of Concerned Scientists. (2009, July 16). *How does the sun affect our climate?* https://www.ucsusa.org/resources/how-does-sun-affect-our-climate
Vetter, D. (2023, December 20). Why 'doing your own research' may make you believe fake news. *Forbes*. https://www.forbes.com/sites/davidrvetter/2023/12/20/why-doing-your-own-research-may-make-you-believe-fake-news/?sh=3c121a951233
Vosoughi, S., Roy, D., & Aral, S. (2018). The spread of true and false news online. *Science, 359*(6380), 1146–1151. https://doi.org/10.1126/science.aap9559
Wardle, C., & Derakhshan, H. (2017). *Information disorder: Toward an interdisciplinary framework for research and policymaking* (Vol. 27). Council of Europe Strasbourg.
Wineburg, S., & McGrew, S. (2019). Lateral Reading and the nature of expertise: Reading less and learning more when evaluating digital information. *Teachers College Record: The Voice of Scholarship in Education, 121*(11), 1–40.
Wineburg, S., Breakstone, J., McGrew, S., Smith, M., & Ortega, T. (2022). Lateral Reading on the open internet: A district-wide field study in high school government classes. *Journal of Educational Psychology, 114*, 893. https://doi.org/10.1037/edu0000740
Wynne, B. (1991). Knowledges in context. *Science, Technology, & Human Values, 16*, 111–121.
Zhang, L., Powell, J. J. W., & Baker, D. P. (2015., July 4). Exponential growth and the shifting global center of gravity of science production, 1900–2011. *Change: The Magazine of Higher Learning, 47*(4), 46–49. https://doi.org/10.1080/00091383.2015.1053777
Zhou, Y., & Shen, L. (2021., June 1). Confirmation bias and the persistence of misinformation on climate change. *Communication Research, 49*(4), 500–523. https://doi.org/10.1177/00936502211028049

Daniel Pimentel is an assistant professor of science education at the University of Alabama. He received his PhD from the Stanford Graduate School of Education. He studies teaching and learning about science media literacy.

Jonathan Osborne is an emeritus professor of science education at Stanford University. His research focuses on policy and pedagogy in the teaching and learning of science.

Chapter 3
A Cognitive and Epistemic Account of the Nature of Engineering

Tamar Ginzburg, Miri Barak, and Sibel Erduran

3.1 Introduction

In recent years, there has been a notable increase in interest and research focus on pre-college engineering education within the realms of science and engineering pedagogy (Cunningham et al., 2020; NASEM, 2020; Purzer et al., 2022). This growing attention highlights a shared recognition among educators and policymakers of the necessity to cultivate an understanding of engineering and its societal impact within K-12 education (AE3 & ASEE, 2020; Antink-Meyer & Brown, 2019; McComas & Burgin, 2020; NGSS Lead States, 2013; Parrish et al., 2024; Purzer et al., 2022).

The significance of engineering education can be attributed to three main dimensions: Firstly, engineering literacy is essential in navigating a world increasingly driven by technology (Carr et al., 2012; NAGB, 2018; NGSS Lead States, 2013; NRC, 2012). Secondly, engineering encapsulates critical skills such as problem-solving, systems thinking, and innovation, which are vital for the sustainable development of contemporary society (Barak & Yuan, 2021; NASEM, 2020; Purzer

T. Ginzburg (✉)
Faculty of Education in Science and Technology, Technion IIT, Haifa, Israel
e-mail: tamar.gin@campus.technion.ac.il

M. Barak
Faculty of Education in Science and Technology, Technion IIT, Haifa, Israel

Department of Education, University of Oxford, Oxford, UK
e-mail: bmiriam@technion.ac.il

S. Erduran
Department of Education, University of Oxford, Oxford, UK

Maison de la Création et de l'Innovation, Université Grenoble Alpes, Grenoble, France
e-mail: sibel.erduran@education.ox.ac.uk

G. Cakmakci, M. F. Tasar (eds.), *Connecting Science Education with Cultural Heritage*, Contributions from Science Education Research 15,
https://doi.org/10.1007/978-3-031-90944-3_3

et al., 2022). Thirdly, persistent shortages in engineering talent, coupled with the designation of engineering occupations as priority fields, underscore the pressing need for robust engineering education (NAGB, 2018; NASEM, 2020).

Nevertheless, debates persist around key inquiries such as: What should comprise the core concepts and skills of engineering education in schools? Which pedagogical methodologies should be employed? Where within the K-12 curriculum should engineering be integrated? What qualifications are requisite for teaching engineering effectively? These questions remain pertinent, fueling ongoing discussions among science and engineering scholars (e.g., Cunningham et al., 2020; Deniz et al., 2020; McComas & Burgin, 2020; Parrish et al., 2024; Purzer et al., 2022).

From a historical point of view, engineering has been an intrinsic facet of human civilization since ancient times, as evidenced by the invention of tools to facilitate daily life. The term 'Engine' originates from the Latin 'Ingenium', denoting mental prowess or clever invention. Derived from this is 'Engineer', a term dating back to the fourteenth century, signifying an individual involved in constructing and operating military machinery (i.e., engines). Across epochs, engineering has been synonymous with the creation and refinement of artifacts to enhance human functionality (Kroes, 2012; Mitcham & Schatzberg, 2009). However, it is imperative to acknowledge that while engineering innovations may enrich our lives, they can also pose risks, exemplified by military machinery or industrial processes that contribute to environmental degradation.

Over the past decade, there has been a gradual integration of engineering education into school curricula, aimed at fostering students' comprehension of engineering design and practice. The publication of the Next Generation Science Standards (NGSS) in the United States in 2013 marked a significant milestone, amplifying references to engineering practices within STEM education discourse (e.g., Kelly & Green, 2018). Nevertheless, despite engineering's longstanding societal relevance, its introduction into school settings has been relatively recent, prompting questions about its delayed inclusion. The incorporation of engineering topics into curricula has been guided by experts from diverse disciplinary backgrounds and epistemological perspectives, resulting in divergent pedagogical approaches (e.g., Daugherty & Carter, 2018; Purzer et al., 2022; Violante & Vezzetti, 2017). These disparities give rise to significant questions such as: Should engineering education be treated as a distinct discipline or integrated into the science curriculum? Are engineering degrees essential for teachers? Can science educators be trained to teach engineering effectively, and if so, how? These questions remain unresolved, reflecting a scarcity of scholarship on a fundamental question: "What constitutes the nature of engineering?" In this chapter, we endeavor to address this pivotal question by exploring the cognitive and epistemic foundations of engineering as a broad domain, alongside the distinctive attributes of various engineering sub-disciplines.

3.2 The Nature of Engineering (NOE)

Over the past two decades, various reports have outlined standards for engineering and technology education, reflecting a concerted effort to integrate these disciplines into educational frameworks (e.g., ITEA, 2007; NAGB, 2018; NGSS Lead States, 2013; NRC, 2009). Notably, the Standards for Technological Literacy (STL) underscored the importance of engineering design by delineating key elements for student engagement (ITEA, 2007). These standards advocate for students to apply a systematic design process to address problems both within and beyond the classroom, emphasizing criteria specification, constraint identification, prototyping, testing, evaluation, and documentation (ITEA, 2007).

Similarly, the Next Generation Science Standards (NGSS) advocate for the integration of engineering concepts and practices to enrich students' understanding of and interest in science (NGSS Lead States, 2013). These standards prompt students to define problem criteria and constraints, evaluate competing solutions, analyze data to inform optimal designs, and develop models of proposed objects, tools, or processes (NGSS Lead States, 2013). The National Assessment Governing Board (NAGB, 2018) has also contributed to the discourse by presenting education standards that encompass technology and engineering literacy. These standards encompass a spectrum of competencies, including systems thinking, maintenance, troubleshooting, idea exchange, solution construction, information research, and more.

Central to these standards is the recognition of the engineering design process and its interconnectedness with science and mathematics education. Consequently, research endeavors have explored students' engagement with design practices through diverse pedagogical approaches. For instance, Mehalik et al. (2008) facilitated student involvement in constructing electrical alarm systems using authentic engineering design practices, revealing greater understanding and knowledge retention, particularly among low-achieving students. Similarly, Berland (2013) investigated the translation of science lessons into engineering contexts, exploring the associated challenges and offering insights for curriculum development.

Building upon these foundations, Cunningham et al. (2020) introduced the 'Engineering is Elementary' program, aiming to integrate engineering and science practices, resulting in significant gains in students' understanding of engineering, technology, and science concepts. Furthermore, interdisciplinary pedagogical approaches advocated by Daugherty and Carter (2018) underscore the interconnectedness of science, technology, engineering, and mathematics (STEM), fostering problem-solving skills and cooperative learning. Recent studies, such as that by Purzer et al. (2022) have advanced frameworks like the honeycomb of engineering, illustrating the versatility of design methodology across multiple domains. This highlights the nuanced goals of engineering education while cautioning against a monolithic definition of engineering (Purzer et al., 2022).

Despite the emphasis on engineering practices within national standards and the promising outcomes showcased in contemporary studies (e.g., Cunningham et al.,

2020; Daugherty & Carter, 2018; Purzer et al., 2022), the integration of engineering concepts remains limited in science, technology, and mathematics curricula, with sporadic implementation in classrooms (Carr et al., 2012; Ekiz-Kiran & Aydin-Gunbatar, 2021). To this day, many students complete K-12 education with little or no exposure to engineering education (Hammack & Ivey, 2017; Pleasants & Olson, 2019; Purzer et al., 2022). Several factors contribute to this phenomenon. Firstly, there is a dearth of schoolteachers with an engineering background, as individuals with engineering degrees seldom pursue careers in teaching (Bybee, 2014; Purzer et al., 2022). Consequently, educators lacking proper engineering education may struggle to meet the demands of teaching this subject effectively (Hammack & Ivey, 2017; Purzer et al., 2022). Secondly, there is a limited number of engineering education training programs for both in-service and pre-service teachers, who often possess limited knowledge of design-based practices (Ekiz-Kiran & Aydin-Gunbatar, 2021; NASEM, 2020; NRC, 2009). Lastly, the multifaceted nature of engineering, necessitating the application of advanced science and mathematics, complicates instructional approaches (Barak & Usher, 2022; Honey et al., 2014; Pleasants & Olson, 2019).

3.2.1 Integration of NOE into the Science Curriculum

The scarcity of schoolteachers with engineering backgrounds, coupled with an already densely packed curriculum, has necessitated the integration of engineering topics into existing science courses (Ekiz-Kiran & Aydin-Gunbatar, 2021; Johnston et al., 2019). Advocates of this integrative and interdisciplinary approach argue that it enriches science education by grounding it in real-world applications (Barak, 2017; Honey et al., 2014; Park et al., 2020). However, without adequate preparation, science teachers may hold misconceptions about the engineering discipline, leading to superficial instruction and limited student learning (Honey et al., 2014; NASEM, 2020; Pleasants & Olson, 2019). Given that most science teachers possess degrees in the sciences, concerns have been raised regarding their lack of engineering background and understanding (Bybee, 2014; Purzer et al., 2022).

Of particular concern is science teachers' ability to distinguish between scientific inquiry and engineering design (Bybee, 2014; Mangiante & Gabriele-Black, 2020). While scientific inquiry seeks to understand natural phenomena by asking questions and testing hypotheses, engineering design aims to solve problems by creating and optimizing practical solutions. Hammack and Ivey (2017) found that many science teachers reported minimal experience in engineering education and struggled to differentiate between engineering and science activities. Other studies have revealed that science teachers, in general, possess limited familiarity with engineering concepts and often hold stereotypical misconceptions about the work of engineers (Cunningham et al., 2006; Purzer et al., 2022). Teachers were found to have limited awareness of the significance of engineering activities and a partial understanding of how to tackle engineering problems. Some erroneously perceived engineers as manual workers in fields like auto mechanics or construction (Hammack & Ivey, 2017; Pleasants & Olson, 2019). For instance, Cunningham et al. (2006) found that teachers were more inclined to view engineers as individuals involved in building construction rather than those overseeing the construction process. Overall, the literature suggests that science teachers' perceptions of engineering and engineering design often do not align with the practices and core ideas outlined in reports such as those by the NRC (2009) or the NGSS Lead States (2013). Given that teachers' conceptions (or misconceptions) can significantly impact students' learning outcomes, there is an increasing need for teacher development programs in engineering education (AE3 & ASEE, 2020; Deniz et al., 2020; Hammack & Ivey, 2017).

Numerous educational studies have proposed pedagogical frameworks for engineering education in schools. Moore et al. (2014) suggested a framework aimed at informing the structure of STEM education standards and initiatives. This framework, focused on quality K–12 engineering education, encompasses twelve key indicators, including 'processes of design' and 'teamwork.' Although distinctions are made between these indicators, there is notable overlap in their application (Moore et al., 2014). Another example is the P–12 engineering learning framework (AE3 & ASEE, 2020), developed to identify learning objectives for engineering literacy. This framework presents a taxonomy of engineering contents derived from a modified Delphi study but places minimal emphasis on connections to science education curricula (AE3 & ASEE, 2020). A more recent framework, the honeycomb of engineering framework, offers a philosophical perspective on pre-college engineering education (Purzer et al., 2022). Categorizing the diverse goals of engineering, this framework emphasizes the centrality of 'design' across various disciplines. However, while the framework can inform lesson design, it does not prescribe effective teaching methods and does not explicitly prioritize teaching the NOE as a learning objective.

To address the scarcity of engineering education programs for science teachers, the National Academies of Sciences, Engineering, and Medicine (NASEM) report (2020) suggests fostering collaborative dialogues among experts from colleges of education, colleges of engineering, and other stakeholders. The report emphasizes the need for teacher education programs to identify and implement actions that highlight the engineering components within the vision for school science education (NASEM, 2020). Such programs should support the development of deep pedagogical content knowledge in engineering to prevent the conflation and confusion of science and engineering (Antink-Meyer & Brown, 2019; Johnston et al., 2019; NASEM, 2020). Authentic representations of science and engineering in classrooms necessitate a well-developed understanding of the nature of science and the nature of engineering (Erduran, 2020; Pleasants & Olson, 2019). Accordingly, the NASEM (2020) report underscores the importance of illuminating the essential qualities of engineering, the design process, and core engineering concepts. It calls for a nuanced conceptualization of 'engineering' and its distinction from 'science' and 'technology,' themes further explored in the subsequent section.

3.2.2 Distinguishing Engineering from Science and Technology

While science, technology, and engineering are interconnected, they possess distinct cognitive and epistemic characteristics that necessitate clear definitions and differentiations (Antink-Meyer & Brown, 2019; McComas & Burgin, 2020; NASEM, 2020). When engineering design tasks are introduced in science classrooms, the boundaries between these disciplines can become blurred, resulting in conflation (Johnston et al., 2019; NASEM, 2020). Given their interdependence, it is crucial to carefully examine the relationships among science, engineering, and technology within the context of school education.

The distinction between science and engineering or technology is relatively clear. Science entails a systematic exploration of the natural and physical world through methods such as observation, experimentation, modeling, and classification (NSTA & ASTE, 2019; NGSS Lead States, 2013). It seeks to answer questions about the natural world, while engineering and technology focus on designing and developing human-made artifacts. Distinguishing between engineering and technology presents a greater challenge, as they are often used interchangeably to describe aspects of human-developed products. Nonetheless, understanding each discipline's theoretical and practical aspects is essential for their integration into the science curriculum (NAGB, 2018; NASEM, 2020).

According to the National Assessment Governing Board (2018), engineering is defined as "a systematic and often iterative approach to designing objects, processes, and systems to meet human needs and wants" (NAGB, 2018, p. 5), while technology encompasses "products, processes, and systems created by people to meet human needs and desires" (NAGB, 2018, p. 23). Pleasants and Olson (2019) assert that engineers work with nature and its laws, as revealed by science, focusing on creative problem-solving processes involving testing and revisions. In contrast, technologists emphasize the construction aspect, viewing technology as both the means to solve problems and the end product. In summary, the connections and distinctions among the three disciplines can be encapsulated as follows: Guided by scientific principles, engineering involves the knowledge and practices necessary for designing new or improved technologies (NAGB, 2018; NASEM, 2020). With appropriate support and guidance, science teachers with limited formal engineering education can develop an understanding of engineering (e.g., Deniz et al., 2020; Pleasants & Olson, 2019). This underscores the importance of developing a conceptual framework of the nature of engineering from both cognitive and epistemic perspectives (Pleasants & Olson, 2019; Purzer et al., 2022). Consequently, it is imperative to highlight the key aspects of various engineering fields.

3.2.3 Domains within Engineering

While engineering education within the context of science education is gaining momentum, engineering is often presented broadly without adequate differentiation across its various sub-fields (Purzer et al., 2022). To understand the nuances of engineering, it is essential to distinguish between its different aspects and fields. Engineering, fundamentally, focuses on designing and creating new or improved artifacts—a term derived from Latin meaning "by skill" and "to make." These artifacts encompass not only tangible machinery and tools but also intangible entities like computer software, design documents, written patents, and even subcellular nanorobots. Over time, engineering has evolved into distinct fields and sub-fields, each with its area of specialization. Some of these include civil, environmental, chemical, mechanical, computer, aerospace, and more (ABET, 2019; ITEA, 2007; NASEM, 2020). Based on the created artifacts, engineering can be broadly categorized into four main categories: Structures, Machines, Materials, and Data.

Structures engineering focuses on planning and constructing buildings and infrastructures to meet human needs and address societal challenges (ABET, 2019; NASEM, 2020). Dating back to ancient times with the construction of pyramids and aqueducts, civil and architectural engineering disciplines are typically involved in this area.

Machines engineering deals with the development of complex devices and tools assembled from mechanical parts to perform specific tasks. It spans various transmission methods such as mechanical, electrical, hydraulic, and pneumatic, and is prominent in mechanical, biomedical, aerospace, and naval engineering (ABET, 2019; ITEA, 2007).

Materials engineering involves the manipulation of materials—solids, liquids, gases, and other condensed phases—to develop useful products. This includes everything from chemical processes to the manufacturing of polymers, ceramics, metals, and advanced materials like semiconductors and nanomaterials (ITEA, 2007; NASEM, 2020).

Data engineering encompasses the development and use of computer-based data, involving creating and manipulating electrical signals to process and store information. It focuses on data analysis, design techniques, and the synthesis of digital signals, playing a crucial role in computer, electrical, and systems engineering (ABET, 2019; ITEA, 2007).

Despite the growing interest in engineering education, existing frameworks often lack a holistic conceptualization of the nature of engineering. While educational frameworks provide pedagogical components and indicators, they fall short in fully defining and instructing the NOE, which is essential for developing engineering literacy (Antink-Meyer & Arias, 2022; Pleasants & Olson, 2019). Defining and teaching the NOE remains a challenge for educational systems worldwide (Antink-Meyer & Arias, 2022; NASEM, 2020), emphasizing the need for comprehensive frameworks and reliable tools for instruction and assessment. Additionally, while there is recognition of the pedagogical benefits of engineering learning, less attention has been given to the cognitive and epistemic aspects of engineering practices (Purzer et al., 2022). Drawing upon previous studies, this paper utilizes the Family Resemblance Approach (FRA) as a methodological framework to analyze and develop the NOE framework (Erduran et al., 2019; Irzik & Nola, 2011; Kaya & Erduran, 2016).

3.3 The NOE from the Perspective of the Family Resemblance Approach

Emerging research suggests that the FRA offers a promising framework for cultivating an inclusive understanding of both the cognitive-epistemic and social-institutional dimensions of NOE (Barak et al., 2024; Aydın-Günbatar & Roehrig, 2023). This is based on the literature that presents the FRA as a framework for holistic and nuanced conceptualization of the nature of science (NOS) (Erduran & Kaya, 2018, 2019). In this study, the FRA was utilized for its unique capacity to incorporate domain-general and domain-specific features, making it particularly useful for examining how science and engineering education is conceptualized and understood (Barak et al., 2024; Erduran & Kaya, 2018, 2019; Park et al., 2020).

Drawing from recommendations made by the literature (e.g., Barak, 2023; Cheung & Erduran, 2023; Erduran et al., 2021; Okan & Kaya, 2023; Park & Brock, 2023; Reinisch & Fricke, 2022), we situated each FRA category within the context of educational studies and policy reports, such as NGSS Lead states (2013), the National Academies of Sciences, Engineering, and Medicine (NASEM, 2020), and the National Assessment Governing Board (NAGB, 2018). Due to space limitations, the following sections provide selected examples of educational applications.

Aims & values: Engineers, similar to scientists, are guided by aims and values such as precision, reasoning, logic, and creativity (AE3 & ASEE, 2020; NASEM, 2020). Table 3.1 shows different perspectives of the aims and values of science and engineering education with examples of educational applications.

Practices: Engineering practices are described in the Framework for K-12 Science Education (NGSS Lead States, 2013) and other policy reports (e.g., NAGB, 2018; AE3 & ASEE, 2020). Accordingly, engineering practices involve a complex process of defining human-related problems, developing and using engineering models, and designing solutions, as detailed by the Practices part of the NGSS Framework (2013): P1. Defining human-related problems; P2. Developing and using (engineering) models; P3. Planning and carrying out investigations; P4. Analyzing and interpreting (engineering) data; P5. Using mathematics and computational thinking; P6. Designing solutions; P7. Engaging in argument from evidence; P8. Obtaining, evaluating, and communicating information. These practices require a meta-level of understanding, which aligns with the FRA framework. Adapting the scientific heuristics (Erduran & Dagher, 2014; Erduran et al., 2021) to engineering practices, Fig. 3.1 shows a holistic representation of the engineering

Table 3.1 Aims and values of science and engineering education

Aims and values	Science applications (from Erduran & Dagher, 2014)	Engineering applications
Objectivity	Seeking neutrality and avoiding bias.	Evaluating design alternatives, emphasizing objectivity with a clear and rational mindset.
Novelty	Searching for new explanations.	Creating new and innovative solutions to complex challenges through creative thinking and critical evaluation.
Accuracy	Ensuring that explanations are accurate.	Ensuring that designs, measurements, and calculations are reliable and free from errors.
Critical examination	Giving reasons to justify claims.	Evaluating the designs, processes, and solutions through systematic analysis, questioning assumptions, testing hypotheses, and scrutinizing the validity of solutions.
Addressing anomalies	Recognizing opposite ideas and responding to objections	Systematically evaluating the feasibility, effectiveness, and trade-offs to recognize the most promising solution.

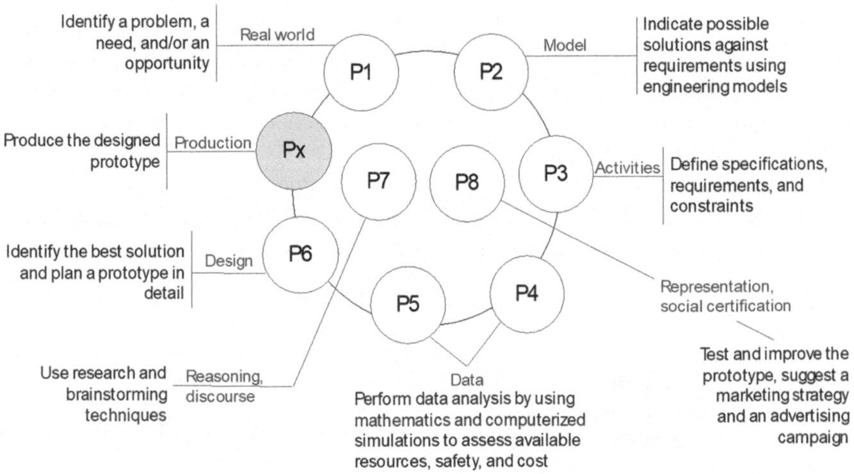

Fig. 3.1 Engineering practices, heuristic features, and example learning assignments

practices with their corresponding heuristic features and examples for learning assignments. P7 and P8 are integral to and interacting with the other six engineering practices, and therefore are in the middle of the model. The engineering Production feature, which replaces the Prediction feature within the scientific practices (Erduran & Dagher, 2014) is not mentioned explicitly as part of the Practices (NGSS Lead States, 2013). It is therefore denoted as Px in the model.

Methods & methodological rules: Methods and methodological rules in engineering encompass a diverse range of systematic guidelines and approaches employed to guarantee that investigative outcomes yield accurate and reliable knowledge (Erduran & Dagher, 2014; Irzik & Nola, 2011). Epistemic ideas relevant to engineering education may include design methods (DM's) grouped into four types of artifacts (4 A's): Materials, Machines, Structures, and Data (Fig. 3.2). In fields such as chemical engineering (*Materials*) and mechanical engineering (*Machines*), Biomimicry emerges as a prominent engineering design approach. Rooted in the emulation of natural materials, mechanisms, and processes, Biomimicry offers innovative solutions to human design challenges. For example, the biomechanics of the gecko lizard's toes have inspired a host of climbing materials, and the hydrodynamics of the whales' swimming has inspired the design of submarines. In fields such as architecture and urban planning (*Structures*), the *Mixed-Use Design* is employed, uniting three or more functions within a single structure, including transportation, residential, hotel, retail, entertainment, and parking. This design method prioritizes the preservation of quality of life while mitigating environmental impact. Other design methods include the *Design for Assembly* and the *Reverse Engineering* method. These four design methods were chosen because they represent a diverse spectrum of systematic approaches that address different aspects of engineering problem-solving. These methods transcend specific disciplines and are adaptable to different contexts, ensuring solutions are accurate and reliable. Of course, these are not the only design methods available, and other approaches may be equally valuable depending on the context and objectives of the project at hand. Regardless of the engineering methods chosen, they—like scientific methods—should be practiced in schools as a creative process involving more than one approach.

Engineering knowledge: While scientific knowledge is typically considered the 'end-product' of scientific endeavors (Erduran & Dagher, 2014; Irzik & Nola, 2011), engineering knowledge emerges from the processes of design and manufacturing (NAGB, 2018; NASEM, 2020). This type of knowledge pertains to the

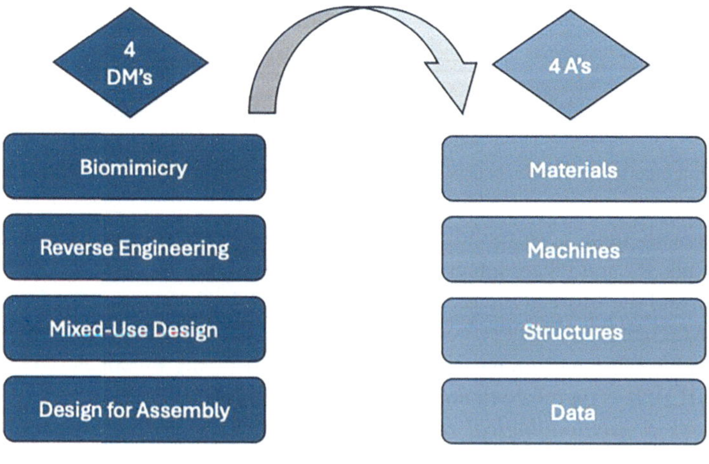

Fig. 3.2 The four design methods (4 DM's), grouped into four types of artifacts (4 A's)

functioning of artifacts and the application of analytical models across various technologies (Pleasants & Olson, 2019). When examining the disciplinary aspects of engineering for K-12 education, scholars inquire into engineering knowledge by posing questions such as: What types of knowledge are inherent to the engineering discipline? How does engineering science differ from and resemble natural science? And to what extent do models of the design process accurately depict designers' actual work? (Pleasants & Olson, 2019). Building upon this foundation, the literature delineates various facets of the nature of engineering knowledge crucial for K-12 educators with limited engineering backgrounds, encompassing empirical, contextually responsive, social, personal, societal and cultural, interdisciplinary, and solution-oriented dimensions (Antink-Meyer & Brown, 2019). Beyond mere content knowledge, engineering knowledge comprises a fusion of scientific and mathematical theories, laws, and models that underpin all engineering practices (NAGB, 2018). It encompasses the practical applications of the engineering enterprise, validating and generating engineering concepts. Providing logical and systematic explanations of the workings of human-made artifacts, engineering knowledge fosters innovation and continued development (Barak et al., 2024).

3.4 Discussion

In recent years, the importance of engineering literacy in navigating a technology-driven society has garnered significant attention, leading to discussions about pre-college engineering education in policy reports and academic studies (AE3 & ASEE, 2020; Antink-Meyer & Brown, 2019; English & Moore, 2018). However, despite this growing recognition, teaching the NOE remains in its nascent stages (Aydın-Günbatar & Roehrig, 2023; Barak et al., 2024). The literature underscores the necessity of conceptualizing and integrating NOE into the science curriculum (Antink-Meyer & Arias, 2022; Deniz et al., 2020; Parrish et al., 2024; Pleasants & Olson, 2019). Drawing parallels with the deeper understanding fostered by exploring the NOS (e.g., Antink-Meyer & Brown, 2019; Barak et al., 2023; Erduran et al., 2019; Lederman, 1992; Okan & Kaya, 2023), there is a clear imperative to establish an ontological and pedagogical framework for engineering that delves into the cognitive and epistemic approaches utilized by engineers across various domains.

This chapter contributes a novel perspective to the ongoing discourse surrounding NOE heuristics and applications by likening it to the conceptualization of NOS. Theorizing NOE presents inherent complexities, as engineering is perceived as both a process, a discipline, and an approach (Antink-Meyer & Brown, 2019). Moreover, delineating the cognitive and epistemic characteristics that both differentiate and unite engineering and science poses significant challenges (Antink-Meyer & Brown, 2019; McComas & Burgin, 2020; NASEM, 2020). Nevertheless, advancing the theoretical understanding of NOE is paramount for fostering scientific and engineering literacies, as integrated teaching can deepen the understanding of modern society (Antink-Meyer & Arias, 2022; Pleasants & Olson, 2019).

Building upon the family resemblance approach (FRA) as framed by Erduran and Dagher (2014), this chapter explains NOE through four cognitive and epistemic categories: Aims & Values, Engineering Practices, Methods & Methodological Rules, and Engineering Knowledge. Leveraging the FRA as a methodological lens can inform the teaching and learning of engineering in educational settings. In teacher development programs, this approach can facilitate the introduction of the multifaceted NOE and aid educators in selecting and delivering NOE-related learning activities aligned with educational objectives. The four discussed FRA cognitive-epistemic categories furnish a comprehensive and cohesive conceptualization of NOE, fostering a shared understanding of the engineering enterprise among educational stakeholders. This cognitive-epistemic dimension of NOE is pivotal for cultivating a unified, holistic comprehension of engineering and its theoretical and practical interplay with science, thereby providing a framework for the development of engineering education in schools.

With the aim of promoting engineering literacy as a guiding principle, future research endeavors should explore additional dimensions of the NOE, aiming to construct a comprehensive conceptualization within the framework of school science education. This study suggests that the FRA may provide a comprehensive, holistic approach to the NOE, allowing for a nuanced understanding of it. Building upon this conceptualization, future research should utilize empirical data to examine the understanding of the NOE using the FRA as a guiding framework. The FRA categories could serve as both theoretical and methodological lenses for identifying gaps in the understanding of NOE and for offering heuristics to enhance this understanding. There is a pressing need to identify further heuristics that contribute to the formation of NOE understanding and facilitate its seamless integration into the science curriculum. By delving deeper into these aspects, researchers can enrich our understanding of engineering education and pave the way for more effective pedagogical strategies in fostering engineering literacy among students.

Acknowledgments This research had partial financial support from the Technion—IIT.

References

ABET. (2019). *2020–2021 Criteria for Accrediting Engineering Programs*. Retrieved from https://www.abet.org/wp-content/uploads/2020/09/EAC-Criteria-2020-2021.pdf

Advancing Excellence in P-12 Engineering Education, & American Society for Engineering Education [AE3 & ASEE]. (2020). *Framework for P-12 engineering learning: A defined and cohesive educational foundation for P-12 engineering*.

Antink-Meyer, A., & Arias, A. M. (2022). Teachers' incorporation of epistemic practices in K-8 engineering and their views about the nature of engineering knowledge. *Science & Education, 31*(2), 357–382.

Antink-Meyer, A., & Brown, R. A. (2019). Nature of engineering knowledge. *Science & Education, 28*(3–5), 539–559.

Aydın-Günbatar, S., & Roehrig, G. H. (2023). Elaborating nature of engineering through family resemblance approach. *Science & Education, 34*, 95–127.

Barak, M. (2017). Science teacher education in the twenty-first century: A pedagogical framework for technology-integrated social constructivism. *Research in Science Education, 47*(2), 283–303.

Barak, M. (2023). Editorial: Family resemblance approach in science education. *Science & Education, 32*, 1–5.

Barak, M., & Usher, M. (2022). The innovation level of engineering students' team projects in hybrid and MOOC environments. *European Journal of Engineering Education, 47*(2), 299–313.

Barak, M., & Yuan, S. (2021). A cultural perspective to project-based learning and the cultivation of innovative thinking. *Thinking Skills and Creativity, 39*, 100766.

Barak, M., Yachin, T., & Erduran, S. (2023). Tracing preservice teachers' understanding of nature of science through their drawings and writing. *Research in Science Education, 53*(3), 507–523.

Barak, M., Ginzburg, T., & Erduran, S. (2024). Nature of engineering: A cognitive and epistemic account with implications for engineering education. *Science & Education, 33*(3), 679–697.

Berland, L. K. (2013). Designing for STEM education. *Journal of Pre-College Engineering Education Research, 3*(1), 22–31.

Bybee, R. W. (2014). NGSS and the next generation of science teachers. *Journal of Science Teacher Education, 25*(2), 211–221.

Carr, R. L., Bennett, L. D., & Strobel, J. (2012). Engineering in the K-12 STEM standards of the 50 U.S. states: An analysis of presence and extent. *Journal of Engineering Education, 101*(3), 539–564.

Cheung, K. K. C., & Erduran, S. (2023). A systematic review of research on family resemblance approach to nature of science in science education. *Science & Education, 32*(5), 1637–1673.

Cunningham, C. M., Lachapelle, C., & Lindgren-Streicher, A. (2006). *Elementary teachers understanding of engineering and technology* (Proceedings of the American Society for Engineering Education American conference and exposition) (Vol. 113). ASEE.

Cunningham, C. M., Lachapelle, C. P., Brennan, R. T., Kelly, G. J., Tunis, C. S. A., & Gentry, C. A. (2020). The impact of engineering curriculum design principles on elementary students' engineering and science learning. *Journal of Research in Science Teaching, 57*(3), 423–453.

Daugherty, M. K., & Carter, V. (2018). The nature of interdisciplinary STEM education. In M. J. de Vries (Ed.), *Handbook of technology education* (pp. 159–172). Springer.

Deniz, H., Kaya, E., Yesilyurt, E., & Trabia, M. (2020). The influence of an engineering design experience on elementary teachers' nature of engineering views. *International Journal of Technology and Design Education, 30*, 635–656.

Ekiz-Kiran, B., & Aydin-Gunbatar, S. (2021). Analysis of engineering elements of K-12 science standards in seven countries engaged in STEM education reform. *Science & Education, 30*(4), 849–882.

English, L. D., & Moore, T. J. (2018). *Early engineering learning*. Springer Nature.

Erduran, S. (2020). Nature of "STEM"? Epistemic underpinnings of integrated science, technology, engineering, and mathematics in education. *Science & Education, 29*, 781–784.

Erduran, S., & Dagher, Z. R. (2014). *Reconceptualizing the nature of science for science education*. Springer.

Erduran, S., & Kaya, E. (2018). Drawing nature of science in pre-service science teacher education: Epistemic insight through visual representations. *Research in Science Education, 48*(6), 1133–1149.

Erduran, S., & Kaya, E. (2019). *Transforming teacher education through the epistemic core of chemistry*. Springer International Publishing.

Erduran, S., Dagher, Z. R., & McDonald, C. V. (2019). Contributions of the family resemblance approach to nature of science in science education. *Science & Education, 28*(3), 311–328.

Erduran, S., Kaya, E., Cilekrenkli, A., Akgun, S., & Aksoz, B. (2021). Perceptions of nature of science emerging in group discussions: A comparative account of pre-service teachers from Turkey and England. *International Journal of Science and Mathematics Education, 19*, 1375–1396.

Hammack, R., & Ivey, T. (2017). Elementary teachers' perceptions of engineering and engineering design. *Journal of Research in STEM Education, 3*(1/2), 48–68.
Honey, M. A., Pearson, G., & Schweingruber, H. (2014). *STEM integration in K-12 education: Status, prospects, and an agenda for research*. National Academies Press.
International Technology Education Association [ITEA]. (2007). *Standards for technological literacy: Content for the study of technology* (3rd ed.). International Technology Education Association. Retrieved May 5, 2022, from https://www.iteea.org/File.aspx?id=67767&v=b26b7852
Irzik, G., & Nola, R. (2011). A family resemblance approach to the nature of science for science education. *Science & Education, 20*, 591–607.
Johnston, A. C., Akarsu, M., Moore, T. J., & Guzey, S. S. (2019). Engineering as the integrator: A case study of one middle school science teacher's talk. *Journal of Engineering Education, 108*(3), 418–440.
Kaya, E., & Erduran, S. (2016). From FRA to RFN, or how the family resemblance approach can be transformed for science curriculum analysis on nature of science. *Science & Education, 25*, 1115–1133.
Kelly, G., & Green, J. (Eds.). (2018). *Framing issues of theory and methods for the study of science and engineering education*. Routledge.
Kroes, P. (2012). *Technical artefacts: Creations of mind and matter: A philosophy of engineering design*. Springer.
Lederman, N. G. (1992). Students' and teachers' conceptions of the nature of science: A review of the research. *Journal of Research in Science Teaching, 29*(4), 331–359.
Mangiante, E. S., & Gabriele-Black, K. A. (2020). Supporting elementary teachers' collective inquiry into the "E" in STEM. *Science & Education, 29*(4), 1007–1034.
McComas, W. F., & Burgin, S. R. (2020). A critique of "STEM" education. *Science & Education*, 29, 805–829.
Mehalik, M. M., Doppelt, Y., & Schuun, C. D. (2008). Middle-school science through design-based learning versus scripted inquiry: Better overall science concept learning and equity gap reduction. *Journal of Engineering Education, 97*(1), 71–85.
Mitcham, C., & Schatzberg, E. (2009). Defining technology and the engineering sciences. In A. Meijers (Ed.), *Philosophy of technology and engineering sciences* (pp. 27–63). Elsevier.
Moore, T. J., Glancy, A. W., Tank, K. M., Kersten, J. A., Smith, K. A., & Stohlmann, M. S. (2014). A framework for quality K–12 engineering education: Research and development. *Journal of Pre-College Engineering Education Research, 4*(1), 2.
National Academies of Sciences, Engineering, and Medicine [NASEM]. (2020). *Building capacity for teaching engineering in K-12 education*. The National Academies Press.
National Assessment Governing Board [NAGB]. (2018). *Technology & Engineering Literacy Framework for the 21st century*. The 2018 National Assessment of Educational Progress. NAGB.
National Research Council [NRC]. (2009). *Engineering in K-12 education: Understanding the status and improving the prospects*. The National Academies Press.
National Research Council [NRC]. (2012). *A framework for K-12 science education: Practices, crosscutting concepts, and core ideas* (H. Quinn, H. Schweingruber, & T. Keller, Eds.).
National Science Teaching Association [NSTA] and Association for Science Teacher Education [ASTE] (2019). *2020 NSTA/ASTE Standards for Science Teacher Preparation*. Retrieved May 5, 2022, from http://static.nsta.org/pdfs/2020NSTAStandards.pdf.
Next Generation Science Standards [NGSS] Lead States. (2013). *Next generation science standards: For states, by states*. The National Academies Press.
Okan, B., & Kaya, E. (2023). Exploring the inclusion of nature of science in Turkish middle school science textbooks. *Science & Education, 32*(5), 1515–1535.
Park, W., & Brock, R. (2023). Is there a limit to resemblances? Teaching about science and pseudoscience from a family resemblance perspective. *Science & Education, 32*(5), 1265–1286.

Park, W., Wu, J. Y., & Erduran, S. (2020). The nature of STEM disciplines in the science education standards documents from the USA, Korea and Taiwan: Focusing on disciplinary aims, values and practices. *Science & Education, 29*(4), 899–927.

Parrish, J. C., Pleasants, J., Reid, J. W., Mulvey, B. K., Peters-Burton, E. E., & Recker, A. (2024). Using card sort epistemic network analysis to explore preservice teachers' ideas about the nature of engineering. *Science & Education, 33*, 301–326.

Pleasants, J., & Olson, J. K. (2019). What is engineering? Elaborating the nature of engineering for K-12 education. *Science Education, 103*, 145–166.

Purzer, Ş., Quintana-Cifuentes, J., & Menekse, M. (2022). The honeycomb of engineering framework: Philosophy of engineering guiding pre-college engineering education. *Journal of Engineering Education, 111*, 19–39.

Reinisch, B., & Fricke, K. (2022). Broadening a nature of science conceptualization: Using school biology textbooks to differentiate the family resemblance approach. *Science Education, 106*(6), 1375–1407.

Violante, M. G., & Vezzetti, E. (2017). Guidelines to design engineering education in the twenty-first century for supporting innovative product development. *European Journal of Engineering Education, 42*(6), 1344–1364.

Tamar Ginzburg is a PhD student at the Faculty of Education in Science and Technology, Technion - Israel Institute of Technology. Her research is in science and engineering education and involves conceptualizing the Nature of Engineering. Ginzburg is a Beta Gama Sigma graduate of the Owen Graduate School of Management at Vanderbilt University in Nashville, TN (MBA); and a Summa Cum Laude graduate of the Technion's Faculty of Education in Science and Technology (MSc). Her academic work is presented in peer-reviewed papers in leading international and national publications.

Miri Barak (PhD) is a Professor of Science and Engineering Education. She is the Dean of the Faculty of Education in Science and Technology, at the Technion - Israel Institute of Technology and an Honorary Research Fellow at the University of Oxford, UK. Her primary research interests include digital learning, AI Ethics Education, and the conceptualization of the Nature of Engineering.

Sibel Erduran is a Professor of Science Education at University of Oxford, United Kingdom. She was the President of the European Science Education Research Association and Editor-in-Chief of Science & Education and currently serves as an Editor for International Journal of Science. She is a Fellow of the Academy of Social Sciences and American Educational Research Association. Her research interests focus on the infusion of epistemic practices of science in science education. Her work on argumentation has received international recognition through awards from NARST and EASE.

Chapter 4
How Has Science Education Research (SER) Evolved in The Last Two Decades?

Mustafa Sözbilir and Dilek Teke

4.1 Introduction

In the early nineteenth century, the ability to read, write and keep simple accounts separated the educated from the uneducated. By the end of the century, however, a broad and practical education was increasingly seen as essential for life in a world increasingly dominated by science, technology and industry (DeBoer, 1991, p.2). Many prominent nineteenth century scientists (e.g., John Tyndall, Thomas Huxley, James Paget, Claude Bernard, Michael Faraday, Herbert Spencer) campaigned for the inclusion of science in school curricula in both Europe and the USA (DeBoer, 1991, p.8). The fundamental changes that took place in the education system during the nineteenth century were the gradual but relentless introduction of the sciences and other modern studies into the curriculum at all levels (DeBoer, 1991, p.2). Science had been part of the school curriculum since the mid-nineteenth century.

Science education has little history as a research field compared to the history of science. The establishment of science education as an independent discipline has been the subject of many studies (i.e., Dahncke et al., 2001; Fensham, 2004; Jenkins, 2001; Keeves, 1998; Kempa, 1991). Development of science education research (SER) is influenced by the development of the social and economic status of countries. There are many examples of research undertaken in the first half of the twentieth century that can be categorized as science education, but much of it has been the work of individuals or quasi-governmental committees (Jenkins, 2001). Before the 1960s, the USA was the only country to establish science education as an academic discipline, with the possibility of research study and doctoral degrees (Fensham, 2004). The development of research in science education was strongly

M. Sözbilir (✉) · D. Teke
Department of Mathematics and Science Education, Atatürk University, Erzurum, Türkiye
e-mail: sozbilir@atauni.edu.tr; dilekteke@atauni.edu.tr

G. Cakmakci, M. F. Tasar (eds.), *Connecting Science Education with Cultural Heritage*, Contributions from Science Education Research 15,
https://doi.org/10.1007/978-3-031-90944-3_4

affected by big-budget, discipline-based curriculum development projects during the 1960s such as Chemical Education Material Study (CHEM Study), Physical Sciences Study Committee (PSSC), Biological Sciences Curriculum Study (BSCS), Chemical Bond Approach (CBA) in the USA and followed by Nuffield Science (UK), and the Australian Science Education Project (ASEP) (Sozbilir et al., 2012).

The beginning of the twentieth century witnessed new trends in science education research around the world such as the nature of science, argumentation, inquiry-based learning and teaching practices together with the integration of information and communication technology into the teaching and learning research. Other trends were the growth in interest in studies of laboratory work, especially (open) inquiry, the implementation and use of problem-solving strategies etc. Very recently, although the Covid-19 Pandemic has had a negative influence on daily life, it opened new doors into education. Emergency remote teaching has become mandatory. It brought immense integration of distance education methods and technologies into all educational processes. Thanks to distance education, teachers' and students' skills in the use of technology have also improved (Öztürk, 2019; Bilgiç & Tuzun, 2020; Tosun, 2024). Today the world is discussing and working immensely on how to use computational thinking, artificial intelligence and machine learning in education.

The goals of science education are intellectual goals of thinking and reasoning, personal goals of appreciation and understanding, practical goals that will help us in our life's work and in our role as intelligent citizens, and futuristic goals of innovation and creativity (DeBoer, 1991, p.xiii). Science education allows students to gain a better understanding of how and why things work. Science teaching improves children's comprehension of the world around them. Everything from the human body to technologies we are surrounded by, science can reveal the mechanisms and the reasons for complicated systems we encounter day to day. Therefore, children growing up in this technologically and scientifically developed world, need to be equipped with scientific literacy and scientific methods to think, learn, solve problems and make informed decisions as the public policymakers of the future.

It was argued that science education should be based on direct experience of the physical world rather than on the words of teachers and textbooks, and there was an objection to the rote learning of meaningless facts. This was a reaction to outdated teaching practices that saw the development of memory and will as the primary goals of education. The famous American educationalist John Dewey joined the debate in 1910, arguing in science for much less teaching of "science as subject matter" (the "facts" discovered by scientists) and much more teaching of "science as method" (Dewey, 1910, p.127).

In order to find future research topics related to science education research [SER], it is necessary to know which topics have been previously studied in this field (Chang et al., 2010). If this requirement is fulfilled, it guides future research (Osborne, 2007). In addition, this trend study also helps the planning of science education policies. It is important to determine the trend to determine in which direction the increasing number of SER articles are going, and which needs they fulfil or which needs they have. Determining these, that is, revealing the current situation, will create sustainable policies in science education (Tosun, 2024).

4.1.1 Research Aims and Questions

This study aims to identify and visualize the trends in science education research (SER) via bibliometric analysis such as the distribution of the number of documents and citations based on country, organization, authors, journals, and keywords of the papers published in the last 22 years in the journals indexed in WoS via VOSviewer ® & RStudio Bibliometrix ®. The research questions of this study are as follows.

- What is the distribution of the number of SER articles by country by year?
- Which are the most cited countries in SER articles?
- Which are the most cited organizations in SER articles?
- Which are the most cited journals in SER articles?
- Who are the most prolific authors of SER articles?
- What is the distribution of the most frequently used words in the abstracts of SER articles?
- What are the most popular research topics in SER articles?

4.2 Research Method

Bibliometric analysis is utilized as the research method in this study. Bibliometric is defined as the use of statistical methods to analyze the bibliometric publications data such as peer-reviewed journal articles, books, conference proceedings, periodicals, reviews, reports, and related documents (Aria & Cuccurullo, 2017; Donthu et al., 2021; Pinto et al., 2019). Bibliometric analysis is a quantitative analysis of scientific literature based on metrics such as evolution over time, number of citations, keywords, most prolific authors, etc. In addition, this analysis is a quantitative method used to examine the structure of knowledge and the development of research fields based on the analysis of related publications (Aria & Cuccurullo, 2017; Donthu et al., 2021).

4.2.1 Data for the Bibliometric Analysis

In the data collection process, "science education" is identified as the keyword. The online search was pursued on August 8, 2023 by conducting the search query in the "topic" field of the Web of Science database. The search was limited to Social Science Citation Index (SSCI) and Science Citation Index Expanded (SCIE); Web of Science Categories: Education Educational Research or Education Scientific Disciplines; Timespan: 2000-01-01 to 2022-12-31; Document type = Articles or Review articles; Citation Topic Meso: &.11 Education and Educational Research. As a result of the search, a total of 14,680 articles from 318 sources were accessed for the analysis. The tools used for analysis in this study include the RStudio, the Biblioshiny© web app (version 4.3.1), and the VOSviewer© software (version

1.6.19). All languages are included as the meta data provided by WoS for all papers in English.

4.3 Findings

This section provides a comprehensive analysis of the publication and citation trends of the SER studies from 2000 to 2022. Figure 4.1 shows the number of SER papers published and cited over the period studied. In general, there has been a gradual increase in the SER studies over the years. However, there has been a decrease in the number of SER papers after 2020 with the effect of Covid-19 pandemic.

Over the last 20 years, a total of 132 countries have contributed to the SER literature. The 20 countries with the highest number of publications are listed in Table 4.1 and the highly cited top 20 countries are given in Fig. 4.2. With a total of 5966 publications, most of the research is concentrated in the USA. England is the second most prolific country with 1108 publications, Türkiye is the third country with 992 papers and Australia follows with 934 publications. However, the country with the highest number of citations per article is Belgium (40 citations per paper) although is not in the top 20 prolific countries. Israel and Scotland (34 citations per paper) come second, the Netherlands (32 citations per paper) is third, and England and USA follow with 31 citations per paper.

Table 4.2 and Fig. 4.5 show the network visualization map of the countries that publish the SER literature, with different node sizes and colors. The size of the node refers to the number of published documents from a country, where the larger the node, the more documents are published from that country. These countries were grouped into ten major clusters represented by ten colors.

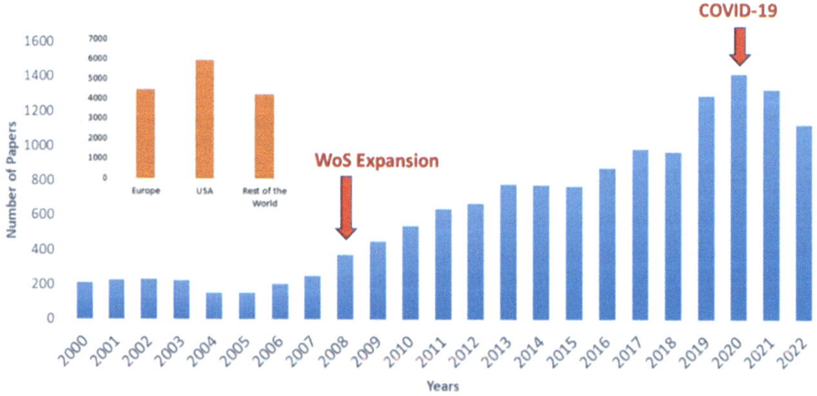

Fig. 4.1 Distribution of the Number of SER Papers Published by Years (2000–2022)

Table 4.1 Top 20 Most Productive Countries in SER

Country	Number of papers	Citations per paper
USA	5966	**31**
England	1108	**32**
Türkiye	992	12
Australia	944	24
Spain	663	14
P.R. China	608	13
Germany	598	23
Taiwan	555	25
Canada	546	27
Netherlands	453	**31**
Sweden	385	14
Israel	375	**34**
South Korea	313	15
South Africa	313	15
Finland	243	23
Brazil	202	16
Greece	194	24
New Zealand	147	26
Norway	146	20
Scotland	140	**34**

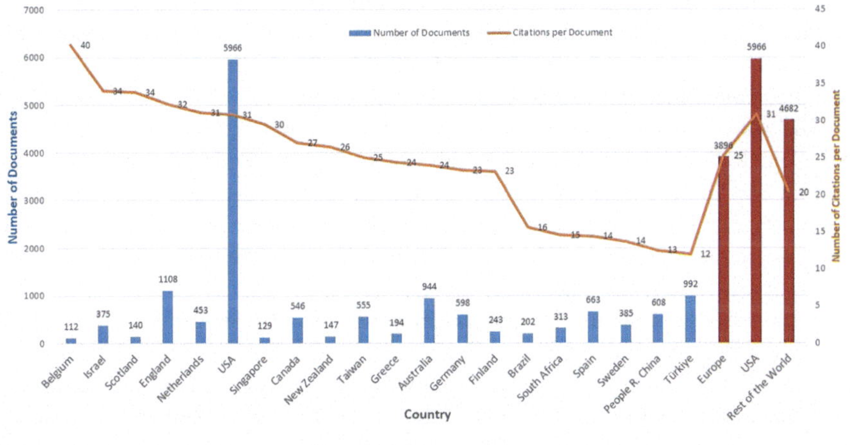

Fig. 4.2 Highly Cited Top 20 Countries in SER

Based on the node sizes shown in Fig. 4.3 and Table 4.2, Germany (cluster 1), USA (Cluster 2), Spain (Cluster 3), England (Cluster 4), Türkiye (Cluster 5), Sweden (Cluster 6), South Korea (Cluster 7), Taiwan (Cluster 8 and 9), and Israel (Cluster 10) were found to be the most influential countries for every cluster in terms of publications in SER.

Table 4.2 Major Clustering Countries in SER

Cluster Number	*Cluster Color*	*Countries*
1	Red	Austria, Botswana, Croatia, Czech Republic, Estonia, Finland, **Germany**, Italy, Lebanon, Lithuania, Malaysia, Poland, Serbia, Slovakia, Slovenia, Switzerland, Venezuela
2	Green	Belgium, Canada, Egypt, Ghana, India, Iran, Jordon, Nigeria, Pakistan, Philippines, Qatar, Saudi Arabia, South Africa, Tanzania, U Arab Emirates, **USA**
3	Dark Blue	Argentina, Brazil, Chile, Colombia, France, Greece, Japan, Luxembourg, Mexico, **Spain**
4	Dark Yellow	Australia, **England**, Ireland, New Zealand, North Ireland, Oman, People R. China, Portugal, Vietnam, Wales
5	Purple	Indonesia, Kazakhstan, Scotland, Thailand, **Turkey**
6	Light Blue	Denmark, Norway, **Sweden**
7	Orange	Russia, **South Korea**
8	Brown	Cyprus, **Netherlands**
9	Pink	Singapore, **Taiwan**
10	Light Red	**Israel**

As a result of the analysis of 14,680 articles, the total number of organizations was 5379. Due to the high number of organizations, the minimum number of citations to be included per organization was limited to 50. With this limitation, 94 organizations were included in the final analysis. According to Fig. 4.4, these organizations formed three clusters among themselves. The colors of these clusters are red, green and blue.

The analysis results showed that the papers were published in 318 different journals. Due to the high number of Journals, the minimum number documents of in a journal was limited to 50. With this limitation, 62 journals (out of 318 journals) were included in the analysis. According to Fig. 4.5, these journals formed seven clusters among themselves. The colors of these clusters are blue, red, green, dark yellow, purple, light blue and pink.

Table 4.3 shows the top 20 journals in which the SER papers were published. While the International Journal of Science Education has the highest number of papers (1228 documents) published, the most cited journal is the Journal of Research in Science Teaching (37,613 citations). However, Journal of Research in Science Teaching, Science Education and Computers & Education are the three journals that are on top in terms of the number of citations per paper as seen in Table 4.3.

As a result of the analysis of 14,680 articles, the total number of authors was 28,900. Due to the high number of Journals, the minimum number of documents in a journal was limited to 20. With this limitation, 23 authors were included out of 28,900 authors. According to Fig. 4.6, these authors formed five clusters among themselves.

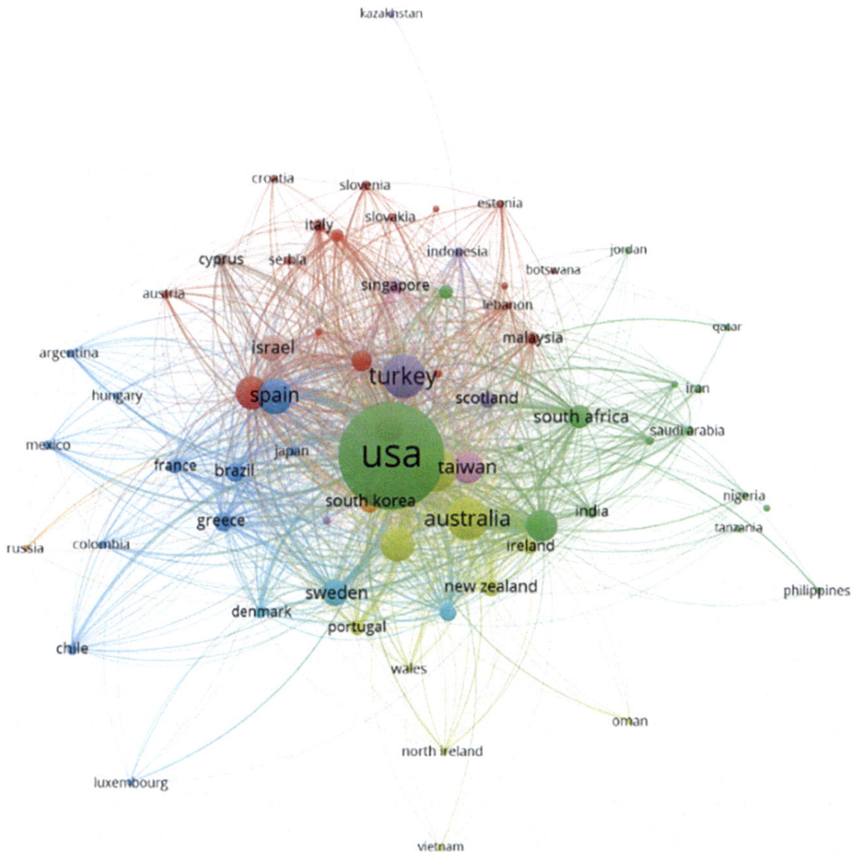

Fig. 4.3 Citations: Countries Network Visualization (2000–2022)*. *In the network visualizations, the size of the spheres indicates the number of documents, and the links between the spheres indicate the strength of the relationship

Table 4.4 shows the most productive authors in SER literature. Among them Tsai, C.C. has the highest number of documents published followed by Erduran, S. and Eilks, I. Table 4.4 shows the top 20 authors with the highest number of citations per document. Among them, Lederman, N. is the first and he is followed by Zeidler, D. and Taber, K.S. It is also visible from Tables 4.4 and 4.5 that 10 of these top 20 highly productive and highly cited authors are the same authors.

In Fig. 4.7, STEM education, professional development and professional development were the trends author keywords. Computational thinking and computer science education were also emerging trends in keywords. When the keywords in the SER articles in Fig. 4.8 were analysed, it was seen that teacher education, high school education and conceptual change keywords were common in the 2000s. while they were replaced by keywords such as nature of science, professional development, curriculum, STEM education and COVID-19 in recent years.

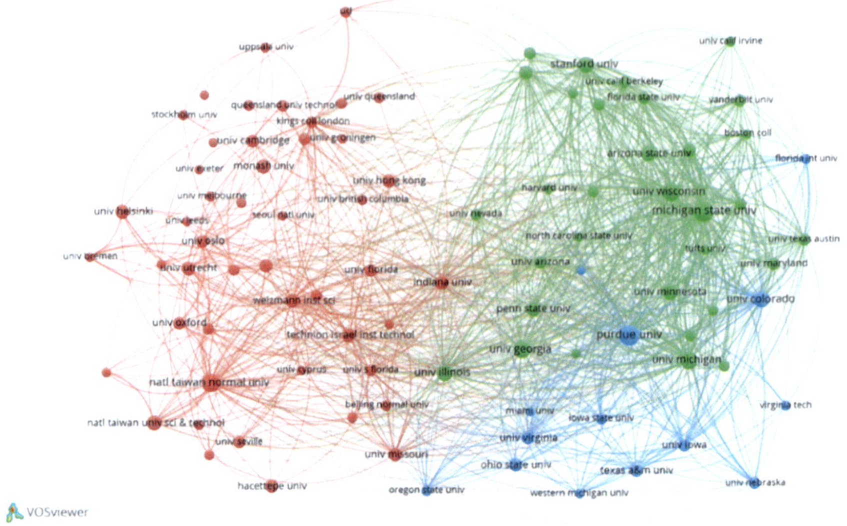

Fig. 4.4 Citations: Organizations Network Visualization (2000–2022)

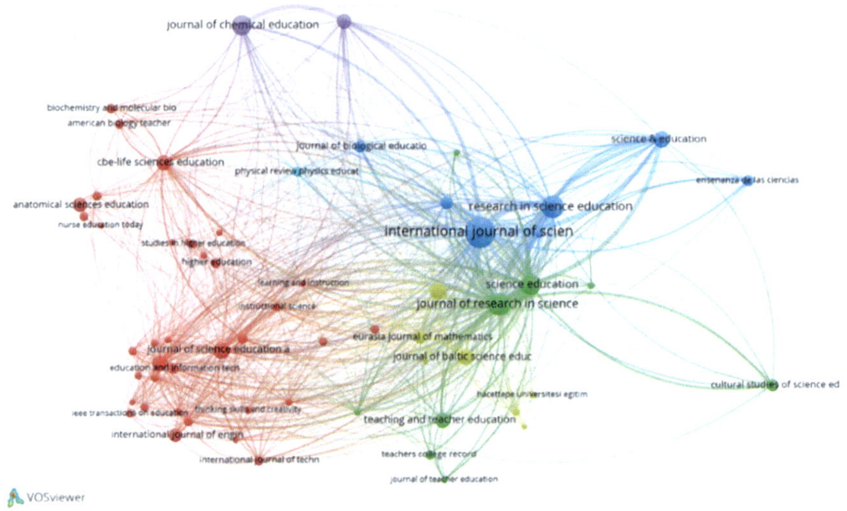

Fig. 4.5 Citations: Journals Network Visualization (2000–2022)

Table 4.3 Citations: Journals (2000–2022)

Journals	Number of papers	Citations per paper
International Journal of Science Education	1228	27
Journal of Research in Science Teaching	689	**55**
Research in Science Education	656	22
Science Education	561	**55**
Journal of Chemical Education	527	15
Science & Education	374	12
Journal of Science Education and Technology	369	20
International Journal of Science and Mathematics Education	366	13
Computers & Education	333	**55**
Journal of Baltic Science Education	322	5
CBE-Life Science Education	307	29
Teaching and Teacher Education	299	43
Chemistry Education Research and Practice	292	18
Anatomical Sciences of Education	285	26
International Journal of Engineering Education	247	7
Research in Science and Technological Education	210	7
Journal of Biological Education	206	11
Cultural Studies of Science Education	199	5
Eurasia Journal of Math. Sci. and Tech. Education	190	10
Ensenanza De Las Ciencias	163	7

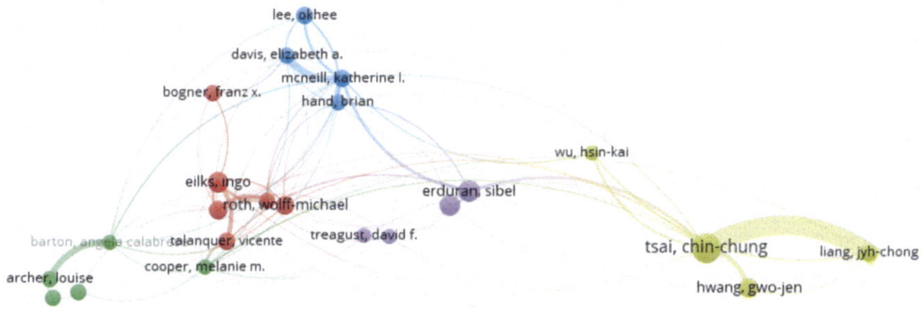

Fig. 4.6 Citations: Authors Network Visualization

Table 4.4 Highly Productive Top 20 Authors (2000–2022)

Authors	Number of Documents	Citations per Document
Tsai, CC.	79	33
Erduran, S.	44	30
Eilks, I.	41	29
Garcia-Carmona, A.	38	10
Hwang, G.J.	34	50
Roth, W.M.	34	32
Lee, O.	34	49
Wang, J.	32	10
Liang, J.C.	30	22
Sadler, T.D.	30	61
Talanquer, V.	27	36
McNeill, K.	26	65
Hand, B.	25	41
Calik, M.	25	26
Bogner, F.X.	25	15
Archer, L.	24	45
van Driel, J.H.	24	35
Davis, E.A.	24	57
Borrego, M.	24	55
Krajcik, J.	23	96

Table 4.5 Highly Cited Top 20 Authors (2000–2022)

Authors	Number of Documents	Citations per Document
Lederman, N.	5	293
Zeidler, D.	5	252
Taber, K.S.	15	234
Osborne, J.	18	199
Windschitl, M.	13	132
Abd-el-khalick, F.	18	124
Hofstein, A.	18	114
Penuel, W.R.	17	103
Krajcik, J.	23	96
Barton, A.C.	20	83
McNeill, K.	26	65
Sadler, T.D.	30	61
Davis, E.A.	24	57
Borrego, M.	24	55
Hwang, G.J.	34	50
Lee, O.	34	49
Hand, B.	25	41
van Driel, J.H.	24	35
Tsai, CC.	79	33
Erduran, S.	44	30

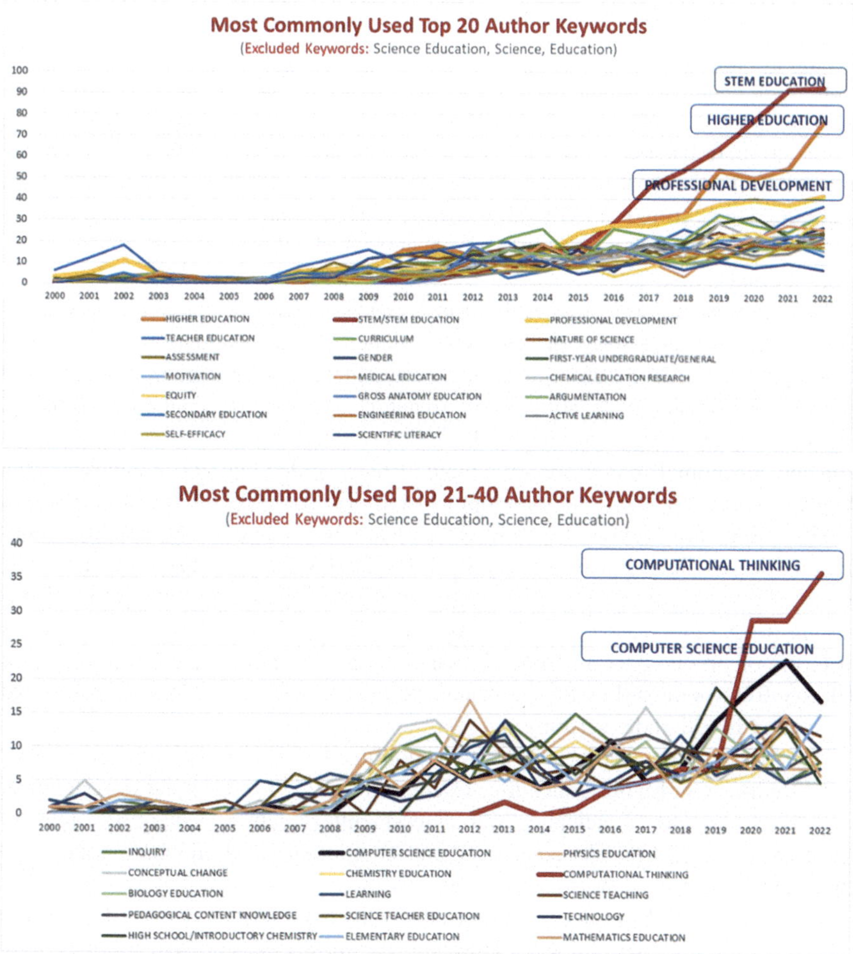

Fig. 4.7 Commonly Used Author Keywords (2000–2022)

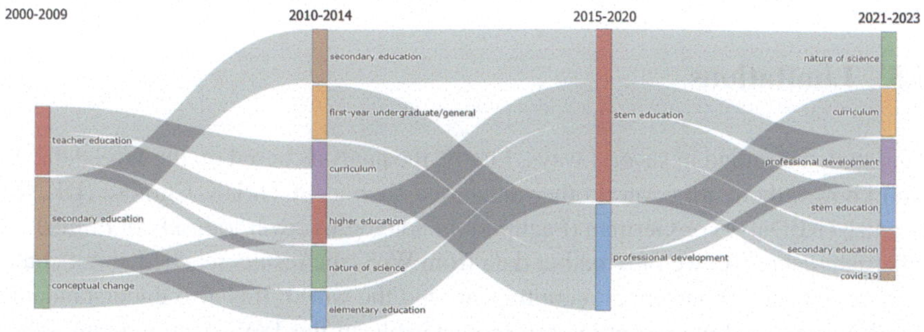

Fig. 4.8 Thematic Evaluation (Rstudio-Bibliometrix®)

4.4 Conclusions and Discussion

SER is an established discipline based on Fensham's (2004) criteria (structural, intra-research & outcome) and SER publications are increasing systematically. However, there has been a decline since 2020 perhaps due to COVID-19 outbreak which interfered education and research around the world. The highest number SER literature comes from the USA (40%), while Europe contributes (30%) and the rest of the World's contribution is 30%. WoS employed an expansion policy to cover more journals since 2004.

Our findings are in parallel with Tosun's (2024) study results that the most commonly studied topics were the nature of science (NOS), evaluation, professional development, STEM/STEM education, science literacy, argumentation, gender and conceptual change. SER studies are dynamic. Trends are changing regularly. Figure 4.8 shows that beginning of the 2000s teacher education, secondary education and conceptual change were the main themes, while later STEM education, the nature of science, professional development and COVID-19 became the main themes.

USA, England, Türkiye, Australia & Spain are the top five most productive countries. However, Belgium, Israel, Scotland, England & Netherlands are listed in the top five in the highest average citation per paper. Türkiye's contribution to SER literature (6.75%) increased gradually. Although Türkiye is in the top three countries in terms of productivity in SER, the citations received by these papers are low. Türkiye follows the policy of «publish or perish» which facilitates the quantity but the quality is still under scrutiny.

According to the author's keywords, dominant areas of research are STEM/STEM education, higher education, teacher education, professional development, curriculum, nature of science, and assessment/evaluation. Emerging areas of research are STEM/STEM education, and computational thinking/computer science education.

The majority of the papers are published by the main journals in the SER such as IJSE, JRST, RISE, SE and JCE. However, the most cited top five journals were *Review of Educational Research, Educational Researcher, Learning and Instruction, American Educational Research Journal, Computers & Education.*

4.5 Limitations

This study is limited in several ways. Firstly, the analysis based on the WoS data. As well known, there are several other databased such as SCOPUS, Chinese (DKNI), Spanish (REDalyc) or German (Fachportal Pädagogik) cover journals in English or other languages. As we only gather data from WoS, the results reached about countries do not fully represent all countries around the world. It is also understandable that English speaking countries are more visible in the WoS data as many other countries value publishing in national languages to address the practitioners and teacher education communities. Apart from the language, the number of papers is closely related with the country populations too. Therefore, it is recommended to read the results by considering the populations and size of the countries too. Taking the language and the population of the countries into consideration, it is clear that some countries would be considered even more productive than others in SER.

Bibliometric analysis is strong in terms of quantitative measurements and evaluations, however, it provides limited information about the content and the quality of the studies. Therefore, these results should be evaluated with caution. Data downloaded from the databases has missing data as well as raw data. It requires data cleaning. Impacts are based on citations. It is known that many papers published in journals may get very low citations or even not get any. Some popular authors or papers may significantly change the quantitative indicators. On the other hand, journal impact factors and citations are also biased by several factors (i.e. Online first publishing, authors' collaborations, author's popularity etc.). Nevertheless, bibliometric analysis is found a useful visualization of research trends in any particular research area or topic to identify the changes.

References

Aria, M., & Cuccurullo, C. (2017). Bibliometrix: An R-tool for comprehensive science mapping analysis. *Journal of Informetrics, 11*(4), 959–975. https://doi.org/10.1016/j.joi.2017.08.007

Bilgiç, H. G., & Tuzun, H. (2020). Issues and challenges with web-based distance education programs in Turkish higher education institutes. *Turkish Online Journal of Distance Education, 21*(1), 143–164. https://doi.org/10.17718/tojde.690385

Chang, Y. H., Chang, C. Y., & Tseng, Y. H. (2010). Trends of science education research: An automatic content analysis. *Journal of Science Education and Technology, 19*, 315–331. https://doi.org/10.1007/s10956-009-9202-2

Dahncke, H., Duit, R., Gilbert, J., Östman, L., Psillos, D., & Pushkin, D. B. (2001). Science education versus science in the academy: Questions – discussion – perspectives. In H. Behrendt, H. Dahncke, R. Duit, W. Graber, M. Komorek, A. Kross, & P. Reiska (Eds.), *Research in science education – Past, present and future* (pp. 43–48). Kluwer Academic Publishers.

DeBoer, G. E. (1991). *A history of ideas in science education: Implications for practice*. Teachers College Press.

Dewey, J. (1910). Science as subject-matter and as method. *Science, 31*(787), 121–127.

Donthu, K., Mukherjee, P., & Lim, W. M. (2021). How to conduct a bibliometric analysis: An overview and guidelines? *Journal of Business Research, 133*, 285–296. https://doi.org/10.1016/j.jbusres.2021.04.070

Fensham, P. J. (2004). *Defining an identity: The evolution of science education as a field of research*. Kluwer Academic.

Jenkins, E. W. (2001). Research in science education in Europe: Retrospect and prospect. In H. Behrendt, H. Dahncke, R. Duit, W. Gräber, M. Komorek, A. Kross, & P. Reiska (Eds.), *Research in science education – Past, present and future* (pp. 17–26). Kluwer Academic Publishers.

Keeves, J. P. (1998). Methods and processes in research in science education. In B. J. Fraser & K. G. Tobin (Eds.), *International handbook of science education* (pp. 1127–1153). Kluwer Academic Publishers.

Kempa, R. F. (1991). Research in chemical education: Past-present and future. In R. Maskill (Ed.), *Research in assessment* (pp. 6–31). Royal Society of Chemistry.

Osborne, J. (2007). In praise of armchair science education. *E-NARST News, 50*(2), 8–11. Retrieved January 16, 2010 from https://narst.org/news/e-narstnews_july2007.pdf

Öztürk, O. (2019). A logistic regression analysis of factors affecting enrolment decisions of prospective students of distance education programs in Anadolu University. *Turkish Online Journal of Distance Education, 20*(1), 145–160. https://doi.org/10.17718/tojde.522459

Pinto, M., Fernández-Pascual, R., Caballero-Mariscal, D., Sales, D., Guerrero, D., & Uribe, A. (2019). Scientific production on mobile information literacy in higher education: A bibliometric analysis (2006–2017). *Scientometrics, 120*(1), 57–85. https://doi.org/10.1007/s11192-019-03115-x

Sozbilir, M., Kutu, H., & Yasar, M. D. (2012). 14. Science education research in Turkey: A content analysis of selected features of published papers. In D. Jorde & J. Dillon (Eds.), *Science education research and practice in Europe: Retrospective and prospective* (pp. 341–374). Sense Publishers.

Tosun, C. (2024). Analysis of the last 40 years of science education research via bibliometric methods. *Science & Education, 33*(2), 451–480. https://doi.org/10.1007/s11191-022-00400-9

Mustafa Sözbilir holds a PhD in educational studies from the University of York, UK. Currently, he is a professor at Atatürk University in Türkiye. His primary research area is science and chemistry education. His research interests include, teaching and learning in thermodynamics at tertiary level, studies on implementation of active learning techniques in science/chemistry education, trends in science/chemistry education research, and teaching science/chemistry to visually impaired students, higher education studies.

Dilek Teke holds a PhD in chemistry education from Atatürk University, Türkiye. Currently, she is an assistant professor at the same university. Her primary research area is chemistry education. Her research interest includes integration of technology into chemistry education, use of AR and AI in chemistry teaching and learning, teaching chemistry to visually impaired students.

Chapter 5
Identity-based Research in Science Education: Past, Present, and Possible Futures

Emily MacLeod and Lucy Avraamidou

5.1 The Story (So Far) of ESERA's Science Identities SIG

The field of identity-based research in science education has grown significantly in a relatively short time. As Carlone (2022) puts it; 'identity, about two decades ago, quickly catapulted to one of [science education]'s hot topics' (p. 3). In their recent methodological review, Danielsson et al. (2023) found that the first papers examining science identities were published as recently as 1998, and that since then this research area has rapidly expanded. Before mapping the findings of this body of literature, we want to recognise and celebrate the role of the ESERA community in the growth of this field.

Launched in a crowded room at the ESERA 2017 conference in Dublin, Ireland, ESERA's Science Identities Special Interest Group (SIG), SIG 5, is one of the organisation's newest but biggest and fastest-growing communities. Having noticed the need for a structural setting for this growing community at the 2015 conference in Helsinki, Henriette Tolstrup Holmegaard (University of Copenhagen), who co-founded the SIG along with Louise Archer (University College London), wanted to create a forum for science identity researchers to discuss and develop methodologies, theories, and results. The intention of the SIG was threefold; 1) to improve the visibility of this new research field, 2) to create a network for researchers, particularly Early Career Researchers, within the field, and 3) to provide a forum to discuss upcoming challenges in science identities research.

E. MacLeod (✉)
Institute of Education, University College London, London, UK
e-mail: e.macleod@ucl.ac.uk

L. Avraamidou
Centre for Learning and Teaching, University of Groningen, Groningen, The Netherlands
e-mail: l.avraamidou@rug.nl

G. Cakmakci, M. F. Tasar (eds.), *Connecting Science Education with Cultural Heritage*, Contributions from Science Education Research 15,
https://doi.org/10.1007/978-3-031-90944-3_5

Since its launch, ESERA's SIG5 has become an active core of the international science identities community and provides a friendly professional network for the increasing number of researchers whose work includes examining science identities in different settings. SIG5 was the first of ESERA's Special Interest Groups to include Early Career Researchers in their leadership, and it holds regular online events to engage its growing community; including writing clubs, reading groups, book launches, and webinars focussing on innovative methods and theoretical tools. In addition, SIG5 members have participated in invited symposia at four ESERA conferences, and have contributed to an edited volume entitled *Science Identities: Theory, Method and Research* (Archer & Holmegaard, 2022). We recognise the strength of this inclusive community and wish to credit SIG5's past and present leadership, and members, in their work to support the growth of the young field of science identities; with regard to both its people and its evidence.

5.2 What Are 'Science Identities' and Why Are They Important?

Increasingly the concept of 'identity', i.e., how people identify and who, or what, people identify as, is a hotly debated topic in media, literature, and modern society. But what do we mean by identities in relation to science education, and why do we and others research them?

For too long science has been dominated by the same groups. Historically, particularly in the Western world, those who are most likely to feel "sciencey", pursue science, and succeed in science are more likely to identify as male, White, and come from privileged social backgrounds. Put briefly, in response to this inequitable patterning of who 'does' science, identity-based research in science education seeks to understand how our differing and multiple identities intersect with our experiences of science, and what practices might improve these experiences; especially for those marginalised from science. In this way, identity-based research in science aims to work towards making science more equitable.

But what do we mean by 'science'? In this chapter, we approach the definition of science in its broadest sense. We acknowledge that people's attitudes towards, interests in, and experiences of science differ within, between, and beyond the disciplines of biology, chemistry and physics. We also know through our own work and the work of colleagues that science identity development is thus differently shaped and informed in different science subjects (e.g., Archer et al., 2020; Avraamidou, 2020; Günter et al., 2021; Madsen & Malm, 2022). As this chapter is intended as an introduction to the conceptual understandings and emerging evidence surrounding science identities, however, we approach the topic of identity-based research in science education as a whole rather than focussing specifically on, for example, biology (see e.g., Günter et al., 2022; Wong et al., 2022), chemistry (see e.g., Archer et al., 2022; Nielsen & Madsen, 2023), or physics (see e.g., Gonsalves, 2020; Wang & Hazari, 2018).

There are many different approaches to conceptualising identity (Côté, 2006), including within science education (Shanahan, 2009). Some researchers use the lens of identity to focus on the experiences of science students, whilst others consider the identity development of science teachers or, to a lesser extent, those working in science. As will be highlighted in more depth later in this chapter, there is no one (accepted) way to understand or study science identities.

What most applications of the lens of identity in science education do have in common is their acknowledgement of both social (re)production and individual agency upon one's science identity development. In other words, science identity researchers recognise that whether and why people see themselves as (and become) "sciencey" is informed by both imposed structural norms and expectations within society, as well as people's own ability to shape themselves and the world around them. In order to better understand this dual approach taken to science identity it is first necessary to briefly outline the different approaches, or stances, which background this interpretation.

The first aspect of the modern approach to (science) identity, the structural stance, has tended to position individuals as puppets-like figures with predetermined or prescribed options and possibilities. The degree of predeterminism in structural applications and understandings of identity is, however, a matter for debate. For example, although Bourdieu often took a structural approach to identity in his landmark contributions to the sociology of education (e.g., 1977), this notion that people have little or no freedom in their life, or in their educational choices, has been strongly criticised (e.g., Nash, 1990). Alas, Sewell (1992) referred to structure not as predeterminism, but as 'the tendency of patterns of relations to be reproduced even when actors engaging in the relations are unaware of the patterns or do not desire their reproduction' (p. 3). Such an interpretation of science identity thus pays close attention to the norms and expectations imposed on individuals within and in relation to science (Shanahan, 2009).

The second, agentic, approach to identity understands that individuals can make completely free choices (Shilling, 1992). This perspective sees the individual as in total control of their thoughts and actions, and thus able to shape the world around them at their free will. In contrast to a structural stance, this agentic interpretation thus understands science identity as entirely, or fully, constructed by the individual according to their own personal choices.

Perhaps unsurprisingly, these two contrasting approaches to identity inform a historic structure/agency debate about the degree to which people can, and do, have their identities formed by the world around them or whether people construct their own identities (e.g., Sewell, 1992; Shilling, 1992; Wilmott, 1999). Today, informed by a growing body of evidence—some of which will be discussed later in this chapter—most researchers (both in science education and beyond) take a middle-ground in their interpretations of identity; one that views structure and agency not as binary opposites but as interrelated and interdependent. As such, structure and agency are often viewed 'as a dialectic where neither exists without the other and one is constantly created and recreated by the other' (Shanahan, 2009, p.46). In the words of Roth and Tobin (2007), who were some of the first to articulate the construct of identity in relation to science, a person appears to have a core identity which undergoes developments that are articulated in autobiographical narrations of the self, yet 'from one setting to the next, our identities, as revealed by our transactions with others, change' (p. 1). In this way, although the term 'identity' can sometimes be understood as an individual's trait(s) or defining feature(s), the lens of identity as applied in science education broadly refers to who we choose to be(come) within science, as well as who we are framed to be(come) by different people and in different contexts within science. As articulated by Carlone and Johnson (2007), who are often credited with developing the first conceptual model of science identity;

> Identity is not just something an individual feels; it is not even what an individual does, although both feelings and actions are components of identity. A science identity is accessible when, as a result of an individual's competence and performance, she is recognized by meaningful others, people whose acceptance of her matters to her, as a science person. (p. 1192)

Our science identities, or the extent to which we 'are' or 'feel' "sciencey", are thus not simply built or defined by what some may call our internal or 'natural' interest or skill in science; but they are co-constructed by the people, places and spaces around us. This is where the concept of recognition becomes central to science identities research. According to Gee (2000), a leading identity theorist; 'when any human being acts and interacts in a given context, others recognize that person as acting and interacting as a certain "kind of person" or even as several different "kinds" at once' (p. 99). For example, we may be seen as a smart person, a feminist, or indeed as a "sciencey" person because our actions and interactions align with how these 'kinds of people' are expected to look or act. Therefore, how we are (not) recognised as talented in or suited to science by scientific experts, science teachers, our friends and families, and other social communities informs the extent to which we develop an identification with science (e.g., Dou et al., 2019).

Importantly, who and what are likely to be recognised as "sciencey" depends upon multiple factors and is not neutral. Empirical evidence has demonstrated over and over again that minorities in science, particularly women and people of colour, are less likely to be seen as 'fitting' within science (e.g., Carlone & Johnson, 2007; Hughes et al., 2021). Science identities are thus informed and shaped by individuals' existing intersectional identities (Feser & Haak, 2022), and draw upon social structures and expected norms (Rushton & Reiss, 2021). In this way, past societal representations of what it means to be, look, or act "sciencey" can cloud who and what is accepted or welcomed within science in the future. To give an example; if men dominate science (as they do), then it may be inadvertently assumed that men are better scientists, meaning that women may struggle to be recognised as viable scientists. Put another way, recognition can enforce power structures and further marginalise those who are already in vulnerable positions (Oliver, 2015). In this way, science identities are political, and the lens of identity enables us to explore how issues related to social, cultural, and political issues support, or hinder, science participation. It allows us to 'stop asking questions about what is wrong with girls (and other students) and start asking what is wrong with science and how that can be changed' (Shanahan, 2009, p. 44). Using identity, we can challenge the taken-for-granted image of science; of who is or can be a scientist, and what it means to become, or un-become a science person. Identity affords us a lens to observe and address forms of oppression and exclusion in science and science-related subjects including, but not limited to, power, sexism, racism, classism, homophobia, transphobia, antisemitism and Islamophobia (Avraamidou, 2020).

In examining the nature of science identities, evidence points to the fact that science identities are not fixed. Science identities are complex and multifaceted processes that are always ongoing (e.g., Archer et al., 2020; Avraamidou, 2014; MacLeod, 2023), rather than 'set' or concrete facets of our personality. Identities can be considered not as 'things' that we have or are, but as 'work' that we conduct or negotiate in order to become who we want to be (e.g., Watson, 2008). This is because 'the "kind of person" one is recognized as "being," at a given time and place, can change from moment to moment in the interaction, can change from context to context, and, of course, can be ambiguous or unstable' (Gee, 2000, p. 99). For this reason, we concur with what Carlone (2022) helpfully terms the 'verbing' of identity (e.g., referring to identity 'work', 'performance' or 'enactment'), rather than treating identities as designated structural constraints. By understanding science identities as processes that we do, author, develop, and construct—rather than have—we acknowledge the ongoing agentic negotiations which make up who we are and can be within science, as well as the spaces around us which shape them. As Holland, Lachicotte, Skinner and Cain helpfully articulate in their book *Identity and Agency in Cultural Worlds* (1998), our identities, including within science, are 'improvised—in the flow of activity with specific social situations—from the cultural resources at hand' (p. 4).

Using this framework and applying the construct of identity to the field of science education leads us to ask research questions such as: What is the relationship between science identity work and STEM study and career choices?; What contexts influence preservice teachers' identity development?; How do young women author science identities in out-of-school contexts?; How do girls of colour develop STEM identities?; How do different kinds of experiences shape biology students' identity production? It is these questions, and questions like these, that ESERA's Science Identities Special Interest Group has been engaging with in recent years.

5.3 What Do We Know from Existing Identity-based Research in Science Education?

Danielsson et al. (2023) found that, since 1998, close to 200 papers have been published on the topic of science identities. To provide a brief introduction to this developing research field, we here share some of what we consider to be the key learnings from the six known review studies into science identities research. Due to word limitations, this introduction is not designed as a systematic or in-depth review, but a brief state-of-the-field assessment and presentation of significant findings using the existing empirical research into science identities.

Research into science identities has consistently found that although formal schooling can provide young learners with the potential to develop science identities, the traditional norms typically associated with science in formal education settings can restrict some young people from positioning themselves, and/or being positioned by others, as "sciencey". For example, the first two known review studies into science identities research were conducted by Shanahan (2009) and Lee (2012), both of whom considered some of the first articles focussing on young people's science identities; including leading studies by Brickhouse and Potter (2001), Calabrese Barton (1998), and Mensah (2008). Specifically, Shanahan (2009) concluded that the extant research indicated that existing structures within classrooms can prevent or discourage students from participating in science, whilst Lee (2012) highlighted evidence that informal science learning environments can provide more positive support for science identity development than formal schooling. Indeed, why and how informal learning environments offer new identity development opportunities for youth continues to be a key focus of science identities research (e.g., Rahm et al., 2022).

According to Shanahan (2009) this finding that some young people's science identity development is restricted by, or as a result of, the norms which are often ingrained in formal education settings was itself a consequence of research which focused upon individuals' personal interactions with/in science, rather than the settings within which these interactions take place. As a result, Shanahan (2009) argued that the field still lacked a detailed understanding of whether and how the social and educational settings within which science identities develop impact students' science identity development;

> There is yet to be broad attention to the actions of students, teachers, parents and administrators that create and make real these norms. We understand that these norms exist but have not thoroughly explored how and why they emerge from the communication and interactions of school science. (p.61)

In response to the shift in science identity studies that followed, the third (Avraamidou, 2014), fourth (Rushton & Reiss, 2021), and fifth (Feser & Haak, 2022) known reviews into science identities research all looked specifically at identity development amongst science teachers. As such, Avraamidou (2014) was the first to define the construct of science teacher identity; through reviewing 29 studies published in international science education peer-review journals over a period of 12 years (2001–2013), Avraamidou (2014) found that the concept of identity had been used as a lens to explore how science teachers see themselves and are seen by others. It was concluded that science teacher identity is 1) socially constructed, 2) dynamic and fluid, and 3) multifaceted, consisting of interrelated sub-identities (Avraamidou, 2014). Likewise, in applying a social identity approach (see Haslam et al., 2011) to the now 79 publications on science teacher identity published between 2000 and 2019, Rushton and Reiss (2021) concluded that science teachers' identities are rooted in both their subject identity and their science community identity. In this way, Rushton and Reiss (2021) argue, the importance of social connectedness and integration to science teacher identity development cannot be underestimated. Feser and Haak (2022) used their meta review of 24 teacher identity review studies to summarise the key features of science teacher identity as; one's characteristics (e.g., how teachers see themselves), one's professional knowledge and abilities (e.g., content knowledge, and pedagogical content knowledge), one's social characteristics (e.g., a sense of belonging, or not, in the school's social environment), as well as wider institutional cultures (e.g., what it means to be a 'science teacher' or a 'good science teacher' within an institution). The authors also highlighted the recurring finding that the identity of science teachers significantly differs from the identity of teachers teaching other subjects (Feser & Haak, 2022). As such, it follows that we need an identity approach specific to science teacher education, research, and practice; rather than a broad perspective applicable to all subject areas. A notable conclusion in these reviews refers to the gaps in the knowledge base of science identities research. Both Avraamidou (2014) and Rushton and Reiss (2021) identified the dominance of small-scale and one-off studies in the field, whilst Feser and Haak (2022) noted the lack of consensus or definition of what is meant by the term 'science teacher identity'.

To perhaps best understand the state of, and current gaps in, the field of identity-based research in science education we turn to the methodological review conducted by Danielsson et al. (2023). By specifically reviewing the methods used by science identities researchers, this sixth and most recent review quantifies some of the trends first identified in earlier reviews. For example, Danielsson et al. (2023) found that 55% of the 198 science identities studies published between 1998 and 2018 used solely qualitative methods (such as observations and/or interviews), whilst only 17% used solely quantitative methods (such as surveys). The authors argue that although small-scale studies contribute detailed and nuanced examinations of science identity development, the lack of consideration of how such research can be scaled-up means that the application of learnings from existing studies remains a challenge (Danielsson et al., 2023). Danielsson et al. (2023) also found that English-language scholars, particularly those based in the US and UK, dominated the field. In addition, Danielsson et al. (2023) noted a lack of consideration of how intersectional identities beyond gender, ethnicity, and social class interact with science identity development. As a result, they call for more consideration of how different axes of power might intersect within the domain of science.

One of the most striking findings from the review by Danielsson et al. (2023) is that, despite the finding that this field has grown significantly in recent years, what we mean by 'science identities' is taken for granted and under-conceptualised. According to the authors;

> Such taken-for-grantedness or under-conceptualisation assumes that identity essentially is a phenomenon that [can] be captured or measured rather than a perspective that, firstly, can be conceptualised in different ways (as, for example, performative or socially constructed), and, secondly, used to understand events, (for example, inclusionary/exclusionary practices

> or aspirations) rather than necessarily being the focus of the investigation in itself. (Danielsson et al., 2023, pp. 713-714)

To this end we begin to ask; (how) can we progress this growing field further without clear definitions or framing of what science identities are, do, or mean? Whilst all science identity researchers note the complex and multifaceted nature of science identity construction (as a process which encompasses, among other things, the personal, social, and collective aspects of how we see ourselves and are seen by others), without a consensus on what we study the field of identity based research in science education remains vulnerable.

The extant empirical research shows that different factors at multiple levels influence science identity development. Individuals' science identities are enacted at the individual level (which refers to personal factors), the institutional level (which refers to schools or universities), the societal level (which refers to influences that we experience as members of different social communities), and the political level (which refers to the socio-political realities of specific contexts). It is precisely these socio-political realities, directly rooted in goals related to social change and justice, that research capitalising on science identity promises to explore, yet in certain areas has so far fallen short.

5.4 Exploring Gaps in the Knowledge Base and Looking towards Recommendations for Future Science Identities Research

Despite its usefulness in framing research explorations related to power, inclusion, and belonging, we note that there remain gaps in our collective understandings and applications of identity-based research in science education. In this penultimate chapter section, we consider the gaps in the knowledge base of science identity research, and use these to propose recommendations for future science identities research. The following is not intended as an exhaustive list of gaps or recommendations, nor a critique of previous research; but as a discussion to help promote the advancement of our field.

First, and perhaps most importantly, the construct of 'science identities' still lacks an operational definition. Whilst we agree that no single lens can fully explain all aspects of science identities (Carlone, 2022), the under-conceptualisation of this construct leads to what Danielsson et al. (2023) describe as 'black spots' in our understanding. As Sfard and Prusak (2005) argued more than 15 years ago, then, 'in the absence of a definition, the reader is led to believe that identity is one of those self-evident notions that, whether reflectively or instinctively, arise from one's first-hand, un-mediated experience' (p. 15). On the contrary, the existing research has demonstrated the multitude of factors that can, and do, contribute to an individual's science identity development. We therefore join Danielsson et al. (2023) in calling for increased consensus or agreement as to the notions or concepts that are included within our definition of science identities.

A second limitation of identity-based research today is connected to the question of where this research is located and, consequently, who conducts identity research in science education. As has already been highlighted, at the time of writing our work is situated mostly in English-speaking and Eurocentric contexts (Danielsson et al., 2023). This practice provides us with a fragmented and limited understanding of science identity work, especially about how different socio-political spaces and contexts might shape science identity. Our second recommendation is therefore for more research into science identity development to be conducted in contexts beyond the US and the UK.

Third, one significant methodological limitation of current research is the dominance of small-scale qualitative studies which, according to analyses by Danielsson et al. (2023), make up over half of current science identity research. Whilst such approaches have enabled us to develop in-depth understandings of how individuals enact their science identities these 'micro', often ethnographic, studies limit our current understandings to specific settings and prevent recommendations from being scaled-up. The few large-scale studies that do exist (e.g., Archer et al., 2023; Hazari et al., 2017) evidence the great potential for such studies to generate knowledge transfer and inform policy and practice improvements. We thus call for more large-scale and quantitative research into the enactment and progression of science identities; so that findings might be understood and applied more widely.

Fourth, multiple theoretical gaps in the realm of science identities research could, if addressed, significantly advance our knowledge of how and why some people are more (or less) likely to identify with/in science. For example, the social axes of gender, race, and, to a lesser degree, social class, have dominated existing research about who identifies with science (Danielsson et al., 2023). Whilst there are exceptions to this narrow approach (e.g., Avraamidou, 2022), how other aspects of our intersectional identities (such as our religion, sexuality, disability) remain under-considered. Whilst existing research has evidenced that science is not neutral and reproduces systems of oppression; how this is enacted is currently not well understood. As we have argued elsewhere (e.g., Avraamidou, 2020), intersectionality calls for an emphasis on the systemic power dynamics that arise as multiple dimensions of social difference interact across individual, institutional, cultural, and societal spheres of influence in relation to our participation in science. We therefore recommend that researchers engage more thoroughly with intersectionality as it can provide us with a lens to re-conceptualise science identity in conjunction, and not in isolation, with other identities; and to examine how these identities intersect with each other and either support or hinder the process of becoming a science person.

Our final two recommendations develop from our knowledge of identity understandings and research in fields beyond science education research. Initially, we wish to recognise that science identity work is not detached from emotions. Whilst affective identity lenses tend to interpret identity as a constant process of becoming informed by both power and agency in a way that mirrors applications of science identities, the application of affective lenses in science identities research is lacking.

This is despite acknowledgment that our emotions—positive, negative, or anywhere in between—might serve as inclusionary or exclusionary resources for becoming a 'certain type' of a person, including in science (Zembylas, 2018). We thus recommend that researchers turn their attention to exploring questions related to how emotions entangle with identity authoring and identity development. For example; how does marginalisation impact people's emotional trajectories as they attempt to form science identities? Whose emotions are (over or under) recognised and valued in specific science places?

Likewise, many existing approaches to understanding science identity work take a deconstructionist approach, meaning that they do not consider material aspects of identity performance. In other words, our current focus on the spoken dimensions of identity as represented through qualitative fieldwork may not account for the relationship between discursive practices and material phenomena that can exist in (science) identity work (e.g., Barad, 2006). Specifically, the final theoretical limitation of current identity-based research in science education that we wish to raise is that it has suffered from the dualism between the self and the body. Whilst we know that identity work more broadly can be viewed as an embodied event by recognising that our physical body and its actions can shape our mental experiences (e.g., Sullivan, 2018), perhaps especially so for young people (e.g., Budgeon, 2003), identity-based research in science education has conceptualised the mind and body relationship as binary and has tended to privilege the mind over the body. The danger in this approach, we argue, is that we treat bodies as a-political, or neutral, instead of as political projects. Finally, then, to explore the ways in which individuals make meaning out of their bodily experiences with/in science we call for researchers to use an embodiment lens when exploring science identities.

5.5 Looking Forward: Three Invitations

In this chapter we have briefly introduced and explored the past and present state of identity-based research in science education. To end the chapter, we wish to welcome others to extend and expand our understandings of science identities through studies carried out in under-served contexts and with underrepresented communities. To encourage the myriad possible futures for our community and research we end by extending three invitations. These invitations are not intended as an exhaustive list of one-way instructions from us to you. Instead, we acknowledge our privileged positionalities within this research field and use this platform to pose these invitations as thought-provoking challenges that can, we hope, prompt the development and expansion of the field of science identities.

1. First, we invite researchers and collectives with diverse identities to engage with science identity, because their own identities matter and can help shape this research field.

2. Second, we invite engagement with science identity research in more diverse geo-sociopolitical contexts and various disciplines.
3. Third and finally, we welcome engagement with research published in different languages to gain a truly international understanding of the field of science identity research.

With these invitations, we aspire to build knowledge and connections that can help to progress current understandings of science identities, and in turn, support a more inclusive world of science for all.

References

Archer, L., & Holmegaard, H. T. (Eds.). (2022). *Science identities: Theory, method and research.* Springer. https://doi.org/10.1007/978-3-031-17642-5

Archer, L., MacLeod, E., & Moote, J. (2020). Going, going, gone: A feminist Bourdieusian analysis of young women's trajectories in, through and out of physics, age 10-19. In A. J. Gonsalves & A. Danielsson (Eds.), *Physics education and gender: Identity as an analytic lens for research.* Springer. https://doi.org/10.1007/978-3-030-41933-2_2

Archer, L., Francis, B., Moote, J., Watson, E., Henderson, M., Holmegaard, H., & MacLeod, E. (2022). Reasons for not/choosing chemistry: Why Advanced Level chemistry students in England do/not pursue chemistry undergraduate degrees. *Journal of Research in Science Teaching, 60*(5), 978–1013. https://doi.org/10.1002/tea.21822

Archer, L., DeWitt, J., Godec, S., Henderson, M., Holmegaard, H., Liu, Q., MacLeod, E., Mendick, H., Moote, J., & Watson, E. (2023). *ASPIRES3 main report: Young people's STEM trajectories, age 10–22.* UCL Institute of Education. https://discovery.ucl.ac.uk/id/eprint/10181968/1/ASPIRES3%20Main%20Report.pdf

Avraamidou, L. (2014). Studying science teacher identity: Current insights and future research directions. *Studies in Science Education, 50*(2), 145–179. https://doi.org/10.1080/03057267.2014.937171

Avraamidou, L. (2020). Science identity as a landscape of becoming: Rethinking recognition and emotions through an intersectionality lens. *Cultural Studies of Science Education, 15*(2), 323–345. https://doi.org/10.1007/s11422-019-09954-7

Avraamidou, L. (2022). Identities in/out of physics and the politics of recognition. *Journal of Research in Science Teaching, 59*(1), 58–94. https://doi.org/10.1002/tea.21721

Barad, K. (2006). *Meeting the universe halfway: Quantum physics and the entanglement of matter and meaning.* Duke University Press. https://doi.org/10.1215/9780822388128

Bourdieu, P. (1977). Cultural reproduction and social reproduction. In J. Karabel & A. H. Halsey (Eds.), *Power and ideology in education.* Open University Press.

Brickhouse, N. W., & Potter, J. T. (2001). Young women's scientific identity formation in an urban context. *Journal of Research in Science Teaching, 38*(8), 965–980. https://doi.org/10.1002/tea.1041

Budgeon, S. (2003). Identity as an embodied event. *Body & Society, 9*(1), 35–55. https://doi.org/10.1177/1357034x030091003

Calabrese Barton, A. (1998). Teaching science with homeless children: Pedagogy, representation and identity. *Journal of Research in Science Teaching, 35*(4), 379–394. https://doi.org/10.1002/(SICI)1098-2736(199804)35:4<379::AID-TEA8>3.0.CO;2-N

Carlone, H. (2022). Understanding and contextualizing the field of science identity research. In H. Tolstrup Holmegaard & L. Archer (Eds.), *Science identities: Theory, method and research.* Springer. https://doi.org/10.1007/978-3-031-17642-5

Carlone, H. B., & Johnson, A. (2007). Understanding the science experiences of successful women of color: Science identity as an analytic lens. *Journal of Research in Science Teaching, 44*(8), 1187–1218. https://doi.org/10.1002/tea.20237

Côté, J. (2006). Identity studies: How close are we to developing a social science of identity?—An appraisal of the field. *Identity, 6*(1), 3–25. https://doi.org/10.1207/s1532706xid0601_2

Danielsson, A. T., King, H., Godec, S., & Nyström, A.-S. (2023). The identity turn in science education research: A critical review of methodologies in a consolidating field. *Cultural Studies of Science Education, 18*, 695–754. https://doi.org/10.1007/s11422-022-10130-7

Dou, R., Hazari, Z., Dabney, K., Sonnert, G., & Sadler, P. (2019). Early informal STEM experiences and STEM identity: The importance of talking science. *Science Education, 103*(3), 623–637. https://doi.org/10.1002/sce.21499

Feser, M. S., & Haak, I. (2022). Key features of teacher identity: A systematic meta-review study with special focus on teachers of science or science-related subjects. *Studies in Science Education, 1-34*. https://doi.org/10.1080/03057267.2022.2108644

Gee, J. P. (2000). Identity as an analytic lens for research in education. *Review of Research in Education, 25*(1), 99–125. https://doi.org/10.3102/0091732x025001099

Gonsalves, A. J. (2020). Constructing inside-ness to physics: How matter comes to matter in physics identity work. *Cultural Studies of Science Education, 15*(4), 911–921. https://doi.org/10.1007/s11422-020-09999-z

Günter, K. P., Gullberg, A., & Ahnesjö, I. (2021). "Quite ironic that even I became a natural scientist": Students' imagined identity trajectories in the figured world of higher education biology. *Science Education, 105*(5), 837–854. https://doi.org/10.1002/sce.21673

Günter, K. P., Ahnesjö, I., & Gullberg, A. (2022). "I try to encourage my students to think, read, and talk science" intelligible identities in university teachers' figured worlds of higher education biology. *Journal of Research in Science Teaching*. https://doi.org/10.1002/tea.21829

Haslam, S. A., Reicher, S. D., & Platow, M. J. (2011). *The new psychology of leadership: Identity, influence and power*. Psychology Press.

Hazari, Z., Brewe, E., Goertzen, R. M., & Hodapp, T. (2017). The importance of high school physics teachers for female students' physics identity and persistence. *The Physics Teacher, 55*, 96–99. https://doi.org/10.1119/1.4974122

Holland, D., Lachicotte, W., Skinner, D., & Cain, C. (1998). *Identity and Agency in Cultural Worlds*. Harvard University Press. https://www.infoamerica.org/documentos_pdf/holland02.pdf

Hughes, R., Schellinger, J., & Roberts, K. (2021). The role of recognition in disciplinary identity for girls. *Journal of Research in Science Teaching, 58*(3), 420–455. https://doi.org/10.1002/tea.21665

Lee, Y.-J. (2012). Identity-based research in science education. In B. J. Fraser, K. Tobin, & C. J. McRobbie (Eds.), *Second international handbook of science education* (pp. 35–45). Springer. https://doi.org/10.1007/978-1-4020-9041-7_3

MacLeod, E. (2023). *The status and safety of teaching: A longitudinal study of why some young people in England become teachers, and why others do not* (Publication No. 10178871) [Doctoral Dissertation, IOE, UCL's Faculty of Education and Society]. UCL Discovery. https://discovery.ucl.ac.uk/id/eprint/10178871/

Madsen, L. M., & Malm, R. H. (2022). Doing geoscience: Negotiations of science identity among university students when learning in the field. In H. Tolstrup Holmegaard & L. Archer (Eds.), *Science identities: Theory, method and research*. Springer. https://doi.org/10.1007/978-3-031-17642-5

Mensah, F. M. (2008). Positional identity and science teacher professional development. *Journal of Research in Science Teaching, 45*(6), 684–710. https://doi.org/10.1002/tea.20258

Nash, R. (1990). Bourdieu on education and social and cultural reproduction. *British Journal of Sociology of Education, 11*(4), 431–447. https://doi.org/10.1080/0142569900110405

Nielsen, K. B., & Madsen, L. M. (2023). Choosing (not) to be a chemistry teacher: Students' negotiations of science identities at a research-intensive university. *European Educational Research Journal*. https://doi.org/10.1177/14749041231157919

Oliver, K. (2015). Witnessing, recognition, and response ethics. *Philosophy & Rhetoric, 48*(4), 473–493. https://doi.org/10.5325/philrhet.48.4.0473

Rahm, J., Gonsalves, A. J., & Lachaîne, A. (2022). Young women of color figuring science and identity within and beyond an afterschool science program. *Journal of the Learning Sciences, 31*(2), 199–236. https://doi.org/10.1080/10508406.2021.1977646

Roth, W.-M., & Tobin, K. (Eds.). (2007). *Science, learning, identity: Sociocultural and cultural-historical perspectives*. Sense Publishers. https://brill.com/display/title/37523

Rushton, E. A. C., & Reiss, M. J. (2021). Middle and high school science teacher identity considered through the lens of the social identity approach: A systematic review of the literature. *Studies in Science Education, 57*(2), 141–203. https://doi.org/10.1080/03057267.2020.1799621

Sewell, W. H. (1992). A theory of structure: Duality, agency, and transformation. *American Journal of Sociology, 98*(1), 1–29. https://doi.org/10.1086/229967

Sfard, A., & Prusak, A. (2005). Telling identities: In search of an analytic tool for investigating learning as a culturally shaped activity. *Educational Researcher, 34*(4), 14–22. https://doi.org/10.3102/0013189x034004014

Shanahan, M. C. (2009). Identity in science learning: Exploring the attention given to agency and structure in studies of identity. *Studies in Science Education, 45*(1), 43–64. https://doi.org/10.1080/03057260802681847

Shilling, C. (1992). Reconceptualising structure and agency in the sociology of education: Structuration theory and schooling. *British Journal of Sociology of Education, 13*(1), 69–87. https://www.jstor.org/stable/1392858

Sullivan, J. V. (2018). Learning and embodied cognition: A review and proposal. *Psychology Learning & Teaching, 17*(2), 128–143. https://doi.org/10.1177/1475725717752550

Wang, J., & Hazari, Z. (2018). Promoting high school students' physics identity through explicit and implicit recognition. *Physical Review Physics Education Research, 14*(2) 020111-020111-020111-020112. https://doi.org/10.1103/PhysRevPhysEducRes.14.020111

Watson, T. J. (2008). Managing identity: Identity work, personal predicaments and structural circumstances. *Organization, 15*(1), 121–143. https://doi.org/10.1177/1350508407084488

Wilmott, R. (1999). Structure, agency and the sociology of education: Rescuing analytical dualism. *British Journal of Sociology of Education, 20*(1), 5–21. https://www.jstor.org/stable/1393391

Wong, B., Chiu, Y.-L. T., Murray, Ó. M., Horsburgh, J., & Copsey-Blake, M. (2022). 'Biology is easy, physics is hard': Student perceptions of the ideal and the typical student across STEM higher education. *International Studies in Sociology of Education, 32*, 1–22. https://doi.org/10.1080/09620214.2022.2122532

Zembylas, M. (2018). Affect, race, and white discomfort in schooling: Decolonial strategies for 'pedagogies of discomfort'. *Ethics and Education, 13*(1), 86–104. https://doi.org/10.1080/17449642.2018.1428714

Emily MacLeod is an Honorary Senior Research Fellow at University College London, UK, and a Postdoctoral Scholar at McGill University, Canada. Her research interests include science identity studies, educational inequalities, young people's career aspirations and trajectories, and (science) teacher shortages.

Lucy Avraamidou is a Professor of Science Education and Director of the Center for Learning and Teaching at University of Groningen, the Netherlands. Her research is associated with theoretical and empirical explorations of what it means to widen and diversify STEM participation in school and out-of-school settings through the lens of intersectionality.

Chapter 6
Comparing Textual Data Analysis Methods in Science Education Research

Martina Caramaschi, Andrea Zanellati, and Olivia Levrini

6.1 Introduction

In Science Education Research (SER), being able to listen to students' voices is crucial to collecting information about the experience of science learning, students' difficulties, and ideas. This focus has become increasingly significant since the shift in SER during the 1960s and 1970s when the emphasis moved from theory verification and generalized outcomes to understanding phenomena in their situated and specific contexts (Szyjka, 2012). As interests have moved toward the study of classroom experiences and related programs, qualitative text-based data have become a fundamental resource for educational inquiry, providing critical access to understanding learning processes.

In this chapter, we focus on specific qualitative textual data, daily collected and analysed in SER. Examples include oral interviews on certain topics made by a researcher that are transformed into text, written comments that are collected together with quantitative data through questionnaires, and students' written tasks or essays about specific scientific topics. Also, open answers written by students in a school test and textual contents of textbooks or educational material provided by the teacher are possible textual data for research.

These textual data in SER have two characteristic traits when they are collected in datasets and employed for research purposes:

M. Caramaschi (✉) · O. Levrini
Department of Physics and Astronomy "A. Righi", University of Bologna, Bologna, Italy
e-mail: martina.caramaschi2@unibo.it; olivia.levrini2@unibo.it

A. Zanellati
Department of Computer Science and Engineering, University of Bologna, Bologna, Italy
e-mail: andrea.zanellati2@unibo.it

G. Cakmakci, M. F. Tasar (eds.), *Connecting Science Education with Cultural Heritage*, Contributions from Science Education Research 15,
https://doi.org/10.1007/978-3-031-90944-3_6

- The amount of available data (e.g., number of texts or their length) for a study case may be moderately high. It is not enough to qualify as *big data*, but substantial enough to render traditional qualitative analysis time-consuming.
- Textual data is collected in a specific context or with a specific task. Therefore, text can display topics, explicitly or implicitly connected with the context. Sentences are moderately articulated in terms of semantics, and specific language is used.

These features, together with the complex nature of the language itself, are what challenge researchers who deal with textual data in SER. For example, Stanley and Robertson (2024) reviewed current literature about qualitative SER, with a focus on methodological aspects. From their organization and analysis of the literature, it emerges that high volumes of textual data are collected in case studies, reaching dozens of audio recording hours and hundreds of interview minutes. Another example of a methodological challenge in textual data analysis arises from the richness and specificity of language. Decisions such as how to reduce the dimensionality of the text (Mariegaard et al., 2022) or what coding method to use (Wulff et al., 2023) become non-trivial, as these significantly affect the interpretability, reliability and reproducibility of results.

In these conditions, methodological questions are likely to arise. For example, you might wonder "how to properly analyze textual data, considering the nuances of context-specific language?", "what volume of textual data can be realistically analysed?", "what method allows for a profitable analysis that answers the research question?". These methodological issues do not concern the technicalities of analytical methods, but rather their connection with epistemology. They push reflection on the type of knowledge textual data offers, how analytical methods process them, and the insights that can be extracted.

In this article, we develop a methodological reflection on different approaches for textual data analysis in the context of SER, with a focus on medium-sized, context-specific datasets. First, we will show the features of the traditional qualitative approach to textual data in SER, implemented in a bottom-up *Thematic Analysis*. Then, space will be given to present main changes occurring in the society, that opened the way to another approach to text-based data analysis: the computational data-driven approaches. As for these approaches, we will focus on *Latent Semantic Analysis*. We will show a case study in which the same dataset has been analysed in parallel by the two methods. Specifically, students' essays were investigated in the context of a Future-Oriented Science Education project in 2021. The chapter concludes with a comparison of the two methods. Epistemological and methodological aspects will be investigated, to critically evaluate where the methods enclose limits and potentialities.

6.2 Different Approaches for Textual Data Analysis

6.2.1 The Tradition: Qualitative Approaches

Traditionally, textual data in SER have been analysed using qualitative approaches. Although not always explicitly stated in articles, qualitative approaches have been historically generated within a specific paradigm, the interpretivist one (Treagust & Won, 2023). A paradigm is based on ontological, epistemological, and axiological assumptions, about the reality under study, its knowability, the type of research problems, the values and philosophical influences, the research design and methods, and the role and practice of researchers. The paradigm plays a crucial role because it "*guide[s] the researchers throughout the empirical research process: from setting the research purpose to selecting data-collection methods to analyzing the data to reporting the findings*" (Treagust & Won, 2023, p.3). In the opening chapter of a recent methodological handbook, Treagust and Won argue that the interpretivist paradigm stems from relativist ontology and constructivist epistemology (Treagust & Won, 2023), where reality is not considered objective per se, but knowable through the meaning attributed by individuals, admitting different interpretations. The research within this paradigm foresees an exploratory design. The usual research topics addressed with this approach are lived experiences of teachers and students. The quality of the interpretation process is ensured by a prolonged meaningful interaction in the context and repeated re-analysis and reflections on data. Additional elements that guarantee the process validity are the researcher's skills, sensitivity, and integrity, member checking, audit trail, thick descriptions, peer debriefing, triangulation, and negative case analysis (Anfara et al., 2002).

6.2.1.1 Thematic Analysis

Thematic Analysis (TA) falls under the qualitative analytical methods and is widely used in many fields such as psychology and social sciences (Braun & Clarke, 2006). It is not considered as a single method but as an umbrella term for a set of approaches, that are all interested in analysing qualitative data, and developing themes from them. The themes, also called *patterns of shared meaning*, consist of terms able to capture something important about the data regarding the research question (Braun & Clarke, 2006).

The general TA procedure for identifying themes does not consist of a predefined set of steps to follow as in the case of a recipe. However, Braun and Clarke have defined a series of phases, that may guide the analyser, which are *familiarising yourself with your data*, in which the analyser immerses in the data, reading multiple times the data available. Then, the *coding* part, in which short labels into the data are found, captures features that might be significant for contributing to the research questions. Afterwards, it is time to start *generating initial themes* to group some codes and develop initial themes that are improved and revised in the next phase called *developing and reviewing themes*. Finally, it is necessary to make sense of the results, *refining, defining and naming themes* and *writing up* the entire process, describing how it has gone.

TA is characterised by its theory-free nature, making it suitable for the theoretical approach preferred by the researcher. Moreover, it is considered flexible because every thematic analysis is characterised by different research questions (i.e. the objective to which the analysis wants to contribute), analysers, and datasets (Braun & Clarke, 2019). To take advantage of this flexibility, keeping a rigorous TA, there are four key elements to enact:

1. *Defining clear guiding questions:* A dataset can be full of potential themes, but "the chosen themes" are the ones that respond to an interest related to RQs.
2. *Putting into play the analyser's sensitivity:* There are no recipes for identifying themes, analyser intuition, and sensitivity are fundamental drivers.
3. *Following practices to make the process trustworthy:* Clear, transparent, and reliable procedures, such as triangulation, are essential to gain credible themes.
4. *Keeping the focus on dataset:* The themes should be meaningful within the dataset and not necessarily counted as the most occurred or frequent.

6.2.2 Epochal Changes Impacting the Methodological Debate

For about twenty years, we have witnessed an epochal shift toward data collection and analysis impacting nearly all research fields. These changes are driven by several factors, primarily the so-called *Big Data* revolution. Data has become available in large amounts and variety, and technological infrastructures and tools allowed us to collect and store it much faster than in the past and in digital format (Klašnja-Milićević et al., 2017). Another process to be considered is the process of *datafication* that is reshaping contemporary society, including SER (Jarke & Breiter, 2019). Education has evolved to incorporate digital platforms, online environments, and technological tools alongside traditional methods, further contributing to this data-driven transformation.

These technological developments now facilitate the collection of extensive data at various levels (individual, classroom, and national) about potentially all processes of teaching, learning, and school management (Jarke & Breiter, 2019). Fischer et al. (2020) identified three big data typologies coming from different educational contexts:

(a) The *microlevel* (e.g., clickstream data), which includes "*fine-grained interaction data with seconds between actions that can capture individual data from potentially millions of learners*".
(b) The *mesolevel* (e.g., text data), which includes "*computerized student writing artefacts systematically collected during writing activities in a variety of learning environments ranging from course assignments to online discussion forum participation*".

(c) The *macrolevel* (e.g., institutional data), which includes "*student demographic and admission data, campus services data, schedules of classes and course enrollment data*".

Finally, the increasing use of data science techniques in various socio-economic fields, including education, has made data-intensive methods an interesting solution for processing large datasets computationally and rapidly (Fischer et al., 2020). In recent years, two important communities were born at the intersection of data science and education: Educational Data Mining, which held its first conference in 2008, and Learning Analytics, in 2011 (Baker & Siemens, 2014). These novelties have supported the evolution of a new research area, Education Data Science, where statistical and computational techniques are employed to explore educational questions and phenomena (McFarland et al., 2021). Specifically for textual data, the field of Educational Text Mining has been established as a research context for analysing large-scale textual data within educational settings.

6.2.3 The Computational Data-Driven Approaches

The digital shift and technological revolution and in turn, the wide rise of Education Data Science have contributed to open new possibilities for textual data analysis in SER (Fesler et al., 2019). Alongside a more traditional qualitative paradigm, an alternative analytical approach emerged. The computational data-driven approach "*is supposed to be data-driven, strongly inductive, and relatively theory-independent*", as well as largely automated (Pietsch, 2015). It assumes, more or less implicitly, the idea that data in big amounts is a necessary and sufficient container of knowledge and that implicit patterns in the data can be highlighted only by the analysis of the data itself. Thus, a specific theoretical design at the basis of data analysis is no longer needed to orient the extraction of valuable information from data.

Belonging to this approach, we refer to all those methods falling under Artificial Intelligence (AI) techniques for textual data analysis, such as Natural Language Processing (NLP) techniques, Machine Learning (ML), and Network Analysis which are precisely those "*new and often non-traditional*" methods included under the umbrella of Education Data Science (McFarland et al., 2021, p. 2). In social sciences research, the approach consists of data collection and pre-processing, followed by data analysis, in which traditional statistics methods and the most widely employed ML methods are usually combined. Finally, the validation phase is needed to ensure correctness and accuracy (Zhang, et al., 2020).

In education research, articles containing keywords such as ML, AI, Learning Analytics, Data Science, NLP, and Network Analysis have massively grown since 2010 (McFarland et al., 2021). Even in SER, interesting works adopted this kind of approach for dealing with textual data (e.g. Wulff, et al., 2023; Odden, et al., 2021; Bruun, et al., 2019). Although strengths started to be highlighted, there are also criticisms and increased awareness of the intrinsic limitations that characterize this approach (Pietsch, 2015). Therefore, in SER, where the debate is still at its beginning, the deepening of the paradigm behind these methods (i.e. ontological, epistemological, and methodological assumptions) is felt as an urge.

6.2.3.1 Latent Semantic Analysis

We briefly present the Latent Semantic Analysis (LSA), the computational data-driven method that we used for the case study. LSA (Landauer et al., 1998) is a text-mining technique in the family of Topic Modeling (Blei, 2011). This family consists of a group of methods for mathematically identifying latent topics in a large corpus of data, i.e. mapping the co-occurrence of words to retrieve the underlying semantic structure of the corpus. The technique applies an unsupervised learning approach and produces some clusters of words that can be used to define the topics that describe the semantic structure of the corpus. Formally, the topic is defined as "*be[ing] a distribution over a fixed vocabulary. For example, the genetics topic has words about genetics with high probability*" (Blei, 2011). In a given collection of documents, all documents share the same set of topics, but display them in varying proportions (Blei, 2011).

Typically, three main phases are carried out to apply the LSA technique. Firstly, the corpus of the documents is pre-processed with some common steps of text mining applications. Specifically, stopword removal, lemmatization, and special character cleaning are performed. This is made both to reduce the complexity of the language and to remove semantically unneeded words. Secondly, the LSA technique is applied as described by Dumais (2004), to embed words into numeric vectors and uncover latent semantic relationships between terms and documents. This process allows the identification of hidden topics by grouping terms that frequently co-occur in similar contexts. Finally, interpretation is attributed to the extracted topics, i.e. the label attribution to the list of words that describes each topic.

To conclude this section, we highlight key features that underscore the relevance of TA and LSA as good representatives of qualitative and data-driven computational methods, respectively, making them suitable for a comparison in the context of SER. Both methods are appropriate for exploring textual corpora, as both *theme* and *topic* represent semantic elements that are characteristic of the document. However, they differ in that the theme is considered a pattern of shared meaning among codes, while the topic is the interpretation of a group of words. Additionally, both methods enable the generation of themes/topics, whether human or computationally driven, tightly connected to the context. Their meaning is not absolute but interpreted within the framework of the specific context and research questions.

Regarding computational methods, LSA is particularly intriguing for a methodological comparison because it stands out as one of the simplest and most transparent technique within the topic modeling family. It embodies many core assumptions of computational data-driven approaches, such as complexity reduction (i.e., text is seen as its bag of words, thus no order nor grammar is kept).

For completeness, it is worth noting that other computational methods, such as Network Analysis (Kristanto & Padmi, 2020), are also suitable for thematic analysis in SER. However, we have chosen not to focus on it in this article due to the vast methodological literature on the topic. Notably, this method also presents similar methodological challenges.

6.3 Aims and Research Questions

Our goal is to investigate the added value of qualitative or data-intensive computational methods when used to analyse qualitative textual data, such as students' essays, interviews, and open answers, in SER. The features of these data (medium quantity and context specificity) create peculiar conditions that seem appropriate for both methodologies.

In particular, the study aims to unveil the main features and differences that lie at the level of methodological process (e.g. resources needed to develop the analysis, and the kind of approach to data), and on the epistemological level (e.g. quality of the knowledge emerging during the process of analysis and the results).

Therefore, our research questions are:

1. When analyzing a corpus of qualitative textual data that has features as described above (see "Introduction" section), how does the choice of a qualitative or a computational data-intensive method characterize the produced knowledge?
2. Moreover, what intrinsic features of each method enhance or limit the analysis?

6.4 Methodology

To answer the research questions, we articulated the study in the following steps. First, we designed the case study in which the textual dataset in SER was analysed by using two separate methods (the qualitative method and the data-driven computational method). Then, we developed an analysis grid able to re-examine the analytical process of the two methods and compare them through specific lenses:

- **Epistemological lens**: This examines how the method influences the quality of the knowledge produced, in terms of generalizability, theory-orientation, interpretability, scalability, granularity, originality, and applicability of results.
- **Methodological lens:** This explores the tradeoffs introduced by the method, including resource requirements, data handling approaches, and conditions for applicability.

Finally, the limits and strengths of the methods have been evaluated.

6.5 Case Study

We present a case study carried out in 2021, from which our methodological reflection started. At that time, there were conditions whereby the same corpus of textual data (i.e. students' essays collection) could be analysed by two methods, namely TA and LSA. In specific, we had chosen them for their suitability to the research purpose and researchers' confidence in performing them. This case is particularly interesting because we have acknowledged specific moments of "struggle" with both methods that revealed important features of the relationship between the dataset, analytic procedures and analysts. The final results show to what extent the two methods can bring value to research, albeit in a problematic and complementary way. Therefore, below we will describe the highlights of the two analyses emphasising where methodological challenges and epistemological criticalities have emerged.

The case study was situated in the context of a future-oriented science education project, called FEDORA, a project of the European Union's Horizon 2020 Research and Innovation program (fedora-project.eu). FEDORA's purpose was to develop research studies to guide and support young people in dealing with future visions through science education. The case study was performed by our local research group in Physics Education at the University of Bologna, aimed at exploring students' future perceptions (Barelli et al., 2022).

Specifically, the dataset[1] was composed of 223 students' essays written in Italian language. The essays collection was performed from 2018 to 2021 by the research group in Physics Education at the University of Bologna. Each student was asked to write an essay based on the same task. The students were asked to describe their future life in 2040, writing *(i)* where they imagine living; *(ii)* the kind of life they are leading; *(iii)* the types of problems that they, in their daily life, in their community of people and, in general, in their society are discussing and/or facing; *(iv)* the possibilities and new opportunities they can seize; *(v)* the technology, objects, the house, the city and the environment that surround them; *(vi)* the type of social life they lead. The context of data collection was the PLS project, which promotes extracurricular courses for high-school students on scientific topics. Examples of topics of these scientific courses are Climate Change, Simulation Computing, Artificial Intelligence, Seismic Risk, and Quantum Computing.

For the FEDORA project goals, it was important to explore the teenagers' image of the desired future. Thus, the research question (RQc) that oriented the analysis in this case study was: "Which topics are expressed by the students in the essay and how are they articulated?". Essentially, the focus was on exploring which topics students expressed and how the topics were articulated, to observe patterns in students' narratives about their imagination of the future. Consequently, the choice of using these two methods was oriented by the RQc, since they share the common goal of identifying recurring topics in the corpus.

[1] The dataset is open-access at the following link: http://amsacta.unibo.it/6964/.

6.5.1 First Method: Reflexive Thematic Analysis

The analysis of the students' essays is described in detail in Barelli's and colleagues' work (2022). Here we report the salient moments of the analytical process. Future students' essays were studied by using the "reflexive thematic analysis" (Braun & Clarke, 2019) which is one specific type of TA that considers the researcher's subjectivity and reflexivity as a central feature. We conducted a "bottom-up" analysis, trying to let information emerge from the data, with the final goal of recognising the emergent patterns. It is important to note that our research group was sensitive toward a critical issue that emerged in previous science education studies, called "polarization attitude". We expected that the attitude could emerge even in the students' narratives about the future.

After a careful reading of the first five essays, we (a group of nine researchers) individually started extracting the first codings and then collectively built the initial themes. Then four analysers continued the TA process, examining further 44 students' essays, moving back and forth between individual immersion and triangulation phases. In this first analytical phase, we identified nine main themes and several sub-themes, which were more specific expressions of how a certain theme manifested itself (see Fig. 1 in Barelli et al., 2022). We explicitly chose to force a bottom-up approach, staying close and faithful to the data until we realised and collectively agreed that the process did not lead to saturation. The process indeed resulted in a very detailed and gigantic map, with the risk of producing a copy of the data itself. Moreover, this categorisation of data was ineffective in capturing the theoretical idea of "polarization attitude".

This collective recognition moved us to change deliberately the approach, starting the second analytical phase. It was characterised by stepping back from data, which we had by then examined and known in depth, and shifting our focus to the phenomenon that was perceived within it (i.e. axial coding) (Vollstedt & Rezat, 2019). A new analytic lens was identified and used to go back on data with a top-down analysis. The analysis allowed us to identify in the students' essays two different modalities adopted to describe their future, called "complexification" and "polarization" attitudes. Moreover, this second analytical phase led to the development of an analysis grid (see Table 1 in Barelli et al., 2022) able to generate new knowledge as it was an effective instrument to see the two different attitudes that we could not capture within the initial themes.

6.5.2 Second Method: Latent Semantic Analysis

Students' future essays were analysed by following the LSA phases described in Sect. 6.2.3. Here, we describe the analytical process, focusing on the analyser's choices that were needed because of the peculiar context and data. The analysis was carried out using the Google Colaboratory Notebook environment, with Python libraries, such as pandas and scikit-learn.

During pre-processing, the analyser started "cleaning" and preparing the textual data, using "nltk" as a source for Italian stop words, and "treetaggerwrapper" library for lemmatization and tokenization. This produced a vocabulary of 1949 terms, later refined by augmenting the list of words to consider as stopwords (Italian stopwords were added to the "nltk" list). Furthermore, a custom dictionary for lemmatization was created by manually merging words with similar lemma, not considered by the previous library. The final vocabulary had 1543 terms. Lemmatization represented the first key decision, based on the assumption that the semantic meaning of essays was maintained by transforming words to their base form.

As for the LSA implementation, a key step was selecting the number of topics. We opted for using a mathematical criterion (Dumais, 2004), as the researcher did not have the means or a specific reason based on the data to establish it otherwise.

Lastly, the list of words that describe each topic needed to be interpreted. We started applying a purely bottom-up procedure, searching for a unifying semantic name, able to capture the meaning of the words present in each list. However, we decided to consider also contextual information that could offer significant lenses for more reliable attribution, considering a top-down knowledge injection approach. In analyzing students' perceptions of the future as expressed in the essays written during STEM education projects between 2018 and 2021, we decided to consider there were two events with potentially high social and global impact: Greta Thunberg environmental protest in the second semester of 2018, and the COVID-19 pandemic started in March 2020. Therefore, we organised the documents into three groups of years: 2018 (81 documents), 2019–2020 (64 documents), and 2021 (56 documents). Then we relied on the following steps to figure out the interpretation for each topic:

1. *Documents clustering:* the K-means algorithm is applied to determine clusters of documents in a latent space.
2. *Main topic estimation:* the estimation of most charactersing topics per each cluster and each year, through a voting mechanism.
3. *Documents overlapping estimation:* Documents overlapping between the clusters and the year groups were measured by using three common metrics: f1-score, intersection over union, and cosine similarity.
4. *Topics interpretation.* Assigning labels to topics by considering terms and extra information related to the year to which a topic is most linked.

In Fig. 6.1, we show the results of the analysis, consisting of the 5 topics identified, to which we attributed the most accurate interpretation possible (Fig. 6.1). Despite in the preprocessing phase, the corpus of textual data was filtered from words that appeared superfluous for the meaning to allow the LSA algorithm to focus only on students' meaningful expressions, the result obtained was still not easy to interpret and also not very informative.

We hypothesize that the medium size of the dataset and the text uniformization during lemmatization may have caused the loss of subtle nuances in the text, which were not subsequently identified by the algorithm. Therefore, lists of words often

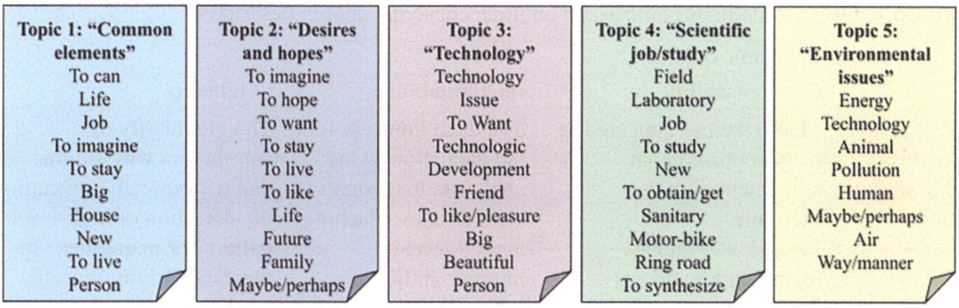

Fig. 6.1 List of most representative words (translated from Italian to English) per each topic found using the LSA method

appear to be too general and thus hard to interpret (e.g., topic 1). Otherwise, some lists of words appear to be too banal to fail to inform in a meaningful way (topic 2). Moreover, the researcher supposes that the words used by the students and their tendency to talk about their "desires" may be influenced by the request to describe their ideal future life. It is possible that the topic was influenced by the task, affecting the results.

6.6 Methods Comparison

In the following tables, the main results of the comparison of the two analytical processes are reported. In Table 6.1 the methods are compared according to the epistemic lens, named "epistemic quality". In Table 6.2 they are compared according to their methodological features.

Regarding the epistemic quality, we focused on the two methods' capacity to produce generalisable, interpretable, and original knowledge. By generality we mean the extent to which analytical results provide insights applicable at least to the whole corpus of data considered (similar to the "saturation" concept in Grounded Theory (Vollstedt & Rezat, 2019). For interpretability, we refer to generating representative and meaningful themes/topics from a set of data in a process that keeps attention to detail and a thorough understanding of context (Kristanto & Padmi, 2020, p. 178). For originality, we focused on the two methods' capacity to provide novel knowledge or at least new information.

As for method features, we compared the two methods in terms of resources needed in the analytical process, such as human resources, specific skills or abilities, and tools. Moreover, we analysed how researchers approached the data and the specific conditions under which the method was applied, shedding light on its strengths and limitations.

From an epistemic perspective, the main value of both methods lies in the interplay between the bottom-up and top-down analysis, where new knowledge is effectively generated. While this is not a novel observation, the close comparison of the

Table 6.1 Grid for methods comparison on three epistemic quality descriptors

	Epistemic quality		
	Generalisability	Interpretability	Originality
Thematic Analysis Bottom-up phase	Macro themes appeared to be generalisable, in agreement with the literature. Instead, sub-themes identification did not saturate, turning into a one-to-one description of the essays, without grasping a generalizable knowledge among all essays.	The deep immersion in the data allowed the analysers to recognise clear themes, reaching a good level of interpretability of macro and sub-themes.	The granularity of sub-themes was too fine and widespread, preventing the detection of high-level patterns or phenomena in the data. Original results were found when we detached from TA, reformulating the sub-themes as dichotomies.
Thematic Analysis top-down phase	The axial coding that led to the development of the analysis grid used to identify "complexified" and "polarized" attitudes, allowed us to grasp the general phenomenon emerging from data. Saturation was reached.	Themes and dichotomies that composed the analysis grid were highly interpretable thanks to a deep understanding of data combined with literature knowledge.	The analysis grid, thanks to internal coherence and operational capacity, leads to identifying patterns in data. Thus, it can be considered an instrument generator of knowledge.
Latent Semantic Analysis	The LSA model is built on the whole corpus of essays. Thus, every essay can be described as a combination of the discovered topics. New essays are likely to be described by the same topics.	Often a topic was described by generic words. Thus, meaning inferring was trivial. Moreover, an overlapping of semantics between the retrieved clusters of words hindered the interpretability of results.	The scarcity in results granularity and topic vagueness made them not very informative. Causes could be identified in the medium-sized corpus and reduction of word complexity, which limited the power of the model.

two methods, and the inner limitations of both bottom-up approaches, makes this epistemic value particularly evident. This interplay is the locus where the new Generative AI can provide further insights and contributions, questioning and exploring the nature of who or what contributes to the generation of knowledge.

The main result of our method comparison is the intrinsic complementarity of their strengths and limitations (e.g., the number of researchers required, the number of analysed texts, or the possibility to work with finer-grained data). However, overcoming the methodological limitations of each method is not achieved simply by applying both. Instead, it requires the thoughtful design of mixed-methods approach that properly integrate one method with some features of the other.

Table 6.2 Grid for methods comparison on three methods features.

	Methods features		
	Analytic resources	Approach to data	Applicability conditions
Thematic Analysis Bottom-up phase	The analytical process involved nine researchers. Several iterations of analysis, triangulation, and refinement were necessary.	Researchers' engagement with data, and sensitivity to capture student voices. Spiral iterative process: a few essays at a time were analyzed to develop the macro and sub-themes. Overall, 49 essays out of 223 were analysed.	When saturation is not reached, the method remains limited not only to the context but also to the subset of the dataset. Disproportion between the human resources available for the analysis and dataset dimension.
Thematic Analysis top-down phase	Theoretical insight is used to look at data, enabling the development of a new analytical lens. A second analytical phase is performed.	Researchers detach from data and then come back to look at them, through the sensitising idea that emerged from the insight.	Since at this point essays had already been read, and the analysis grid was able to synthesize the phenomenon, its application was fast.
Latent Semantic Analysis	The analytical process involved only one researcher, who had computational and science education expertise. It required a scrupulous and customized preprocessing phase.	Iterative refinement of topics, automatically retrieved, improving, in turn, the list of removed and aggregated words. All essays were analysed.	Scalable method to reuse in the future, if additional data will be collected. Issue given by a slight difference in the essay task over years. Borderline quantity of texts for a data-driven approach.

6.7 Conclusions

This study aims to contribute to the methodological debate that arose in this historical moment when epochal changes are occurring, brought by digitalisation and new analytical instruments such as artificial intelligence and data science. In particular text-based inquiry represents a challenge in SER due to the medium-sized volume, specificity, and contextuality of data. Analysing and comparing the analytical process of TA and LSA, in parallel used on the same textual data corpus, we critically observed how a qualitative or a data-driven computational method generates knowledge and manages data.

A contribution of this work is provided by the development of a table for methods analysis and comparison, to examine epistemological and methodological features. A limitation was that only two methods were analysed. However, the use of the table can be extended for analysing other methods.

We observed that topic or theme generation is particularly difficult when only a bottom-up analysis is carried out. In TA the risk is to not saturate, generating a more and more specific and extended description of the data, without creating general patterns of meaning. Instead LSA approach Struggled to recognise original patterns, likely due to the small dataset and the delicate preprocessing choices that reduced the text's complexity and nuanced meanings. However, the creation of less vague topics in LSA deserves further reflection. Certainly, the interplay with top-down knowledge to recognise meaningful clusters or to better interpret results seems necessary for both TA and LSA.

This work is part of a wider methodological reflection that is continuing, particularly focusing on three main aspects. We are investigating to what extent a mixed method approach (Treagust, & Won, 2023) in which AI-based technologies are integrated with qualitative research can help to overcome the limits of interpretability highlighted in the automatic data-driven method, and at the same time reinforce the generalizability of retrieved knowledge, mitigating the human-resources needed. Interesting mixed approaches come from Computational Grounded Theory and its very first application in SER (Tschisgale et al., 2023). Or from the use of Network Analysis for thematic analysis in SER (Mariegaard et al., 2022). Moreover, we are focusing on the importance of including external knowledge into ML pipelines for education, to enrich the model with expert and contextual knowledge. Thus, the concept of Informed Machine Learning has been reviewed and expanded in a recent work by Zanellati et al. (2024). Finally, we feel that the categories traditionally adopted to describe methods (i.e., qualitative, quantitative) struggle to grasp some aspects that characterize the AI-based and data science-based methods (Caramaschi et al., 2023). While AI is changing science nature (Erduran, 2023; Barelli et al., 2024) and its practices (Erduran & Levrini, 2024), it is also questioning foundational aspects of methods. Thus, a new metric for methods comparison is under development, to deepen how AI and data-driven approaches are changing the debate on methodological aspects.

References

Anfara, V. A., Brown, K. M., & Mangione, T. L. (2002). Qualitative analysis on stage: Making the research process more public. *Educational Researcher, 31*(7), 28–38. https://doi.org/10.3102/0013189X031007028

Baker, R., & Siemens, G. (2014). Educational data mining and learning analytics. In R. K. Sawyer (Ed.), *The Cambridge handbook of the learning sciences* (pp. 253–272. (Cambridge Handbooks in Psychology)). Cambridge University Press. https://doi.org/10.1017/CBO9781139519526.016

Barelli, E., Lodi, M., Branchetti, L., & Levrini, O. (2024). Epistemic insights as design principles for a teaching-learning module on artificial intelligence. *Science & Education*. https://doi.org/10.1007/s11191-024-00504-4

Barelli, E., Tasquier, G., Caramaschi, M., Satanassi, S., Fantini, P., Branchetti, L., & Levrini, O. (2022). Making sense of youth futures narratives: Recognition of emerging tensions in students' imagination of the future. *Frontiers in Education, 7*, 1–17. https://doi.org/10.3389/feduc.2022.911052

Blei, D. (2011). Introduction to probabilistic topic models. *Communications of the ACM, 55*, 77–84. https://doi.org/10.1145/2133806.2133826

Braun, V., & Clarke, V. (2006). Using thematic analysis in psychology. *Qualitative Research in Psychology, 3*(2), 77–101. https://doi.org/10.1191/1478088706qp063oa

Braun, V., & Clarke, V. (2019). Reflecting on reflexive thematic analysis. *Qualitative Research in Sport, Exercise and Health, 11(4)*, 589–597. https://doi.org/10.1080/2159676X.2019.1628806

Bruun, J., Lindahl, M., & Linder, C. (2019). Network analysis and qualitative discourse analysis of a classroom group discussion. *International Journal of Research and Method in Education, 42*(3), 317–339. https://doi.org/10.1080/1743727X.2018.1496414

Caramaschi, M., Fantini, P., & Levrini, O. (2023). Do the epistemological bases of qualitative and quantitative paradigms in STEM education still hold today? Oral contribution. *Measurement in STEM Education conference*. https://indico.unina.it/event/60/contributions/638/

Dumais, S. T. (2004). Latent semantic analysis. *Annual Review of Information Science and Technology (ARIST), 38*, 189–230.

Erduran, S. (2023). AI is transforming how science is done. Science education must reflect this change. *Science, 382*, 9788. https://doi.org/10.1126/science.adm9788

Erduran, S., & Levrini, O. (2024). The impact of artificial intelligence on scientific practices: An emergent area of research for science education. *International Journal of Science Education*. https://doi.org/10.1080/09500693.2024.2306604

Fesler, L., Dee, T., Baker, R., & Evans, B. (2019). Text as data methods for education research. *Journal of Research on Educational Effectiveness, 12(4)*, 707–727. https://doi.org/10.1080/19345747.2019.1634168

Fischer, C., Pardos, Z. A., Baker, R. S., et al. (2020). Mining big data in education: Affordances and challenges. *Review of Research in Education, 44*(1), 130–160.

Jarke, J., & Breiter, A. (2019). Editorial: The datafication of education. *Learning, Media and Technology, 44*(1), 1–6. https://doi.org/10.1080/17439884.2019.1573833

Klašnja-Milićević, A., Ivanovic, A., & Budimac, M. (2017). Data science in education: Big data and learning analytics. *Computer Applications in Engineering Education, 25*, 1068–1078. https://doi.org/10.1002/cae.21844

Kristanto, Y. D., & Padmi, R. S. (2020). Using network analysis for Rapid, Transparent, and Rigorous Thematic analysis: A case study of online distance learning. *Jurnal Penelitian dan Evaluasi Pendidikan, 24*(2), 177–189.

Landauer, T., Foltz, P. W., & Laham, D. (1998). Introduction to Latent semantic analysis. *Discourse Processes, 25*, 259–284. https://doi.org/10.1080/01638539809545028

Mariegaard, S., Seidelin, L. D., & Bruun, J. (2022). Identification of positions in literature using thematic network analysis: The case of early childhood inquiry-based science education. *International Journal of Research & Method in Education, 45*(5), 518–534.

McFarland, D. A., Khanna, S., Domingue, B. W., & Pardos, Z. A. (2021). Education data science: Past, present, future. *AERA Open, 7*. https://doi.org/10.1177/23328584211052055

Odden, T. O. B., Marin, A., & Rudolph, J. L. (2021). How has science education changed over the last 100 years? An analysis using natural language processing. *Science Education, 1–28*. https://doi.org/10.1002/sce.21623

Pietsch, W. (2015). Aspects of theory-ladenness in data-intensive science. *Philosophy of Science, 82*(5), 905–916. https://doi.org/10.1086/683328

Stanley, S. D., & Robertson, W. B. (2024). Qualitative research in science education: A literature review of current publications. *European Journal of Science and Mathematics Education, 12*(2), 175–199. https://doi.org/10.30935/scimath/14293

Szyjka, S. (2012). Understanding research paradigms: trends in education research. *Problems of Education in the 21st Century. 43*, 110-118. https://doi.org/10.33225/pec/12.43.110

Treagust, D. F., & Won, M. (2023). Paradigms in science education research. In N. G. Lederman, D. L. Zeidler, & J. S. Lederman (Eds.), *Handbook of research on science education* (Vol. III, 1st ed.). Routledge. https://doi.org/10.4324/9780367855758

Tschisgale, P., Wulff, P., & Kubsch, M. (2023). Integrating artificial intelligence-based methods into qualitative research in physics education research: A case for computational grounded

theory. *Physical Review Physics Education Research, 19*, 2. https://doi.org/10.1103/PhysRevPhysEducRes.19.020123
Vollstedt, M., & Rezat, S. (2019). An introduction to grounded theory with a special focus on axial coding and the coding paradigm. In G. Kaiser & N. Presmeg (Eds.), *Compendium for early career researchers in mathematics education. ICME-13 Monographs*. Springer. https://doi.org/10.1007/978-3-030-15636-7_4
Wulff, P., Westphal, A., Mientus, L., Nowak, A., & Borowski, A. (2023). Enhancing writing analytics in science education research with machine learning and natural language processing—Formative assessment of science and non-science preservice teachers' written reflections. *Frontiers in Education, 7*. https://doi.org/10.3389/feduc.2022.1061461
Zanellati, A., Mitri, D. D., Gabbrielli, M., & Levrini, O. (2024). Hybrid models for knowledge tracing: A systematic literature review. *IEEE Transactions on Learning Technologies, 17*, 1021–1036. https://doi.org/10.1109/tlt.2023.3348690
Zhang, J., Wang, W., Xia, F., Lin, Y., & Tong, H. (2020). Data-driven computational social science: A survey. *Big Data Research, 21*. https://doi.org/10.1016/j.bdr.2020.100145

Martina Caramaschi has an academic background in Physics and is currently a PhD student in Data Science and Computation, affiliated with the Department of Physics and Astronomy "A. Righi" at the University of Bologna, Italy. Her research focuses on the use of various methods for textual data analysis within the field of science education research. She is currently exploring how AI-based approaches are reshaping the methodological debate in science education, particularly the concept of mixed-methods.

Andrea Zanellati is a mathematics and physics teacher and researcher at the University of Bologna. He earned a degree in Mathematics with honors in 2011 and completed his PhD in Data Science and Computation in 2025. His research interests focus on the application of Data Science and Machine Learning methods and techniques in educational contexts.

Olivia Levrini is Full Professor in Physics Education and History of Physics at the Department of Physics and Astronomy "A. Righi" of the University of Bologna, Italy. She coordinated the European Projects I SEE, IDENTITIES, FEDORA and is now coordinating the EU Teacher Academy FEDORAS. Her current research work concerns future-oriented science education, interdisciplinarity in STEM education, cognition and conceptual change, identity and processes of appropriation, instruction design on contemporary STEAM themes (climate change, artificial intelligence, quantum computing). She served as Conference President for the 2019 ESERA conference (www.esera2019.org).

Part II
Teaching, Learning, and Student Engagement in Science Education

Chapter 7
Just Play It: Factors That Support Learning in Videos on Physics

Tanja Mutschler, Stefan Sorge, David Buschhüter, Christoph Kulgemeyer, and Andreas Borowski

7.1 Introduction

With the advent of the Covid-19 pandemic, schools were forced to shut down, leading to a rapid adoption of online learning methods. Yet, this transition exposed the unpreparedness of educational institutions. Students, teachers, and parents were all caught off guard by the abrupt transition to digital education. Positive effects of this shift towards digitally supported learning can also be a reduced teaching load for teachers and enhanced self-learning phases for students. Meta-analyses show that the use of digital media in the context of flipped classroom or Computer-Supported Collaborative Learning constellations has moderate effect sizes, especially in STEM teaching (Jeong et al., 2019; Wagner et al., 2021). To utilize the benefits of digital media in science classroom beyond the Covid-19 pandemic, high-quality materials are required.

T. Mutschler (✉) · A. Borowski
Department of Physics Education, University of Potsdam, Potsdam, Germany
e-mail: tanja.mutschler@uni-potsdam.de; andreas.borowski@uni-potsdam.de

S. Sorge
Department of Physics Education, IPN - Leibniz Institute for Science and Mathematics Education, Kiel, Germany
e-mail: sorge@leibniz-ipn.de

D. Buschhüter
Educational Technology Lab, German Research Center for Artificial Intelligence, Berlin, Germany
e-mail: david.buschhueter@dfki.de

C. Kulgemeyer
Department of Physics Education, University of Bremen, Bremen, Germany
e-mail: kulgemeyer@physik.uni-bremen.de

G. Cakmakci, M. F. Tasar (eds.), *Connecting Science Education with Cultural Heritage*, Contributions from Science Education Research 15,
https://doi.org/10.1007/978-3-031-90944-3_7

In this context, educational videos represent a promising option with a high degree of flexibility (Wagner et al., 2021). They enjoy a high level of acceptance among learners and are already widely used outside of formal lessons (Welbourne & Grant, 2016). A national study on media consumption of teenagers in Germany (mpfs, 2021) shows that up to 30% of all teenagers use YouTube regularly as a source of learning content. While the interest in digital science content can be considered a positive development, the videos themselves pose problems from a didactical perspective. Kulgemeyer and Peters (2016) were able to show that the freely accessible videos differ widely in terms of quality and content-related correctness. In some cases, explanations in the videos reinforce learners' misconceptions (Kulgemeyer & Wittwer, 2022). In addition, they are often not conform to the curriculum.

For videos to be used as high-quality material in the context of digitally supported teaching, they must be oriented towards the students' learning process. It is known from studies focusing on factors influencing students' learning success that the learners' motivation and interest as well as the (in-depth) structuring of the learning unit are highly predictive (Fischer et al., 2014; Seidel et al., 2005; Shulman, 1986). Based on a comprehensive literature analysis, Kulgemeyer (2018) has developed quality criteria (e.g. a level of language and mathematization appropriate to the target group and the adaptation to the learners' pre-knowledge) for explanatory videos that have a measurable effect on comprehension performance when tested subsequently. Regarding structuring, Kulgemeyer (2018) suggests the rule-example approach for learning factual knowledge. Yet, the optimal positioning of the example is subject to contradicting learning theories and competing findings (e.g., Fyfe et al., 2015; Kokkonen et al., 2022). The main aim of this study is, therefore, to further investigate the connection between learning success and structuring within educational videos on physics.

7.2 Theoretical Background

For the structuring of learning processes, there are numerous different approaches. With its strong focus on learning objectives, the Basic Model Theory (Oser & Baeriswyl, 2001) is recognised as a helpful planning tool for designing learning-effective instruction. For concept-building learning processes, this theory advocates approaching a new subject matter by starting with a concrete example which is gradually abstracted and subsequently lead to the general rule or concept. The Concreteness Fading Approach (Fyfe & Nathan, 2019; Fyfe et al., 2014), which originates from mathematics education, similarly follows this structural sequence with its three-step progression from concrete to abstract. Studies show that instruction based on both theories is certainly capable of generating significant learning gains, particularly for lower-performing students (Fyfe et al., 2015; Maurer, 2016; Zander, 2016). Both approaches also build on a broad basis of established learning theories (Bruner, 1966; Renkl et al., 2002; Smith & Medin, 1981), yet stand in contradiction to the rule-example structuring that is advocated by deductive approaches and proposed for educational videos. To decide if positive effects are due to the positioning of the example before the abstract representation of the concept, direct comparisons between the two structural sequences (example-rule vs. rule-example) are required. Although McNeil and Fyfe (2012) and Fyfe et al. (2015) examine different structural sequences in their interventions, they do not directly compare both structural approaches with each other. Besides this, there are only a few other empirical studies that investigate the effect of the positioning of the (concrete) example on students' learning success. Table 7.1 gives an overview.

According to the body of research, none of the studies can show the superiority of the procedure in which a topic is first elaborated using a specific example and then transferred to the general rule for learning declarative knowledge. However, the relevance of a generalized representation both on its own and in a position preceded by the example is confirmed several times (Kaminski et al., 2008; Seidel et al., 2013). The only exception here is the study by Jaakkola and Veermans (2018), which came to the same and in some cases better results with the exclusive use of examples. In total, it seems that the effect of the structural sequence on learning success depends on individual learner prerequisites such as conceptual level, pre-knowledge, and an age-dependent understanding of the topic (Fyfe et al., 2015; Jaakkola & Veermans, 2018; Tomlinson & Hunt, 1971). For physics and physics-related areas, no effect of the structural sequence has been found so far (Kokkonen et al., 2022; Van Hout & Mettes, 1976). Regarding the learners' abilities to use their knowledge in new contexts, however, the importance of an abstract representation or a generalized rule is also emphasized here. Thematically, all physics-related studies are in the context of electrodynamics limiting a transfer to other contexts. Overall, this shows that the study situation regarding the influence of the structural sequence on learning success is inconclusive.

We therefore pose the following research questions:

Q1: What influence does the structural sequence (example-rule | rule-example) have on the learning success within an educational video on physics?

H1: Due to the findings of the physics-related research, we assume that the order of the structural sequence within an educational video on physics will not affect the learning success of the learners if the effect is considered across all participants. However, we expect group differences in terms of pre-knowledge:

Q1.1: To what extent does the learners' pre-knowledge influence the effect between the structural sequence and learning success when learning the law of interaction?

H1.1: Pre-knowledge is assumed to be a moderating variable. Learners with little pre-knowledge are more likely to benefit from the rule-example structure sequence as the rule provides them with a structure in the sense of an advance organizer that enables them to comprehend the concrete example and focus on its relevant features (Tomlinson & Hunt, 1971).

Since, in addition to learning success, student perception of the structure is also an important requirement for the educational videos, the extent to which this differs between the different structural sequences will be examined:

Table 7.1 Body of research on the positioning of the example

Study / Authors	Domain	Sample	Constructs	Structural sequence more effective for student learning[a]				
				E-R	=	R-E	E	R
Kokkonen et al. (2022)	Physics Education	N = 70 high school students	Concreteness fading vs. concreteness introduction		+			
Jaakkola and Veermans (2018)	Science Education	N = 127 primary school students	Concreteness fading vs. constantly concreteness		+		(+)[b]	
Johnson et al. (2014)	Engineering Education	N = 343 high school students	Abstr.-abstr. vs. abstr.-context. vs. context.-abstr. vs. context.-context.		(+)[c]	+[c]		
Van Hout and Mettes (1976)	Engineering Design	N = 235 university students	Ex.-rule-ex. vs. rule-ex. x teacher-centered vs. learner centered		+			
Kaminski et al. (2008)	Mathematics	N = 80 undergraduates	One, two or three concr. examples vs. a generalized presentation		+			(+)[d]
			Concr. example-abstr. representation vs. abstr. representation only					(+)[d]
Seidel et al. (2013)	Educational Science	N = 54 teacher students	Example rule vs. example rule (with adapted instruction forms)			+		
Tomlinson and Hunt (1971)	Psychology Education	N = 160 high school students	Comparison of low, medium & high pre-structuring (examples only, examples-rule, rule-examples)		+[e]	+[e]		

[a]Abbreviations: "E-R" example-rule; "=" both structural sequences were equally effective; "R-E" rule-example, "E" concrete examples only; "R" generalised rule only

[b] Effects in favor of constant concreteness were only found for one specific grade level within the sample

[c] The result in favor of the structural sequence R-E refers to the near transfer and to the investigated target construct of problem-solving ability (and not to the increase in conceptual knowledge); for the far transfer only a (non-significant) tendency could be determined

[d] Significant differences in favor of the exclusively generalized presentation were only found with regard to transfer ability, not general learning success

[e] The effects here are group-dependent: For students with a low conceptual level, the structural sequence R-E proves to be more effective; for students with a high conceptual level, the various structural sequences do not differ in terms of learning success

Q2: To what extent does the structure of the two structural sequences (example-rule | rule-example) perceived by the learners differ within an educational video on physics?

To understand the learning success within educational videos, possible conditional factors should be examined:

Q3: To what extent can learning success within an educational video on physics be predicted by other motivational factors?

7.3 Design

7.3.1 *Instruments*

At the core of the study is the educational video. We decided to use the law of interaction as the context. Newtonian mechanics is a highly sought-after topic on conventional video platforms. In addition, various studies show that learners have specific difficulties in understanding the law of interaction in particular (Hestenes et al., 1992; Savinainen & Scott, 2002). Bao and Fritchman (2021) even show that misconceptions are persistent after instruction.

When creating the video, a close focus was placed on typical student ideas regarding Newton's 3rd Law and the key aspects proposed by Bao and Fritchman (2021) such as the simultaneity and symmetry of the interacting forces. Based on the importance of theory-based structuring, the video was structured according to the basic model theory (Oser & Baeriswyl, 2001). This means that relevant pre-knowledge is activated first, then the concept is explained using the skateboard experiment as a prototypical example before it is generalized. This is followed by an application and transfer phase with examples from different contexts and learning tasks (e.g. transferability to magnetic forces; sports and everyday contexts). A second learning video was then produced identical to the first, but with a different structural order so that the general rule precedes the example. Both videos are about 25 min long. In an interview study with teachers, the target group orientation and conformity with the quality criteria for good explanatory videos according to Kulgemeyer (2018) were checked and adjustments were made based on the feedback.

The educational videos were then embedded in a pre-post test setting. The time required to carry out the study, including instruction, educational videos, and testing, corresponded to exactly one double lesson (90 min), so that teachers, who were provided access to the online test environment and additional material beforehand, could use our unit as an alternative to their own instruction. Assignment to the reference groups then took place on a class-by-class basis.

Since the focus of the study's interest lies in the extent to which a structural sequence is more effective for learning and the extent to which a possible effect is influenced by the students' pre-knowledge, students' pre-knowledge before the educational videos and their knowledge of the interaction law afterwards were assessed.

Table 7.2 shows the content of the individual items of the two constructs and, thus, connects to the idea of the content aspect of construct validity according to Messick (1995). The individual items originated from different sources. The Force Concept Inventory (FCI) (Hestenes et al., 1992) was the primary source for creating the test instrument due to its thematic orientation, its well-established position, and its focus on student conceptions in mechanics. In its original version, it was also designed as an evaluation instrument for mechanics lessons in high school—i.e. comparable to our target sample. All four items of the FCI on the interaction law were included in the post-test. Both the pre-test and the post-test were then supplemented with further items from sources closer to the learners (e.g. students' learning websites) as well as with items developed individually in order to map non-conceptual task areas and various item difficulties. The complete item set can be requested from the corresponding author.

The second research question focuses on possible non-cognitive predictors of learning success within this multimedia learning unit. From the theoretical analysis, learners' interest in the subject and their motivation emerged as possible factors influencing their knowledge. Furthermore, structuring is clearly defined as a characteristic of teaching quality and thus as a conditional factor for learning success (Fischer et al., 2014). For these constructs established instruments were used. For motivation, the sub-constructs interest/enjoyment, pressure/tension, perceived competence, effort/importance of the Intrinsic Motivation Inventory (IMI) (Deci & Ryan, 2003) were included. Items from German studies (e.g., Maurer, 2016) were used for interest and perceived structuring. The reliabilities of the scales were tested using Cronbach's alpha. The values for the content knowledge related items lie between $\alpha = 0.66$ for pre-knowledge and $\alpha = 0.69$ for knowledge on the law of interaction and as such are acceptable according to Cohen (1988). For the non-cognitive constructs, values between $\alpha = 0.71$ and $\alpha = 0.96$ were achieved stating a good to excellent range.

Table 7.2 Items on content knowledge

Construct	Pre-Knowledge	Knowledge on law of interaction (=learning success)
Item Content	*1st Newton's axiom:* conceptual	*3rd Newton's Axiom:* conceptual, calculations
	2nd Newton's axiom: simple terms, formula symbols, calculations, conceptual	Situation-related differentiation between the three axioms (recognizing the law of interaction)
	Equilibrium of forces: conceptual	*2nd Newton's axiom:* calculations
	Acceleration concept: meaning, visualization in diagram	
Item sum:	10	13

7.3.2 *Sample*

To promote enrolment in the study, we contacted all secondary schools in two German federal states by postal mail. A total of 399 pupils from 19 different schools then took part in the study. After removing the individuals who did not complete the instrument, the final sample comprised of $N = 258$ participants (49% female, 34% male, 17% non-binary or not reported). Students which did not complete the test showed no significant difference in their pre-knowledge ($p = 0.85$) compared to those who completed the instrument. Thus, there seems to be no positive selection. The participants were distributed almost equally between the two groups ("example-rule": $N = 135$, "rule-example": $N = 123$).

7.4 Results

The assessment of the normality using the Shapiro-Wilk-Test showed that none of the constructs are normally distributed, $p < 0.05$. The non-parametric Mann-Whitney-U-test was therefore used for the following analyses of differences between the two intervention groups.

Table 7.3 provides information on the central measures of the constructs at the pre-test time and the statistics of the Mann-Whitney U-test comparing both groups. With a median of 6, more than half of the items on pre-knowledge were generally answered correctly. The two groups example-rule and rule-example did not differ significantly in their pre-knowledge, $U = 8219$, $z = -0.14$, $p = 0.89$, $r = 0.01$. With a mean value of 2.76 for interest, the construct is below the expected value of 3. The overall sample thus has a neutral to indifferent attitude towards the subject of physics on average. The two reference groups example-rule and rule-example similarly do not differ in their interest in physics, $U = 7776$, $z = -0.88$, $p = 0.38$, $r = 0.05$. Hence, the comparability of the two groups in terms of pre-knowledge and interest is ensured as a prerequisite for relating any effects that occur to the intervention.

Table 7.3 Central measures and values of the Mann-Whitney U-test for the constructs of the pre-test

Construct	Group	M	SD	Mdn	U	z	p
Pre-knowledge					8219.00	−0.14	0.89
	All	5.89	2.09	6.00			
	E-R	5.93	2.10	6.00			
	R-E	5.85	2.09	6.00			
Interest					7776.00	−0.88	0.38
Likert scale 1–5	All	2.76	0.84	2.75			
	E-R	2.80	0.84	2.88			
	R-E	2.71	0.85	2.63			

Table 7.4 Central measures and values of the Mann-Whitney U-test for the constructs of the post-test

Construct	Group	M	SD	Mdn	U	z	p
Learning success					7314.00	−1.66	0.10
	All	4.91	2.83	5.00			
	E-R	5.12	2.72	5.00			
	R-E	4.68	2.94	4.00			
Perc. structuring Likert scale 1–4					8268.00	−0.06	0.95
	all	2.80	0.79	2.75			
	E-R	2.79	0.78	2.83			
	R-E	2.82	0.81	2.75			

Table 7.4 provides information on the central measures of the constructs at the post-test time and the statistics of the Mann-Whitney U-test. Across the entire sample, the mean score for knowledge of the interaction law is 4.91 and thus less than half of the achievable points. This rather low score indicates a less pronounced understanding of the law of reciprocal action among the students. Complete conceptual knowledge of the interaction law is not yet available. With a mean value of 2.80, the perceived structure is slightly above the mean of 2.5 on the four-point Likert-scale. On average, students rate the video units as structured, goal-oriented and coherent.

The analysis of possible differences between the example-rule and rule-example group using the Mann-Whitney U-test showed no significant effects for either knowledge of the interaction law ($U = 7314$, $z = -1.66$, $p = 0.10$, $r = 0.10$) or the perceived structuring ($U = 8268$, $z = -0.06$, $p = 0.95$, $r < 0.01$). So, the two videos differing in their structural sequence are equally effective for learning the law of interaction and are perceived as equally structured by the students.

To investigate the influence of pre-knowledge on the impact of the structural sequence, a moderation analysis is carried out. The analysis shows that the overall model, $F(3, 254) = 20.46$, $p < 0.001$, becomes significant with a moderate variance explanation of $R^2 = 19\ \%$ according to Cohen (1988). However, the interaction effect of structural sequence and pre-knowledge is neither significant ($p = 0.14$) nor relevant ($\Delta R^2 < 1\%$). A follow-up regression model, in which the interaction term was removed, similarly shows no significant relationship between the structural sequence and the learning success on the interaction law, $B = -0.04$, $p = 0.22$, but one between pre-knowledge and the learning success, $B = 0.57$, $p < 0.001$. These results reinforce the findings from the analyses of differences that neither of the two structural sequences is more effective in terms of learning success. Consequently, only pre-knowledge is predictive for learning, which, however, shows no interaction effect with the structural sequence. Therefore, no learning group classified according to pre-knowledge benefits particularly from a structural sequence.

To identify predictors of the learning success of the video learning unit, a linear regression analysis was carried out. Based on a preceding correlation analysis, only those constructs were selected that correlated significantly with the knowledge of

Table 7.5 Regression model predicting learning success

Construct	Regression coefficient B (*stand.*)	*p*
(constant)	1.73 ()	0.02
Pre-Knowledge	0.48 (*0.36*)	<0.001
Interest	0.28 (*0.08*)	0.13
Perceived Structuring	0.54 (*0.15*)	0.01
Pressure/Tension	−0.85 (*−0.35*)	<0.001
R^2	0.30**	

the interaction law (pre-knowledge, interest, perceived structuring, and pressure/tension). The resulting model (Table 7.5) has a high goodness of fit according to Cohen (1988) with $R^2 = 0.30$. Apart from interest ($p = 0.13$), all other predictors statistically significantly predict the criterion learning success, $F(4, 253) = 27.35$, $p < 0.001$. Learning success within this learning unit on the law of interactions can therefore be predicted to a considerable extent by the students' pre-knowledge, their feeling of structure and the absence of pressure/tension. The standardized regression coefficients show that a change in pre-knowledge has the greatest effect on learning success, closely followed by the feeling of pressure or tension.

7.5 Discussion

In summary, our results show that the structural sequence (example rule vs. rule example) has no influence on learning success in the context of the presented video learning units on the law of interaction. In all analyses, pre-knowledge was found to be the most important construct for learning success. The effect between structural sequence and learning success is not moderated by pre-knowledge indicating that lower-performing students benefit equally from both units. With the help of the perceived structuring, the feeling of pressure and tension as well as the pre-knowledge performance, the learning success can be predicted with a high goodness of fit of $R^2 = 0.30$.

With these outcomes, the present study joins the empirical findings of physics-related (Jaakkola & Veermans, 2018; Johnson et al., 2014; Kokkonen et al., 2022) and non-physics-related (Kaminski et al., 2008; Tomlinson & Hunt, 1971; Van Hout & Mettes, 1976) studies, which also found no (or only a limited) evidence in favor of a specific structural sequence. Our study now demonstrates that these findings can also be applied to educational videos. As an extension to Kokkonen et al. (2022), Jaakkola and Veermans (2018) and Johnson et al. (2014), who investigated the influence of the structural sequence in the teaching of electrodynamic content, we have now been able to apply this result to a different physical context, namely Newtonian mechanics.

Since the positioning of the example within the structural sequence has no effect on either the learning performance or the perception of the students, it provides greater flexibility in the arrangement of the respective phases when creating educational videos. The analysis of the conditional factors also reproduces the findings of Seidel et al. (2005) and Fischer et al. (2014) for educational videos and emphasizes the need to focus particularly on activating pre-knowledge and to create a climate conducive to learning when implementing the videos in the classroom to avoid students feeling tense.

7.6 Limitations

An obstacle that was indeed not insignificant, but which we underestimated, was the insufficient internet coverage at the schools, which presumably caused the high drop-out rate of over a third of participants. To some extent, pupils in the final sample also reported stuttering sequences. Although this difficulty does not compromise the study design per se as both groups were equally affected, it does impair the learners' viewing experience and possibly depresses aspects of the motivational sub-constructs. For subsequent studies, it would therefore be advisable to consider the option of establishing a network by the researchers at the schools to ensure a smoother implementation. This would also lead to further standardization. In our design, we made a conscious decision to allow teachers to use the videos independently in order to reach more schools and enable a more realistic environment for integrating the videos into everyday school life, but at the same time this decision led to more openness: despite the instruction manual and feedback forms, it remains unclear to some extent whether students worked together, or teachers provided support.

Another limitation was the lack of interaction and personal feedback in the application and transfer phase. Since our findings demonstrate that the concrete order of the structural sequence in the introductory part has little relevance for learning success, it is worthwhile for subsequent studies to focus on this application and transfer phase and to investigate what degree of interaction within educational videos is most beneficial for knowledge acquisition and how it can be achieved. In this context, a new study by Pi et al. (2023) shows a promising proposal for how the phase after watching the videos can be designed so that particularly students with little pre-knowledge will benefit.

7.7 Conclusion

This study suggests that when developing a physics learning video, it will probably not influence the students' learning success if the introduction of a new concept starts with the example (followed by the rule) or with the rule first (followed by the example). This speaks for a higher flexibility in creating such videos.

Our results also speak for the transferability of findings from educational research often conducted in studies on face-to-face teaching to learning situations with videos and possibly other kinds of learning material. In this sense, our findings are also a reminder of the theory- and evidence-based development of such materials. This process is resource intensive, but, once produced, the materials can be used an infinite number of times and can be adapted efficiently, thus offering to cushion the shortage of teachers.

Acknowledgments We would like to express our sincere thanks to all the dedicated teachers and their classes who participated in this study. We would also like to thank Helen Börner and Dennys Gahrmann for their contribution to the educational videos.

References

Bao, L., & Fritchman, J. C. (2021). Knowledge integration in student learning of Newton's third law: Addressing the action-reaction language and the implied causality. *Physical Review Physics Education Research, 17*, 020116.

Bruner, J. S. (1966). *Studies in cognitive growth*. Wiley.

Cohen, J. (1988). *Statistical power analysis for the behavioral sciences*. Lawrence Erlbaum.

Deci, E. L., & Ryan, R. M. (2003). *Intrinsic Motivation Inventory*. Retrieved from https://selfdeterminationtheory.org/intrinsic-motivation-inventory/

Fischer, H. E., Labudde, P., Neumann, K., & Viiri, J. (Eds.). (2014). *Quality of instruction in physics comparing Finland, Switzerland and Germany*. Waxmann.

Fyfe, E. R., McNeil, N. M., & Borjas, S. (2015). Benefits of "concreteness fading" for children's mathematics understanding. *Learning and Instruction, 35*, 104–120. https://doi.org/10.1016/j.learninstruc.2014.10.004

Fyfe, E. R., McNeil, N. M., Son, J. Y., & Goldstone, R. L. (2014). Concreteness fading in mathematics and science instruction: A systematic review. *Educational Psychology Review, 26*(1), 9–25.

Fyfe, E. R., & Nathan, M. J. (2019). Making "concreteness fading" more concrete as a theory of instruction for promoting transfer. *Educational Review, 71*(4), 403–422. https://doi.org/10.1080/00131911.2018.1424116

Hestenes, D., Wells, M., & Swackhammer, G. (1992). Force concept inventory. *The Physics Teacher, 30*, 141–158.

Jaakkola, T., & Veermans, K. (2018). Exploring the effects of concreteness fading across grades in elementary school science education. *Instructional Science, 46*, 185–207. https://doi.org/10.1007/s11251-017-9428-y

Jeong, H., Hmelo-Silver, C. E., & Jo, K. (2019). Ten years of computer-supported collaborative learning: A meta-analysis of CSCL in STEM education during 2005–2014. *Educational Research Review, 28*, 100284. https://doi.org/10.1016/j.edurev.2019.100284

Johnson, A. M., Reisslein, J., & Reisslein, M. (2014). Representation sequencing in computer-based engi-neering education. *Computers & Education, 72*, 249–261. https://doi.org/10.1016/j.compedu.2013.11.010

Kaminski, J. A., Sloutsky, V. M., & Heckler, A. F. (2008). The advantage of abstract examples in learning math. *Science, 320*, 454–455. https://doi.org/10.1126/science.1154659

Kokkonen, T., Lichtenberger, A., & Schalk, L. (2022). Concreteness fading in learning secondary school physics concepts. *Learning and Instruction*. https://doi.org/10.1016/j.learninstruc.2021.101524

Kulgemeyer, C. (2018). A framework of effective science explanation videos informed by criteria for instructional explanations. *Research in Science Education, 50*, 2441–2462.

Kulgemeyer, C., & Peters, C. H. (2016). Exploring the explaining quality of physics online explanatory videos. *European Journal of Physics, 37*, 065705.

Kulgemeyer, C., & Wittwer, J. (2022). Misconceptions in physics explainer videos and the illusion of under-standing: An experimental study. *International Journal of Science and Mathematics Education, 21*, 417–437.

Maurer, C. (2016). *Strukturierung von Lehr-Lern-Sequenzen [Structuring of learning sequences]*. Logos.

McNeil, N. M., & Fyfe, E. R. (2012). "Concreteness fading" promotes transfer of mathematical knowledge. *Learning and Instruction, 22*(6), 440–448. https://doi.org/10.1016/j.learninstruc.2012.05.001

Medienpädagogischer Forschungsverbund Südwest [mpfs]. (2021). JIM-Studie 2021: Basisuntersuchung zum Medienumgang 12-19 Jähriger. [JIM Study 2021: Basic investigation into the media consumption of 12-19 year olds]. https://www.mpfs.de/fileadmin/files/Studien/JIM/2021/JIM-Studie_2021_barrierefrei.pdf

Messick, S. (1995). Validity of psychological assessment: Validation of inferences from persons' responses and performances as scientific inquiry into score meaning. *American Psychologist, 50*(9), 741–749.

Oser, F., & Baeriswyl, F. J. (2001). Choreographies of Teaching: Bridging Instruction to Learning. In V. Richard-son (Ed.), *Handbook of Research on Teaching* (pp. 1031–1065). American Educational Research Association.

Pi, Z., Guo, X., Liu, C., & Yang, J. (2023). Students with low prior knowledge learn more when explaining to a peer than to a teacher after viewing educational videos. *Active Learning in Higher Education, 0*(0). https://doi.org/10.1177/14697874231212260

Renkl, A., Atkinson, R., Maier, U., & Staley, R. (2002). From example study to problemsolving: Smooth transitions help learning. *Journal of Experimental Education, 70*, 293–315.

Savinainen, A., & Scott, P. (2002). Using the force concept inventory to monitor student learning and to plan teaching. *Physics Education, 37*(1), 53–58.

Seidel, T., Blomberg, G., & Renkl, A. (2013). Instructional strategies for using video in teacher education. *Teaching and Teacher Education, 34*, 56–65.

Seidel, T., Rimmele, R., & Prenzel, M. (2005). Clarity and coherence of lesson goals as a scaffold for student learning. *Learning and Instruction, 15*(6), 539–556. https://doi.org/10.1016/j.learninstruc.2005.08.004

Shulman, L. S. (1986). Paradigms and research programs in the study of teaching: A contemporary perspective. In M. C. Wittrock (Hrsg.), *Handbook of research on teaching* (3. Aufl.,S. 3–36). Macmillan.

Smith, E. E., & Medin, D. L. (1981). *Categories and concepts*. Harvard University Press.

Tomlinson, P. D., & Hunt, D. E. (1971). Differential effects of rule-example order as a function of learner conceptual level. *Canadian Journal of Behavioural Science, 3*(3), 237e245. https://doi.org/10.1037/h0082265

Van Hout, J. F. M. J., & Mettes, C. T. C. W. (1976). The effect of EGRULEG versus RULEG and teacher-centredness versus student-centredness on pupil gain and satisfaction. *Instructional Science, 5*, 181–187.

Wagner, M., Gegenfurtner, A., & Urhahne, D. (2021). Effectiveness of the flipped classroom on student achievement in secondary education: A meta-analysis. *Zeitschrift für Pädagogische Psychologie, 35*(1), 11–31. https://doi.org/10.1024/1010-0652/a000274

Welbourne, D. J., & Grant, W. J. (2016). Science communication on YouTube: Factors that affect channel and video popularity. *Public Understanding of Science, 25*(6), 706–718. https://doi.org/10.1177/0963662515572068

Tanja Mutschler is a postdoctoral researcher at the University of Potsdam, Germany. Her current work centers on knowledge brokering in a large-scale national initiative aimed at fostering teachers' digital competencies. In her PhD research, she explored how learning processes can be effectively structured in physics education.

Stefan Sorge is a postdoctoral researcher in physics education at the IPN – Leibniz Institute for Science and Mathematics Education in Kiel, Germany. His work focuses on the development of pre-service teachers' professional knowledge, the impact of professional development on in-service teachers and the design of coherent instruction. He has contributed to various national and international research initiatives and currently serves as a scientific advisor for the Alliance for Teacher Education at Kiel University.

David Buschhüter holds a PhD in Physics Education from the University of Potsdam in Germany. His main research interests focus on the use of AI in Education. Currently, he works as senior researcher at the Educational Technology Lab of the German Research Center for Artificial Intelligence (DFKI).

Christoph Kulgemeyer is a professor of physics education at the University of Bremen. He works on physics teachers' professional knowledge, instructional explanations, and reflection skills. His latest work focuses on how to learn effectively using physics explainer videos.

Andreas Borowski is a professor of physics education and the director of the Center for Teacher Education and Education Research (ZeLB) at the University of Potsdam, Germany. His research focuses on the professional knowledge of both preservice and inservice physics teachers. As director of the ZeLB, he is particularly interested in strengthening the link between professional knowledge and practical experience gained during student teaching internships, as well as fostering collaboration between university faculty and schoolteachers.

Chapter 8
Decoding Visual Attention: Eye-Tracking Insights Into Students' Interaction with Images in Physics Textbooks

Martina Kekule and Alžběta Krejčí

8.1 Theoretical Background

Learning from an educational text is an important part of students' learning skills. The constructed scientific knowledge is communicated to the science community through various modalities like text and image (Eilam & Gilbert, 2014) or graphs, schemes, etc. Researchers across the discipline are exploring non-textual explanations in textbooks, particularly graphs, illustrations and photographs, over the use of analogies within illustrations (Vojir & Rusek, 2019). The visual behavior of students when they process such text can provide a deeper insight into their cognitive processes. Significant associations between eye-tracking data and reading comprehension and prior knowledge emerged (Mason et al., 2013). For example, the greater integrative processing of the illustrated text was associated with higher learning performance. And it seems that pictorial presentations appear to convey physics concepts more quickly and efficiently than textual ones (Chen et al., 2014).

Readers' preexisting knowledge also influences cognitive processing of reading (Wolfe & Goldman, 2005). Learning achievement when students deal with text and pictorial visualizations can differ according to cognitive styles, such as a visualizer or also a verbalizer (Koć-Januchta et al., 2017). For example, visualizers spent more time on pictures than verbalizers and vice versa; verbalizers entered the irrelevant areas of pictures sooner than visualizers and visualizers outperformed verbalizers on comprehension. Mason et al. (2013) investigated how long students inspected a picture during their first visit. Students who were the best integrators of a picture and a text, for example, showed a longer inspection time during the first encounter, and students who were the worst integrators showed a shorter inspection time.

M. Kekule (✉) · A. Krejčí
Department of Physics Education, Faculty of Mathematics and Physics, Charles University, Prague, Czech Republic

G. Cakmakci, M. F. Tasar (eds.), *Connecting Science Education with Cultural Heritage*, Contributions from Science Education Research 15,
https://doi.org/10.1007/978-3-031-90944-3_8

Based on Mayer's cognitive theory of multimedia learning (CTML), multimedia learning relies on several processes that must be activated: selection, organization and integration processes (Mayer, 2009). Selection means selection of relevant words or pictures for further processing in working memory.

Textbooks or various educational materials can include various relevant pictures or schemes. There are many classification approaches of pictures in educational materials among scientists. For example, Kress and van Leeuwen (1996) characterize the pictures as narrative and conceptual representation. Dimopoulos et al. (2003) provide in-depth analysis of pictures in three main categories—classification, framing and formality. Korfiatis et al. (2004) conducted content analysis of nature images in primary school textbooks in two main categories—metaphors of nature and depictions of nature. One aspect we can assess pictures from is the relevant point of view. For example, Tangen et al. (2011) examined the effect of congruent images (i.e., images that were relevant to the target information) and incongruent images (i.e., images that were relevant to the narration but not the target information) on interest and cognitive processes during learning from slideware presentations. Williams et al. (2017, 2022) examined the impact of irrelevant visual images in a PowerPoint lecture on attention and information retention performance. In one of the most cited papers in textbook research (Vojir & Rusek, 2019), Pozzer and Roth (2003) classified pictures into four categories according to how decorative or informative they are. Based on eye movements, Slykhuis et al. (2005) identified that students did not pay much attention to only decorative pictures in a presented Power Point presentation. Interesting but irrelevant information, also known as seductive details, included in multimedia educational texts can capture students' attention towards the topic but may also have a harmful effect on the learning process (Rey, 2012). For example, on transfer of knowledge with low prior knowledge (Bender et al., 2021).

Research Questions Like Slykhuis et al. (2005), our research questions focused on a similar problem, but in physics education. This is a stimuli-based experiment where we focused on comparing visual attention for different variations of the visual stimulus (Hahn & Klein, 2022). We created a model situation that students learn from a textbook, for example, during their preparation for a final exam, and we investigated their visual behavior during the process. As textbooks or educational materials can be accompanied by images with varying degrees of information, ranging from purely motivational decorative pictures to highly informative images that augment the text itself, we were interested in how students would respond to this level of decorativeness when learning from such material. In line with Mayer's cognitive theory of multimedia learning we would expect that student selection, particularly, could differ depending on whether a picture is relevant or not within educational text, aiming at relevance from an information viewpoint. Operationally, we focused on the attention that students pay to various pictures included in real educational material. Pictures differed from a decorative point of view similarly to Pozzer and Roth (2003). The most decorative image in the presented material can be considered a so-called seductive detail. The following research questions particularly guided our study:

(a) When do students visit a picture for the first time? Pictures vary according to the decorative level; however, we expect that all the pictures will attract students' attention soon.
(b) To what extent can the degree of decorativeness of images be distinguished based on eye movements? We expect less attention to more decorative pictures. In the case of less decorative images, we expect greater variability of the attention due to additional confounding variables (such as students' interest in physics or their performance).

8.2 Method

Students were observed by the eye-tracking method as they were reading several texts from national textbooks supplemented with both pictures and questions. Texts were presented in pairs. Firstly, students read a text from a textbook with an accompanying picture (see Fig. 8.1). After this first slide, the same slide but with a question followed. So, students first read an educational text with a picture. This is a model of a learning situation when they just try to go through and learn from a textbook. For example, during preparation for a final exam. We then presented the same slide to them, but accompanied by a question to which they should give an answer. We also tracked their eyes during the process. After the eye-tracking session, students filled out an additional questionnaire, freely recalled information

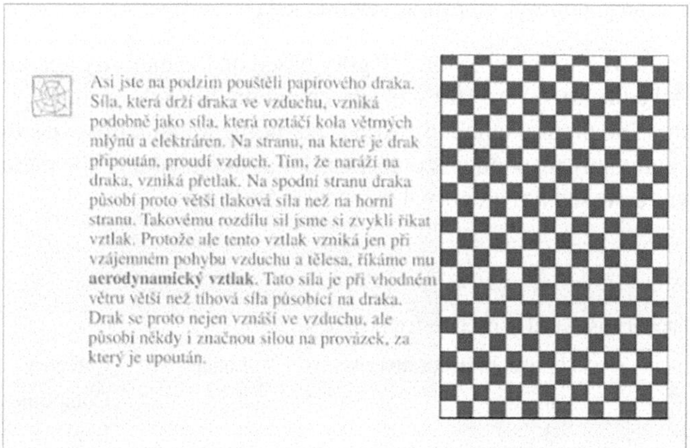

Fig. 8.1 Layout of the tested materials. Text on the left, picture on the right. Created AOIs within each picture: AOI of the whole material area (gray), AOI of the picture (check). The most decorative picture labeled 'Kite'

from the text, answered given questions and they participated in an interview under their eye movements records.

Because we aim to investigate students' attention, we carefully considered what could have an influence on their attention. Firstly, when students read and are eye tracked, they cannot move their head freely, so it is challenging for them. That is why we can only use a few materials for the test, typically we used 10 slides. So altogether they went through 5 pairs of texts. In the paper, we will mostly only focus on dealing with the first paired slide. It means that they also only went through 5 different pictures. As Pozzer and Roth (2003) suggest four categories, we could only work with one picture in each category. So, it is therefore important to follow trends in the students' attention with the increasing degree of decorativeness of the picture. Secondly, we expected that the pictures' order could influence the attention devoted to the picture, so we worked with two test set-ups. In the first set the text with the two most decorative pictures were placed at the end, in the second set these pictures were presented at the beginning. Thirdly, students may be bored reading about the same topics, so we chose texts with different topics.

Materials Each student read 5 pairs of texts from our national textbook (Rauner et al., 2007). Each picture was labeled according to the Pozzer and Roth (2003) classification: how much is a picture only decorative or more informative. They distinguish four coding categories: decorative, illustrative, explanatory and complimentary. The decorative category means a picture least integrated with the text and without a caption, and the complimentary category means a picture with the highest level of integration and with a caption. Our test samples more closely resemble the more modern digital concept of slides in a presentation, so we presented the images without captions. Thus, the categorization of the images was similar to Slykhuis et al. (2005): solely based on the inherent information of the images and their connectivity to the text. Based on the assessments of 15 physics educators, the pictures were assigned into the categories as follows (Table 8.1):

Each picture was assigned to the category based on the majority agreement of the educators. All educators were 100% in agreement on the 'Kite' being a purely decorative picture. The picture labelled 'Moon' about the lunar eclipse has the weakest agreement, and many respondents considered it as explanatory. The picture labeled 'Particle' was assessed as category 4 by several educators.

Table 8.1 Category assigned to each tested picture

Picture	Level of decorativeness	Label	Category
Iron	4	4	Complimentary
Particles	3	3	Explanatory
Microwave oven	2	22	Illustrative
Moon	2	21	Illustrative
Kite	1	1	Decorative

Level 1—the least integrated text and picture, level 4—the highest level of integration

Test Set Up Pictures were on the upper right side of the screen; they took approximately 1/3 of the whole presented area (see Fig. 8.1). After this first slide the same slide, but with a question (on the left down), followed. Texts were focused on the following fields: particles, how an iron works, how a microwave oven works, lunar eclipses and kiting. All students went through the topic during their lower secondary education. We did not want their attention to fade while reading so much text, so we chose different topics. Pictures were presented in two set ups:

1. Test setup 1: level 3 ('Particles'), level 4 ('Iron'), level 2 ('Microwave oven'), level 2 ('Moon') and level 1 ('Kite').
2. Test setup 2: level 1 ('Kite'), level 2 ('Moon'), level 2 ('Microwave oven'), level 4 ('Iron') and level 3 ('Particles')

Participants Thirty-three high school students took part in the study, 16 of which went through test setup 1 and 17 through test setup 2. They were assigned to a test set up randomly. The median age is 16 years (lower quartile is 16 and upper quartile 17 years old). More than half of students (57%) stated that they enjoy physics (15% very much). 72% of students received the best or second-best grade in physics. 21% of students received an average grade in physics.

Apparatus A TX300 eye-tracker was used (Tobii, frequency 300 Hz, visual angle accuracy of less than 0.5°). The infrared camera was placed under the 23-inch screen (1900 × 1280 pixels) of the stimulus PC. Participants had their eyes 38 cm from the screen. A five-point calibration and validation procedure was used before the start of the experiment. Eye movement was classified as a saccade when the eye's velocity exceeded 30°/s. The minimum fixation duration was set to 60 ms. The testing room was a low distraction room with controlled lighting.

Data Analysis For each material we created so called AOIs, Areas Of Interest. The whole material and the picture were marked as a separate AOI for every material (Fig. 8.1). All the images were approximately the same size, yet we made a correction in the total fixation duration so that it respects an area size of the pictures in a simple linear approach—the bigger the picture the more viewing time. We assigned coefficient 1 to the area size of 100,000 px (approximately the area size of the 'Iron' picture). Then for example, an 85,000 px picture area size (approximately the area size of the 'Particle' picture) had the coefficient 0.85. So, if a participant had a total fixation duration on the 'Iron' picture of 14,918 s it stayed the same, and if they had a total fixation duration on the smaller 'Particle' picture of 6742 s, we divided it by 0.85 and we further used this bigger 7932 s value. We used three typical AOI metrics in order to investigate our research question: time to the first fixation/visit of the picture AOI, total fixation duration and count on a picture AOI.

Time to the first fixation operationally indicates the first direct attention devoted to the picture. When the stimuli are presented to students, they usually start in the middle of the stimuli and then move to the left and start to read the text first. Because we read the text from the left to the right, to visit a picture located on the right, one needs to consciously direct their attention to the right.

8.3 Results

First, we investigated how much time the students needed in order to go to the picture. Particularly, we determined the time to the first fixation on the picture area. See histograms for each of the five pictures in Fig. 8.2. The data shows that students in their first focus on a picture differed across pictures. For example, in the case of the more decorative picture (21 'Moon'/level 2) most students looked at the picture within 8 s. Yet in the case of the most complimentary picture (4 'Iron' / level 4), more students visited the picture after 10 s. We did not have identified any special students' strategy, for example, always first read a text and then look at a picture or vice versa.

The most interesting result is the number of students who did not visit a picture at all (Table 8.2). Fifteen of 33 participants did not visit the most decorative picture

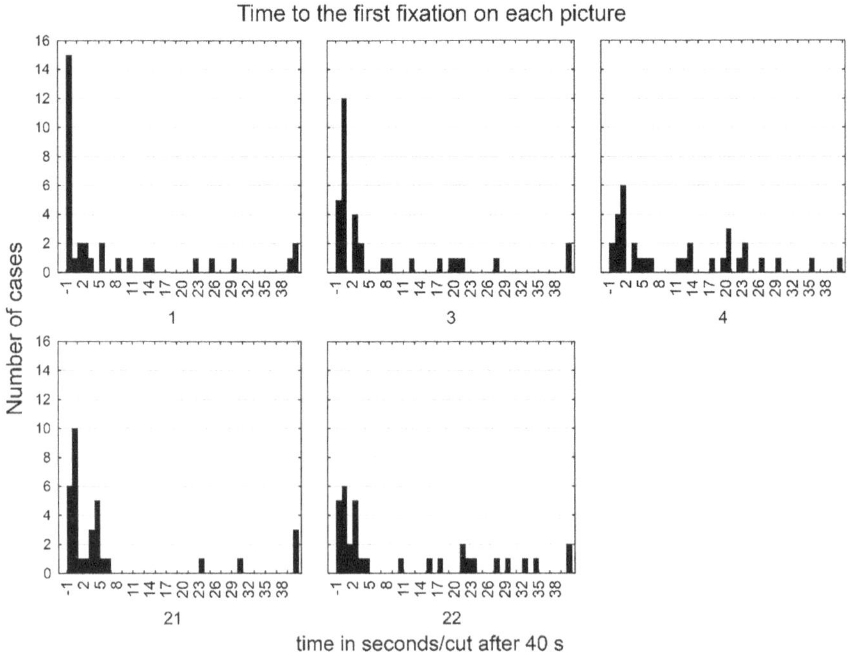

Fig. 8.2 Histogram of time needed to the first fixation on each picture. The histogram is cut after 40 s (outliers) and the 0 (first bar) symbolizes no visit. Numbers 1, 21, 22, 3 and 4 are labels of pictures (Table 8.1): 4—picture 'Iron' is the least decorative, 1—picture 'Kite' is the most decorative

Table 8.2 Number of students who did not visit a picture at all

Picture	Level of decorativeness	Label	No visit No. of students
Iron	4	4	2
Particles	3	3	5
Microwave oven	2	22	5
Moon	2	21	6
Kite	1	1	15

Table 8.3 Total fixation duration mean of students who did not pay any attention to the decorative picture ('Kite') yet paid, i.e., attention to explanatory picture and to pictures on the second slide of the pairs (the slide with a question)

Picture	Level of decorativeness	Total fixation duration mean on a picture AOI (s)
Kite	1	0
Kite with a question	1	0.75
Particle	3	4.40
Particle with a question	3	1.52

('Kite'). Many fewer students did not visit less decorative pictures. Just because students did not fixate on a picture does not mean they did not see it at all. Their brain could access it via an image in their peripheral vision.

All students who did not visit 'Kite' picture spent some time on the other pictures. For example, on average 4.40 s on the 'Particle' picture (Table 8.3). Probably, their strategy is not to ignore pictures, but they recognize the lack of information in the picture and do not waste their energy to attend it. When the same students attended the slide with the 'Kite' picture and the given question, several of them visited the picture (see Table 8.3). In the followed-up interview they mentioned that they just want to take a look at the kite design. Several of them did it after they answered a question, several of them during the process of searching for an answer. In the latter case the picture could serve as a distraction.

Secondly, we focused on the total time that students spent on a picture. Since we can assume that students' attention will decrease with more and more slides, we implemented the test with two different orders of image setups. Rather decorative pictures ('Kite', level 1, and 'Moon', level 2 (21)) were presented in test setup 1 at the end. Rather complementary pictures ('Particles', level 3, and 'Iron', level 4) were presented in test setup 1 at the beginning. The Kolmogorov-Smirnov test rejects the hypothesis about the same distribution only for the 'Moon' picture (Table 8.4). Particularly, students paid more attention to the picture when it was presented at the beginning of the test. However, still they did not pay so much attention to the higher-level picture (level 3). That is why we will further analyze data from both set ups together.

Table 8.4 Kolmogorov-Smirnov test by Statistica, Inc. Comparing the student groups who went through test setup 1 and 2. Marked tests are significant at $p < 0.05000$

Picture	D[a]	Max Neg value	Max Pos value	*p*-value	Test set up 1 Mean ± Std. Dev (s)	Test set up 2 Mean ± Std. Dev (s)
Iron	4	−0.397	0.015	>0.100	5.927 ± 4.3651	8.964 ± 4.6141
Particle	3	−0.118	0.232	>0.100	4.716 ± 2.3856	4.419 ± 3.8056
Moon	2	**−0.631**	**0.000**	**<0.005**	**0.814 ± 0.7479**	**3.091 ± 2.8872**
Kite	1	−0.346	0.000	>0.100	0.200 ± 0.3035	0.463 ± 0.5500

[a]Level of decorativeness

Table 8.5 Total fixation duration/count mean and median on each picture

			Total fixation duration		Total fixation count	
Picture	Level	Label	Mean (s)	Median (s)	Mean (s)	Median (s)
Iron	4	4	7.49	7.07	28	20
Particles	3	3	4.56	4.22	16	16
Microwave oven	2	22	1.36	0.77	6	5
Moon	2	21	2.02	1.41	7	6
Kite	1	1	0.34	0.06	2	1

On average students spent less time on the decorative picture and more time on more complimentary pictures (Table 8.5 and Fig. 8.3). Because of the sample size, we provide nonparametric statistics on the data, but we report mean value as well. As can be seen in Table 8.5, in all cases the median and mean are similar and maintain the pattern in the results. The decorative picture 'Kite' not only received the least attention, but also the variability among students is the smallest in this case. In contrast, for the complimentary image, the variability is the greatest. The same pattern can be also observed for total fixations count.

Nonparametric Kruskal-Wallis test rejects the hypothesis that total fixation duration/count medians are the same (Table 8.6, left/right). So, we could suggest that there is a difference in total fixation duration/count medians spent on the different pictures.

8.4 Discussion

Non-visitation Almost half of the students did not visit the purely decorative picture at all. And about one fifth of students did not visit the less decorative picture either. Images were only perceived through the peripheral image on the retina. So, the results suggest that students are able to determine whether the picture possesses useful information or not based on their peripheral vision. It could indicate their good level of selection process within Mayer's cognitive theory of multimedia learning.

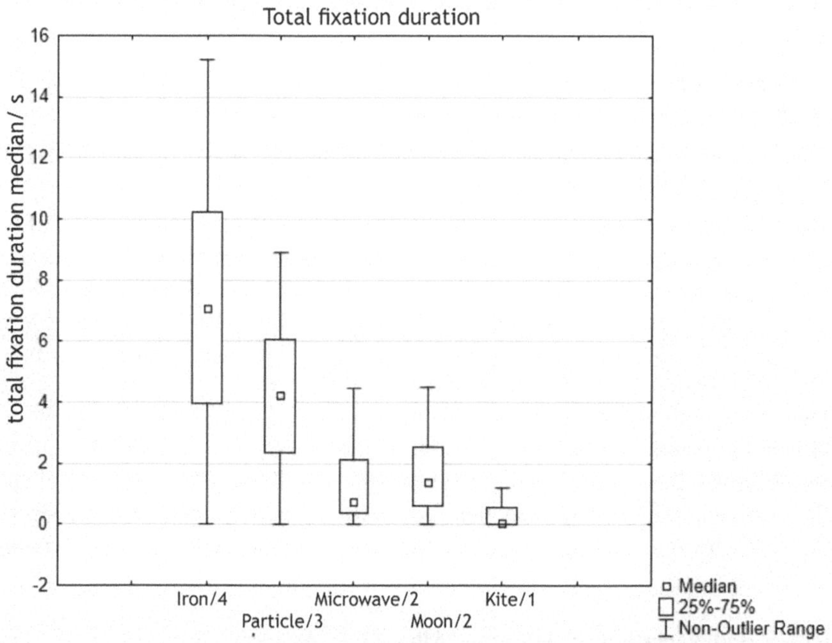

Fig. 8.3 Total fixation duration medians (s) on each picture. Box plots provided by Statistica, Inc. Picture 'Iron'/4 (left) is the least decorative, picture 'Kite'/1 (right) is the most decorative

Table 8.6 Nonparametric Kruskal-Wallis tests by Statistica, Inc. Total fixation duration (left). $H(4, N = 164) = 84.90446$, $p = 0.00$. Reject hypothesis that medians are the same. Total fixation count (right). $H(4, N = 165) = 79.66393$, $p = 0.00$. Reject hypothesis that medians are the same

		Total fixation duration		Total fixation count	
Picture	Level	Sum of ranks	Mean rank	Sum of ranks	Mean rank
Kite	1	1096.500	33.2273	1108.500	33.5758
Particles	3	3645.500	110.4697	3621.500	109.7273
Iron	4	4268.500	129.3485	4236.500	128.3636
Moon	2	2135.000	64.6970	2265.000	68.6364
Microwave	2	2384.500	74.5156	2465.500	74.6970

Seductive Detail The most decorative image could be considered as a so-called "seductive detail." An element that attracts attention but doesn't carry relevant information. Although data suggests that students were not distracted by this image, they may have been disciplined only because of the testing environment, and in reality, they might have paid more attention to the image. As one student mentioned in a subsequent interview, during the second projection of the slide with the image, she felt the need to visit the image, even though she believed it did not contain relevant information. In future research, it would be beneficial to focus on multiple images and conduct the test in a more natural environment, such as presenting the image on a tablet.

Seductive details can have a negative impact on learning, particularly in the case of students with lower final school grades (Bender et al., 2021). Further research could also focus on how other variables (such as students' performance, interest in science or art) influence non-visitation of purely decorative pictures.

Color The image non-visitation could also have been influenced by the color of the image. Particularly, bottom-up attention directed to an image may be influenced mainly by the red-green color contrast (Frey et al., 2008). The images were selected from a single textbook, thus using the same color filter. The image 'Microwave oven' (level 2) featured a prominent orange area, which might have been more attention-grabbing. The most decorative image, 'Kite' (level 1), was multicolored, which could again have drawn more attention compared to the other images, which were predominantly in blue tones. Therefore, the data on non-visitation for images at levels 1 and 2 could potentially be even higher. Williams et al. (2022) included only black-and-white images in their research. To better simulate a real-world context, future research could explore the effects of varying color modifications of images.

Median and Variation of Attention Allocation As research results of Slykhuis et al. (2005) determined, within a biology presentation, students paid less attention to decorative pictures than complementary ones. Students in the research about the educational texts from physics showed the same behavior. Similar to the way physics educators assigned a category to a given image based on their experience, we would categorize images in the same way based on the students' average fixation time. And not only based on the medians of the data, but also based on its variability. According to our picture categorization, we would expect less variance in the case of the image on level 2/21 'Moon'. In the follow-up interview, some students mentioned that they were confused by the color of the moon, wondering if it was a double image. Thus, for some students the picture was clearly interpretable, for some of them it was not, which could then result in more variance than would be expected for the category based on our data. Also, in categorizing the images, the experts agreed the least in this case. Due to the time length of the experiment itself, only one image from each category (two from the middle and the ambiguously identifiable category/level 2) was included in the test and so further research in the area, including more investigated pictures is needed.

Relevant Variables Attention allocation on a picture (total fixation duration / count) varied differently among the pictures. The smallest variation was for the most decorative picture, whilst the largest variation was for the least decorative picture. In previous studies, several differences in time spent on a picture within an educational material were identified and can contribute to variability in more information-rich images. Koć-Januchta et al. (2017) identified differences between visualizers and verbalizers; students' preferred cognitive styles are characteristics that could impact time spent on a picture. The high-ability group made more effort to integrate the textual and pictorial information (Jian & Ko, 2017). Readers with the text with an abstract illustration made a greater effort to integrate verbal and pictorial information than readers with only text or text with a concrete illustration (Mason et al., 2013). Not only cognitive, but also affective variables (such as the graphic design of the image, interest in the physical topic, etc.) can influence the amount of attention students will pay to the image.

Implications for Science Education Textbooks with colorful pictures are attractive for students and, obviously, they will prefer to start to study such a textbook. When creating educational materials, we could take into account that "an important factor in the creation of textbooks is the non-use of excess amounts of stimuli that attract pupils' attention at the expense of more important information" (Drexler et al., 2018) and not overload the material with many only decorative pictures which could serve as seductive details. Although, as our preliminary results determined, students are able (even by peripheral vision) to distinguish the relevance of a picture and not pay almost any attention to a purely decorative picture. However, even if they attend the picture twice or even not at all, their brain still needs to assess the relevance of a picture resulting in an increased cognitive load.

8.5 Conclusion

The paper focused on students' interaction with pictures in physics textbooks that can contain different amounts of information; information ranging from purely decorative pictures (no information) to being important for understanding the text (similarly to Pozzer & Roth, 2003; Slykhuis et al., 2005). As expected, students paid the least attention to the purely decorative picture. Images classified as having a higher degree of complementarity on average received more attention (total fixations duration and total fixations count) from students. The variability in the case of these images was higher than in the case of a purely decorative image. All students clearly understood the very low level of information (in terms of physics) contained in this picture and more or less ignored the picture. Almost half of the students did not pay any attention to this image with their foveal vision. These results suggest that the eye-tracking method is suitable for research in the selection process within Mayer's cognitive theory of multimedia learning and that the amount of picture decorativeness could be another relevant factor that could influence time spent on a picture. Further research focusing on how other relevant variables (such as students' learning outcomes, interest in science or art, their attitudes towards the material, differences in visual strategies, etc.) influence interaction with various more or less decorative pictures is needed.

References

Bender, L., Renkl, A., & Eitel, A. (2021). When and how seductive details harm learning. A study using cued retrospective reporting. *Applied Cognitive Psychology, 35*(4), 948–959.

Chen, S. C., She, H. C., Chuang, M. H., Wu, J. Y., Tsai, J. L., & Jung, T. P. (2014). Eye movements predict students' computer-based assessment performance of physics concepts in different presentation modalities. *Computers & Education, 74*, 61–72. https://doi.org/10.1016/j.compedu.2013.12.012

Dimopoulos, K., Koulaidis, V., & Sklaveniti, S. (2003). Towards an analysis of visual images in school science textbooks and press articles about science and technology. *Research in Science Education, 33*, 189–216.

Drexler, M., Drexler, D., & Sladek, P. (2018). Eye-tracking as a tool for measuring students' attention when working with physics textboks. In L. G. Chova, A. L. Martinez, & I. C. Torres (Eds.), *Edulearn18: 10th Int. Conf. on education and new learning technologies* (pp. 9219–9228). IATED.

Eilam, B., & Gilbert, J. K. (2014). The significance of visual representations in the teaching of science. In B. Eilam & J. K. Gilbert (Eds.), *Science teachers' use of visual representations* (pp. 3–28). Springer International Publishing. https://doi.org/10.1007/978-3-319-06526-71

Frey, H. P., Honey, C., & König, P. (2008). What's color got to do with it? The influence of color on visual attention in different categories. *Journal of Vision, 8*(14), 6. https://doi.org/10.1167/8.14.6

Hahn, L., & Klein, P. (2022). Eye tracking in physics education research: A systematic literature review. *Physical Review Physics Education Research, 18*, 013102. https://doi.org/10.1103/PhysRevPhysEducRes.18.013102

Jian, Y. C., & Ko, H. W. (2017). Influences of text difficulty and reading ability on learning illustrated science texts for children: An eye movement study. *Computers & Education, 113*, 263–279. https://doi.org/10.1016/j.compedu.2017.06.002

Koć-Januchta, M., Höffler, T., Thoma, G. B., Prechtl, H., & Leutner, D. (2017). Visualizers versus verbalizers: Effects of cognitive style on learning with texts and pictures—An eye-tracking study. *Computers in Human Behavior, 68*, 170–179. https://doi.org/10.1016/j.chb.2016.11.028

Korfiatis, K., Stamou, A., & Paraskevopoulos, S. (2004). Images of nature in Greek Primary School Textbooks. *Science Education, 88*, 72–89. https://doi.org/10.1002/sce.10133

Kress, G., & van Leeuwen, T. (1996). *Reading images: The grammar of visual design*. Routledge.

Mason, L., Tornatora, M. C., & Pluchino, P. (2013). Do fourth graders integrate text and picture in processing and learning from an illustrated science text? Evidence from eye-movement patterns. *Computers & Education, 60*(1), 95.

Mayer, R. E. (2009). *Multimedia learning* (2nd ed.). Cambridge University Press.

Pozzer, L. L., & Roth, W. M. (2003). Prevalence, function, and structure of photographs in high school biology textbooks. *Journal of Research in Science Teaching, 40*(10), 1089–1114.

Rauner, K., Havel, V., & Randa, M. (2007). *Textbook of physics for secondary schools*. Fraus Publishing House, Pilsner.

Rey, G. D. (2012). A review of research and a meta-analysis of the seductive detail effect. *Educational Research Review, 7*(3), 216–237. ISSN 1747-938X.

Slykhuis, D. A., Wiebe, E. N., & Annetta, L. A. (2005). Eye-tracking students' attention to PowerPoint photographs in a science education setting. *Journal of Science Education and Technology, 14*(5/6). https://doi.org/10.1007/s10956005-0225-z

Tangen, J., Constable, M., Durrant, E., Teeter, C., Beston, B., & Kim, J. (2011). The role of interest and images in slideware presentations. *Computers & Education, 56*. https://doi.org/10.1016/j.compedu.2010.10.028

Vojir, K., & Rusek, M. (2019). Science education textbook research trends: A systematic literature review. *International Journal of Science Education, 41*(11), 1496–1516.

Williams, J. L., Mccarley, N., Sharpe, J., & Johnson, C. (2017). The ability to discern relevant from irrelevant information on PowerPoint slides: A key ingredient to the efficacy of performance feedback. *North American Journal of Psychology, 19*, 219–236.

Williams, J. L., McCarley, N., Willis, S. D., & Huddleson, A. N. (2022). Images and text on PowerPoint slides: "Tracking" their impact on information retention. *International Journal of Mobile and Blended Learning, 14*(1), 1–14. https://doi.org/10.4018/IJMBL.291979

Wolfe, M. B. W., & Goldman, S. R. (2005). Relations between adolescents' text processing and reasoning. *Cognition and Instruction, 23*(4), 467–502. https://doi.org/10.1207/s1532690xci2304_2

Martina Kekule she holds a PhD in Physics Education from Charles University in Prague. As part of national and European projects, she has focused on students' performance and attitudes toward science, and has participated in large-scale evaluation initiatives (e.g., SMIRC TIMSS 2011). In recent years, her research has centered on eye-tracking, particularly examining how students solve multiple-choice science tests and interpret graphs. She also contributes to the development of interdisciplinary educational resources that bridge science and art, with a special emphasis on 3D photography.

Alžběta Krejčí is a science and mathematics teacher and aspiring psychologist in Czechia. Her research interests are eye-tracking and the effective use of technology in teaching.

Chapter 9
Applying Computational Thinking to Enhance Learning About Electricity: A Norwegian Case Study

Thomas Frågåt, Katarina Pajchel, and Louise Mifsud

9.1 Introduction

Over the last decade, several countries have introduced computational thinking (CT) into their curricula (Hsu et al., 2019; Kafai & Proctor, 2022). Arguments for the introduction of CT into curricula have ranged from viewing CT as a key transversal skill for the twenty-first century (Voogt et al., 2015) to a skill that is necessary for bringing school science and mathematics up to standard (Weintrop et al., 2016). In the Nordic countries, CT has been introduced by integrating it with specific subjects, across subjects or as a stand-alone subject (Pajchel et al., 2024). This article focuses on the Norwegian context, where CT—which is described as a problem-solving strategy—was introduced as part of the 2020 curriculum reform in mathematics, science, music and arts and crafts (Norwegian Directorate for Education and Training [NDET], 2019a). CT is only mentioned once in the mathematics curriculum; the term 'programming' is used throughout the rest of the curriculum. While the relationship between CT and programming is not explicated in the curriculum, the curriculum describes programming as a tool for exploration, problem-solving and design (NDET, 2019a). Kaufmann and Stenseth (2023) argue that the Norwegian curriculum encourages students to become good problem solvers and discover connections within and between the subjects.

T. Frågåt (✉)
Faculty of Education, Department of Mathematics, Natural Sciences, and Physical Education, University of Inland Norway, Hamar, Norway
e-mail: thomas.fragat@inn.no

K. Pajchel · L. Mifsud
Faculty of Teacher Education and International Studies, Department of Primary and Secondary Teacher Education, Oslo Metropolitan University, Oslo, Norway
e-mail: kapaj@oslomet.no; lomi@oslomet.no

G. Cakmakci, M. F. Tasar (eds.), *Connecting Science Education with Cultural Heritage*, Contributions from Science Education Research 15,
https://doi.org/10.1007/978-3-031-90944-3_9

The Norwegian curriculum in mathematics introduces specific programming concepts, such as variables, conditionals and loops, in the mathematics curriculum (NDET, 2019b), while the natural science curriculum, which includes technology as one of its core elements, emphasises that students are expected to use programming to 'explore, make and program technological systems that consist of parts that work together' (competence goal seventh grade) (NDET, 2019c). Thus, the curriculum indicates a progression where mathematics is responsible for introducing different programming competencies that are then used in other subjects (Pajchel et al., 2024). Furthermore, Norwegian curriculum and policy documents indicate that CT is introduced to enhance learning in the subjects (Andersen et al., 2023; Vinnervik & Bungum, 2022).

Kafai and Proctor (2022) argue that the focus of CT in curricula has shifted from a narrow emphasis on skills to a broader approach of literacy. CT has been described as a way of thinking, a problem-solving strategy and programming skills (Brennan & Resnick, 2012; Grover & Pea, 2013; Shute et al., 2017; Weintrop et al., 2016). Shute et al. (2017) argue that CT is 'the conceptual foundation required to solve problems effectively and efficiently (i.e. algorithmically, with or without the assistance of computers) with solutions that are reusable in different contexts' (p. 151). Shute et al.'s (2017) definition highlights six different CT facets—'decomposition, abstraction, algorithms, debugging, iteration, and generalization' (p. 153)—and emphasise inclusion of CT in different instructional settings and disciplines. Weintrop et al.'s (2016) definition takes mathematics and sciences as its point of departure and proposes four CT practices: data practices, modelling and simulation, computational problem-solving and systems thinking. Hence, Weintrop et al. (2016) depart from the more generic approach towards a taxonomy that targets the processes of knowledge development that are characteristic of mathematics and science. Weintrop et al.'s (2016) taxonomy aims to mirror how computing is currently used in scientific exploration and argumentation, thus contributing to the ongoing efforts to engage students in rich and realistic scientific inquiry as practice (Duschl, 2008).

Sengupta et al. (2013) and Nordby et al. (2022) found indications that integrating CT into subject-specific learning is challenging. Tannert et al. (2021) propose that because CT is a new topic in schools, teachers may experience tension between teaching CT as the primary learning objective or integrating it as a means of learning, making it a secondary learning objective. Finding suitable and meaningful ways of using CT in subjects is one of the challenges faced by teachers (Kite & Park, 2023; Li et al., 2020b). Although computational thinking has become a fundamental part of mathematics and science, it is still difficult to translate such methodologies into school instruction. Many curricula aim to integrate CT into the combined science, technology, engineering and mathematics (STEM) field (Bocconi et al., 2022). The presence of technology ('T') in STEM indicates that programming and computer science may have a natural place there (Li et al., 2020b). However, the vison of 'CT for all' emphasising applications and thinking and problem-solving skills (Kafai & Proctor, 2022; Li et al., 2020a; Papert, 1980) encourages CT integration through activities and approaches that are characteristic of and relevant in the context of subject learning.

The categories developed by Weintrop et al. (2016) are particularly interesting, as they are closely related to processes of exploration that are typical of inquiry-based learning as described by for example Crawford (2014). Several other frameworks indicate that CT can be meaningfully integrated into inquiry-based learning activities (Palts & Pedaste, 2020; Sengupta et al., 2013) by mapping it onto the typical stages of inquiry, which start by defining a problem or asking questions and formulating and testing hypothesis. This first stage also involves abstraction, which is frequently represented by modelling and is considered the core practice in scientific inquiry (Windschitl et al., 2008). Next, students engage in problem-solving, which involves the design of algorithms, programmes or other artefacts that facilitate the exploration of a phenomenon through testing, data collection and analysis of the model (Palts & Pedaste, 2020; Sengupta et al., 2013). Thus, integrating CT in inquiry-based learning can potentially support what Duschl (2008) calls conceptual and epistemic practices. This study aims to explore how CT can be integrated into inquiry-based science teaching to enhance subject-specific learning. In this chapter we raise the following research question:

What categories of CT are used when working with an inquiry-based approach in school science?

We explore a teaching unit for seventh grade which is an introduction to electric circuits that involves working with analogue and programmable equipment as mediating embodiments, drawing on Weintrop et al.'s (2016) model as an analytical framework.

9.2 Research Method

This case study is part of a larger design-based research project that aims is to develop an approach to integrating science and CT in course curricula. We employed the methodology of conjecture mapping (Sandoval, 2014). The guiding principle of the intervention is to work with real-world problems and embodiments—or tasks—that engage students in creating artefacts using CT. In this unit students were working unplugged with analogue circuits and plugged using micro:bits (MakeCode, n.a).

The participants were 25 seventh graders and their science teacher in Norway. Informed consent was obtained from teachers and students participating in the study and participants were pseudonymised. The class had been working with CT and programming in previous lessons and had been introduced to block programming in Scratch and micro:bit the previous year. The science project lasted over seven weeks (eight lessons). The aim of the project was to build a model of a smart house with electrical circuits and control systems. In this chapter, we report on an introductory lesson that lasted 1 hour and 18 minutes. This lesson was chosen because it presents the culmination of how circuits are developed and built, covering both unplugged and plugged approaches.

The lesson was divided into two parts, with the first 30 minutes being dedicated to investigating how to connect a light bulb to a battery so that a closed circuit is established. This part was regarded as unplugged, as the students worked on classical electrical wiring with light bulbs and a battery. During the second part, the students used micro:bits that were coded using the MakeCode platform (MakeCode, n.a), which we categorised as a plugged activity. The teacher gave the students readymade code as a starting point for a stepwise exploration of both the code and circuits through the Predict, Run, Investigate, Modify and Make (PRIMM) approach (Sentance et al., 2019). The students worked in fixed pairs throughout the lesson, and this was interrupted only by short whole-class teacher instructions. The teacher created the learning material herself, with support from the researchers.

Data were collected using video recordings. One camera followed the teacher, who had a microphone attached to her to capture any instructions provided and conversations with students. Five chest-mounted GoPro cameras captured the construction of the electrical circuit and the students' screen activities. In this chapter, we present data from the teacher camera, which captured the classroom activity as a whole and included the teacher's conversations with the students.

The qualitative data were analysed deductively, focusing on dialogues between teachers and students and between students while working with artefacts (e.g. electric circuits) and programming code. The codes were derived from the CT framework proposed by Weintrop et al. (2016) and the characteristics of inquiry-based teaching and learning described by Crawford (2014); see Table 9.1. We analysed the video sequences, primarily looking for indicators of CT practices, and looked for indicators of inquiry-based learning – the teacher's chosen pedagogical approach. Although CT and inquiry-based learning are different and distinct concepts, we juxtapose them in Table 9.1. Table 9.1 operationalises our analytical framework, which focuses on the similarities of aspects of CT and inquiry-based learning. The descriptions of inquiry-based learning are written alongside the descriptions of the CT practices. Some of the operationalisations in inquiry-based learning are relevant to several aspects of the CT practices. The codes used for the analysis are in bold fonts.

Coding was done using a spreadsheet. The video recordings were divided into segments of 8 minutes (the first segment is only 4 minutes due to privacy considerations)—long enough to elucidate the context and the codes that co-occurred. Each segment was analysed by looking for indicators of the five categories described in Table 9.1. Each segment was given a weighted value between 0 and 4. A value of 4 indicates a very high level of a given code and the observation of the occurrence of one or more aspects of the category and its presence throughout most of the duration of the segment. A value of 0 indicates that there was no trace of the code. In addition, a rough transcript was made for each segment. The coding of the whole dataset was done independently by the first and second authors to ensure inter-coder reliability and validated by the third author.

Table 9.1 The categories and operationalisations used for the analysis (bold fonts) based on Weintrop et al. (2016) and Crawford (2014)

CT practices	*Description of CT practices*	**Inquiry-based learning**
Data practices	Collecting Data, Creating Data, Manipulating Data, Analysing Data, Visualizing Data	Asking questions, designing and carrying out investigations and interpreting data as evidence
Modelling and simulation practices	Using Computational Models to Understand a Concept, Using Computational Models to Find and Test Solutions, Assessing Computational Models, Designing Computational Models, Constructing Computational Models,	Building models and interpreting data as evidence
Computational problem-solving practices	Preparing Problems for Computational Solutions, Computer Programming, Choosing Effective Computational Tools, Assessing Different Approaches/Solutions to a Problem, Developing Modular Computational Solutions, Creating Computational Abstractions, Troubleshooting and Debugging	Asking questions, designing and carrying out investigations and using logic and evidence
System thinking practices	Investigating a Complex System as a Whole, Understanding the Relationships within a System, Thinking in Levels, Communicating Information about a System, Defining Systems and Managing Complexity	Communicating findings, creating arguments and using logic and evidence

9.3 Results

The study presented in this chapter explores what categories of CT are applied when working with an inquiry-based approach in school science, namely the introduction to electrical circuits. The results are intertwined with a description of the activities that integrate inquiry and CT. Table 9.2 gives an overview of the weights assigned to the segments throughout the teaching unit. The first 30 minutes of the lesson, which were unplugged, were characterised by inquiry-based approaches, while the in second part we observed more prominent presence of CT practices, mirroring the teacher's lesson design.

While working unplugged, the students started by creating a hypothesis based on a drawing of six different possible connections of a light bulb, battery, and set of wires, where only one of the solutions was correct (see Fig. 9.1). The teacher went through each of the drawings, and the students had to show, with their thumbs up or down, whether they thought that the circuit would make the light bulb light up. The students had to explain their hypothesis based on previous knowledge and the teacher ensured that the students could give an argument in support of their hypothesis.

Several of the explanations of the battery-light bulb connections given by the students mentioned the positive and negative poles of the battery and the necessity of a closed circuit. However, the details about the impact of the positive and

Table 9.2 Overview of the results showing the coding of all segments

Segment	Inquiry-based learning	Data practices	Modelling and simulation practices	Computational problem-solving practices	System thinking practices
00:00–04:22	1	0	0	0	0
04:22–12:00	3	0	0	0	1
12:00–20:00	3	0	0	0	1
20:00–28:00	1	0	0	0	1
28:00–36:00	0	0	0	0	0
36:00–44:00	0	0	0	0	0
44:00–52:00	1	0	2	4	1
52:00–60:00	2	0	2	4	2
60:00–68:00	2	0	2	4	2
68:00–71:49	0	0	0	0	0

Fig. 9.1 Which light bulb will light up? The teacher asks about different ways of connecting a light bulb to a battery

negative poles were, at this stage, not clear to all the students, revealing a lack of understanding of the system as a whole. One of the students said, 'Couldn't we just test it?' This student had conceptualised that the simple system consisting of a battery and light bulb had to be a closed circuit and wanted to test the hypothesis. Thereafter, the students were given wires, a battery and a light bulb to test the different connections and check whether their hypothesis was correct. In our analysis, this sequence contained instances of asking questions and carrying out investigations; therefore, it scored high in terms of inquiry-based learning. It also introduced the concept of a closed circuit as a system, which must be understood to carry out troubleshooting and debugging.

The task was then extended by giving the students two (and later three) light bulbs and a different number of wires. Their teacher challenged the students to couple the light bulbs in different ways and prompted them to observe their brightness. Most of the students managed to create a parallel coupling. The teacher then asked them to remove one wire and still get the two light bulbs to light up:

Teacher See if you can manage ... now you have connected like that ... Can you get it to work if I remove a wire? Do it with just one wire...

[*The teacher returns after a little while.*]
Student What did you mean?
Teacher If, for example, we take ... like that, and then we fasten ... and then we fasten a wire on, oops! There!
Student It won't work because one light doesn't get a plus, or whatever it was...
Teacher Yes, but look now...
Student Ah, aha, yes, it was the wrong point [connector] we put it on [coupled it to].
Teacher But, right? But you see that something else happens with the light in these [couplings].
Student They get weak...
Teacher Yes, but why?
Student Because there is a too small current.
Teacher Yes, but it is... but you managed it on the others [*Teacher draws a circle in the air with her hand*]. Why...? What's the difference?
Student Because it is the world's longest wire.

The students were scaffolded by the teacher's challenge to discover serial couplings, which was a new concept to them. A requirement to solve this task was that the students understood how to connect a closed circuit and how the system worked as a whole. Their first attempt did not work, since they had connected to the wrong connector, even though the rules for circuits had already been introduced in the previous task. Our analysis indicates that the students corrected this by applying troubleshooting and using different approaches to assess solutions to the problem. To explore the circuits, especially when students had to remove a wire, they had to apply both problem-solving practices and inquiry-based approaches, such as investigation and using logic. Furthermore, our findings indicate that the students had to apply both logic and system thinking to understand why the light bulbs were dimmer in a serial coupling than in a parallel coupling, despite the argument that 'less current was running through the light bulbs' being based on intuition rather than a scientifically precise or complete understanding.

The unplugged part of the lesson was concluded with a whole-class discussion of the different circuits and an introduction to the notions of parallel and serial circuits to consolidate the learning sequence by letting the students communicate their findings. The aim was to bring the students to the same level of understanding of circuits before they started working with the micro:bits.

In our analysis, the unplugged part of the lesson was coded with a high degree of inquiry-based learning, as the students started with a concrete question, formulated hypotheses and carried out investigations that tested these. They also engaged in some degree of system-thinking practices, as they had to understand the relationship between the components. CT problem-solving practices were only present when the students were troubleshooting.

In the second part of the lesson, the battery and light bulbs were replaced by micro:bits and light-emitting diodes (LEDs). Before starting, the students were introduced to LEDs and how they should be connected in a circuit with their short (-) and long legs (+) to the correct pole. The teacher highlighted the similarities between light bulbs and LEDs to relate the unplugged electrical circuit to the new task in which the micro:bit would be the energy source, as well as a new element in the circuit: a switch.

Throughout this part of the lesson, the students continued working in an inquiry-based way following a progression inspired by the PRIMM approach. The first task was scaffolded with readymade code, which let the students more or less follow a recipe to connect and get an LED to light up by using the micro:bit as a switch. The focus was on understanding and testing the programming code of a switch and combining it with the previously explored concept of electrical circuits.

First, in a whole-class exercise, the teacher read and explained the block-based programme that controls the micro:bit as a switch. The teacher started by asking, 'What are we actually going to find out?' She explained that the aim was to find out how to 'get light in the LEDs', which meant connecting an electrical circuit. This part of the system was familiar for the students from the unplugged part of the lesson. However, now the students were to use the micro:bit as a microcontroller to get the LEDs to light up. Students expressed their initial understanding of the program in comments attached to their code. The students connected the circuit and tested their understanding by running the code. They were also expected to revise their comments if their understanding changed after running the code. As the students gradually completed the first task, they were encouraged to modify both the circuit and the code. They were asked to add LEDs relating to the concepts of parallel and series circuits. This entailed adding code for more switches so that the LEDs could be operated independently. Some students received LEDs in different colours, which they programmed as traffic lights or blinking lights. All modifications implied extending the physical circuit and working with concepts such as closed circuit, positive and negative poles, ground and current, as well as recognising and reusing the basic example code for a switch. Running, investigating and gradually modifying the code and circuit intertwines physics subject content with several of the central CT categories, such as modelling, system thinking and problem-solving.

The teaching unit was intended as a basis for creating a model of a smart house in later lessons. At this stage, viewing the circuit as a small-scale model, the circuit is used to understand the concepts of large-scale circuits working in systems such as smart houses. For example, the teacher explained the concept of a closed circuit to the whole class as follows:

Teacher We talked that it [the wire] had to be connected to the battery to have a circuit. And this will also be very important when we work with micro:bits later on ...If we want to have light in the light bulbs or diodes, which we will work with later, then we need to have a circuit, the current must flow, have a circuit to flow through all the way, which is a closed circuit.

The teacher referred to this notion throughout the lesson, often illustrating it by drawing a circle in the air with her hand. Similarly, as in the unplugged part, the circuit was used to test the students' understanding of both the code and the circuit, which implied new connection points and LEDs that needed to be connected with the correct polarity. Thus, the system clearly functioned as a basis for students' inquiry.

Our findings indicate that the students had to apply system-thinking practices to make the circuit work as intended. The task of getting light in a bulb or LED required an understanding of each component and how to connect them together, noting the basic patterns of series and parallel circuits. After managing to programme the micro:bit as a switch for one LED, the extra tasks involving circuits and switches for additional LEDs were more explorative in nature. Some of the higher-achieving students were asked to make the LED blink regularly, meaning they had to add new programming concepts such as variables and loops to the code. The excerpt below exemplifies how the teacher helped a group expand their system from one to two LEDs:

Teacher Did you manage?

Student We did the same, but there is always only one [LED] that shines.

Teacher Yes, then we must think, maybe it is possible to use one more wire and use some of these patches [micro:bit connection pins] to make the other shine as well. But maybe you need some more blocks; for example, you could add one 'when A and B'. Then, one can write the digital value; then it is possible to change where it says P0 to P1. And then turn it on, for example [*the teacher points to the screen and the micro:bit*].

When students started working on modifying and extending the circuits, system-thinking and inquiry-based practices were the most conspicuous practices, notably those related to communicating information about the system. Communicating information about the system was essential in both the scaffolding of the tasks and troubleshooting.

In the second, plugged, part of the lesson we find a high degree of CT problem-solving practices. After the thorough introduction of the basic code for a switch, students had to recognise it as a basic module that they could modify and reuse when adding more LEDs to the circuit. In this way, exploring circuits by extending them with different combinations of LEDs facilitated discussions about effective computational tools. Reusing the basic code for a switch also gave the students an opportunity to learn implicitly about computational abstraction. However, the most explicit problem-solving practices observed were troubleshooting and debugging, as the following excerpt illustrates:

Student It [micro:bit] smiles and is grumpy…

[The teacher points at the screen.]
Teacher But that means you haven't downloaded your code; you should have these symbols. So, you must try to download one more time.

The excerpt illustrates how the teacher pointed out that the code running on the micro:bit was not the one that the students had programmed. Similar dialogues occurred in other groups, as students needed help downloading the code to the micro:bit. Many students also had to be reminded to ensure that the LED was connected with the correct polarity and that the wires were properly connected to the micro:bit. This troubleshooting was often explained using physics concepts, such as closed circuits. Even though the code used only a few blocks and could be copied from a template, the students needed to adjust it, meaning that they engaged in more classical debugging, which involved many cycles of trial and error, giving them the opportunity to practice systematic debugging.

To summarise, our analysis indicates that CT modelling, as well as system-thinking and problem-solving practices, co-occurred with the inquiry-based activities. Interestingly, it was often difficult to distinguish between the problem-solving practices of CT and the inquiry-based approaches, which highlights the close relationship between CT and inquiry-based learning in the approach of this teaching unit. Even when troubleshooting and investigating the system students were more focused on understanding the underlying scientific content, i.e. electrical circuits, rather than the micro:bits, LEDs and wires. Thus, the main focus was subject specific. Although models with integrated programmable units were the objects of the students' inquiry, most of the computational modelling and simulation practices were less common during the intervention. Data practices were naturally absent, as the focus was on inquiry through the construction of technological artefacts.

9.4 Discussion

The aim of this study was to explore how CT can be integrated into inquiry-based science teaching to enhance subject-specific learning and identify which CT categories are applied when using such an approach. This was accomplished through a case study that followed a lesson in which students used programmable micro:bits to explore the concept of an electrical circuit. The analysis based on the framework that combined CT practices and inquiry-based learning allowed us to investigate the co-occurrence of aspects of both learning approaches. Our findings indicate that the combination of working with analogue components and programable microcontrollers involved both system-thinking and problem-solving practices, two CT categories proposed by Weintrop et al. (2016). Students were introduced to a system consisting of the basic electricity components that were used to create the fundamental model of a closed circuit and served both as a conceptual model and a model for testing different hypotheses.

When the students programmed the micro:bit, programming and CT were integrated into the inquiry. Rather than mere 'add-ons', they constituted tools that allowed the students to express and explore their ideas and interact with physics and technology through the system. Programming, as represented by basic blocks, is a language that encodes the logic of students' ideas. In line with Li et al. (2020a) we argue that CT and programming becomes an expression of students thinking. When CT melts into working methods of inquiry in a science/technology topic, the micro:bit and programme controlling the circuit could be interpreted as what Papert (1980) called 'objects to think with'.

The analysis of the plugged part of the lesson showed that the analogue introduction was not only a warmup but was actively integrated and presented in the plugged part, especially when students encountered problems. Solving problems required both an understanding of the basic principles of an electrical circuit and CT practices. Students were observed working systematically and developing skills while troubleshooting, debugging and communicating about the system with peers and the teacher, which were most prominent CT practices in our analysis. The teacher actively used the notion of a closed circuit when supporting students' debugging. In our analysis, the CT practices were interpreted as tightly linked to the subject content. They cooccurred with and were integrated into practices that are central to inquiry-based approaches as described by Crawford (2014) and Duschl (2008), among others.

Throughout the lesson, we observed students engaging in subject-specific learning through the simple model of a technological system—an electric circuit—when working both with and without the programable micro:bit. In the Norwegian science curriculum, programming is supposed to be used to explore both technology and natural phenomena. It was therefore a conscious choice to include programming and CT not only as technology but rather as a tool to be used in subject-specific injury and problem-solving. Li et al. (2020b) encourage studies of concrete implementations of CT integration in STEM. The analysis of the teaching unit through the lens of our framework that combines CT and inquiry-based learning provides a promising example of CT and science and technology integration. In our analysis, the integration did not simply entail adding new technology; rather, it entailed designing an inquiry and modelling activity in which a programable device facilitated thinking and problem solving, where CT co-occurred in mutually enhancing synergy with inquiry-based learning. The role of CT could be interpreted as what Tannert et al. (2021, pp. 78–79) describe as 'secondary learning objective'. This suggests a promising pathway for using CT to enhance subject-specific learning outcomes. Consequently, the inquiry-based approach exemplified in this case study could help remedy the challenges Sengupta et al. (2013) highlighted, namely those of integrating CT into subject-specific learning. Unlike the findings of Sengupta et al. (2013) and Nordby et al. (2022), our findings indicate that CT neither took the focus away from the subject, nor was it detrimental to the teaching of the subject; rather, CT was integral to the creation of the model of the system.

9.5 Conclusion and Further Directions

This study suggests an approach teachers can use to apply CT in inquiry-based teaching. With regard to CT, the inquiry-based lesson we observed contained several aspects of the practices described by Weintrop et al. (2016) and showed interesting synergies between these two approaches. Previous studies have highlighted the challenges of integrating CT in subjects at the cost of subject content (Nordby et al., 2022; Sengupta et al., 2013). However, our findings indicate that through engaging with artefacts, in this case represented by an analogue circuit and a circuit controlled by a micro:bit, and through the teacher's scaffolding, CT and programming were integrated with the subject content. CT became what Tannert et al. (2021, pp. 78–79) describe as a secondary rather than primary learning objective.

While this chapter reports on a short teaching sequence, it draws on a longitudinal study covering a period of seven weeks. Thus, this study presents a snapshot of the interconnection and possible synergies between CT and inquiry-based learning and suggests avenues for future studies. Focusing on artefacts resembling real-world problems might be one approach that mitigates the challenges of integrating CT into subject-specific learning that warrants further investigation.

Acknowledgements This project was funded by the Norwegian Research Council grant no. 320322

References

Andersen, R., Frågåt, T., Boğar, Y., Jensen, J. J., & Mifsud, L. (2023). Representations of computational thinking in policy documents in an educational context: The cases of Denmark, Finland, and Norway. In *Proceedings of the 17th International Conference of the Learning Sciences-ICLS 2023* (pp. 35–42) https://repository.isls.org//handle/1/10269

Bocconi, S., Chioccariello, A., Kampylis, P., Dagienė, V., Wastiau, P., Engelhardt, K., Earp, J., Horvath, M., Jasutė, E., & Malagoli, C. (2022). *Reviewing computational thinking in compulsory education: State of play and practices from computing education.* (JRC128347). Publications Office of the European Union.

Brennan, K., & Resnick, M. (2012). New frameworks for studying and assessing the development of computational thinking. In *Proceedings of the 2012 annual meeting of the American educational research association, Vancouver, Canada.*

Crawford, B. A. (2014). From inquiry to scientific practices in the science classroom. In *Handbook of research on science education* (Vol. II, pp. 529–556). Routledge.

Duschl, R. (2008). Science education in three-part Harmony: Balancing conceptual, epistemic, and social learning goals. *Review of Research in Education, 32*(1), 268–291. https://doi.org/10.3102/0091732x07309371

Grover, S., & Pea, R. (2013). Computational thinking in K–12: A review of the state of the field. *Educational Researcher, 42*(1), 38–43. https://doi.org/10.3102/0013189X12463051

Hsu, Y.-C., Irie, N. R., & Ching, Y.-H. (2019). Computational thinking educational policy initiatives (CTEPI) across the globe. *TechTrends, 63*(3), 260–270. https://doi.org/10.1007/s11528-019-00384-4

Kafai, Y. B., & Proctor, C. (2022). A revaluation of computational thinking in K–12 education: Moving toward computational literacies. *Educational Researcher, 51*(2), 146–151. https://doi.org/10.3102/0013189x211057904

Kaufmann, O. T., & Stenseth, B. (2023). Programmering som del av matematikkfaget. *Norsk pedagogisk tidsskrift, 107*(2), 109–123. https://doi.org/10.18261/npt.107.2.2

Kite, V., & Park, S. (2023). What's computational thinking?: Secondary science teachers' conceptualizations of computational thinking (CT) and perceived barriers to CT integration. *Journal of Science Teacher Education, 34*(4), 391–414. https://doi.org/10.1080/1046560X.2022.2110068

Li, Y., Schoenfeld, A. H., diSessa, A. A., Graesser, A. C., Benson, L. C., English, L. D., & Duschl, R. A. (2020a). Computational thinking is more about thinking than computing. *Journal for STEM Education Research, 3*(1), 1–18. https://doi.org/10.1007/s41979-020-00030-2

Li, Y., Schoenfeld, A. H., diSessa, A. A., Graesser, A. C., Benson, L. C., English, L. D., & Duschl, R. A. (2020b). On computational thinking and STEM education. *Journal for STEM Education Research, 3*(2), 147–166. https://doi.org/10.1007/s41979-020-00044-w

MakeCode. (n.a.). https://makecode.microbit.org/

Nordby, S. K., Bjerke, A. H., & Mifsud, L. (2022). Primary mathematics teachers' understanding of computational thinking. *KI - Künstliche Intelligenz, 36*(1), 35–46. https://doi.org/10.1007/s13218-021-00750-6

Norwegian Directorate for Education and Training. (2019a). *Curricula in English.* Retrieved from https://www.udir.no/in-english/curricula-in-english/

Norwegian Directorate for Education and Training. (2019b). *Mathematics year 1–10 (MAT01-05), Competence aims after year 4.* Retrieved from https://www.udir.no/lk20/mat01-05/kompetansemaal-og-vurdering/kv18?lang=eng

Norwegian Directorate for Education and Training. (2019c). *Natural science (NAT01-04), Competence aims after year 7.* Retrieved from https://www.udir.no/lk20/nat01-04/kompetansemaal-og-vurdering/kv79?lang=eng

Pajchel, K., Mifsud, L., Frågåt, T., Rehder, M. M., Juuti, K., Bogar, Y., Lavonen, J., Schrøder, V., Aalbergsjø, S. G., & Rognes, A. (2024). Sign of the times: The framing of computational thinking in Danish, Finnish, and Norwegian Curricula. *Nordic Journal of Comparative and International Education (NJCIE), 8*(4). https://doi.org/10.7577/njcie.5744

Palts, T., & Pedaste, M. (2020). A model for developing computational thinking skills. *Informatics in Education, 19*(1), 113–128.

Papert, S. A. (1980). *Mindstorms: Children, computers, and powerful ideas.* Basic books, Inc.

Sandoval, W. (2014). Conjecture mapping: An approach to systematic educational design research. *Journal of the Learning Sciences, 23*(1), 18–36. https://doi.org/10.1080/10508406.2013.778204

Sengupta, P., Kinnebrew, J. S., Basu, S., Biswas, G., & Clark, D. (2013). Integrating computational thinking with K-12 science education using agent-based computation: A theoretical framework. *Education and Information Technologies, 18*(2), 351–380. https://doi.org/10.1007/s10639-012-9240-x

Sentance, S., Waite, J., & Kallia, M. (2019). Teaching computer programming with PRIMM: A sociocultural perspective. *Computer Science Education, 29*(2-3), 136–176. https://doi.org/10.1080/08993408.2019.1608781

Shute, V. J., Sun, C., & Asbell-Clarke, J. (2017). Demystifying computational thinking. *Educational Research Review, 22*, 142–158. https://doi.org/10.1016/j.edurev.2017.09.003

Tannert, M., Lorentzen, R. F., & Berthelsen, U. D. (2021). Computational thinking as subject matter: As an independent subject or integrated across subjects? In *Computational Thinking in Education* (pp. 73–89). Routledge.

Vinnervik, P., & Bungum, B. (2022). Computational thinking as part of compulsory education: How is it represented in Swedish and Norwegian curricula? *Nordic Studies in Science Education, 18*(3), 384–400.
Voogt, J., Fisser, P., Good, J., Mishra, P., & Yadav, A. (2015). Computational thinking in compulsory education: Towards an agenda for research and practice. *Education and Information Technologies, 20*(4), 715–728. https://doi.org/10.1007/s10639-015-9412-6
Weintrop, D., Beheshti, E., Horn, M., Orton, K., Jona, K., Trouille, L., & Wilensky, U. (2016). Defining computational thinking for mathematics and science classrooms. *Journal of Science Education and Technology, 25*(1), 127–147. https://doi.org/10.1007/s10956-015-9581-5
Windschitl, M., Thompson, J., & Braaten, M. (2008). Beyond the scientific method: Model-based inquiry as a new paradigm of preference for school science investigations. *Science Education, 92*(5), 941–967. https://doi.org/10.1002/sce.20259

Thomas Frågåt has been an assistant professor at the University of Inland Norway. His primary research interests are physics education and computational thinking in science education.

Katarina Pajchel is an associate professor at OsloMet – Oslo Metropolitan University in Norway. Her primary research interests include science education and teacher education, focusing on research-based teacher education, programming, and computational thinking within the context of science education.

Louise Mifsud is a professor at OsloMet – Oslo Metropolitan University in Norway. She holds a PhD in Education from the University of Oslo. Her research and teaching interests focus on digital literacy in schools and teacher education, particularly cyber ethics and computational thinking. Currently, she serves as the lead PI on a Norwegian Research Council-funded project regarding Mathematics, Science, and Computational Thinking in Nordic schools and teacher education (MASCOT).

Chapter 10
Students' Multimodal Meaning Making of Nature of Science: From Isolated Scientific Practices to Coherent Scientific Methods

Kason Ka Ching Cheung, Alis Oancea, and Sibel Erduran

10.1 Introduction

The goal of helping students develop an informed understanding of nature of science (NOS) has been emphasized in curriculum reform documents around the world (CDC, 2017; DOE, 2013; NGSS, 2013; NRC, 2012). From a holistic account, NOS is defined as a set of inter-connected cognitive-epistemic and social-institutional characteristics of science (Erduran & Dagher, 2014), including but not limited to how scientific knowledge is generated, how scientific investigation was conducted, as well as how scientific work interacts with broader social and institutional systems. In the book length account by Erduran and Dagher (2014), the cognitive-epistemic system comprises aims and values, scientific methods, scientific practices and scientific knowledge, while the social-institutional system encompasses social certification and dissemination, scientific ethos, social values, professional activities, financial systems, political power structures and social organisations and interactions. Such conceptualisation of NOS adopts a "big approach" (Gandolfi, 2020) which is inclusive of both nature of scientific knowledge and nature of scientific inquiry. In the authentic scientific world, these NOS categories are interconnected (Chan et al., 2023; Cheung, 2020; Cheung et al., 2023). For example, in news

K. K. C. Cheung (✉)
Department of Curriculum and Instruction, The Education University of Hong Kong, Hong Kong, China
e-mail: cheungkac@eduhk.hk

A. Oancea
Department of Education, University of Oxford, Oxford, UK

S. Erduran
Department of Education, University of Oxford, Oxford, UK

Maison de la Création et de L'Innovation, Université Grenoble Alpes, Grenoble, France

G. Cakmakci, M. F. Tasar (eds.), *Connecting Science Education with Cultural Heritage*, Contributions from Science Education Research 15,
https://doi.org/10.1007/978-3-031-90944-3_10

media's communication of science during the COVID-19 pandemic, two or more NOS categories, such as scientific practices and scientific methods, were often co-articulated in representing the works of scientists. Hence, interconnections are far more important than meaning making of seemingly disperse NOS categories.

One type of connections is that a NOS category can foreground students' meaning making of another NOS category. For example, the category of "scientific practices" can form a basis of meaning-making of other NOS categories such as "scientific knowledge" or "scientific methods". Students can engage in a range of scientific practices in the production of scientific knowledge (Osborne, 2014). Moreover, during engagement in scientific practices, students are the active agents of making decisions regarding the steps of scientific methods, as well as what kinds of products (e.g., scientific models or politics and economics of science) need to be created (Berland et al., 2016).

Although our previous work demonstrated that meaning making of NOS is multimodal in nature (Cheung et al., 2025, 2024), it is still unknown how students use different modes of representation to make meaning across aspects of NOS. Emerging works in this field targeted teacher education where pre-service teachers can use the visual mode to show connections among ideas within a single NOS category (Barak et al., 2023; Erduran & Kaya, 2018). To explore the affordances of multimodality in students' meaning making across NOS categories, we co-planned explicit-reflective NOS instruction with secondary science teachers in Hong Kong, while opportunities were given to grade seven students to reflect on their NOS categories using the visual, verbal, and written modes of representation. Data from a recent research project demonstrated that students' multimodal meaning-making of nature of scientific methods foregrounds their understanding of individual scientific practices.

In the following section, we will review the Family Resemlance Approach to NOS, particular the version developed by Erduran and Dagher (2014) and highlight the importance of multimodal connections in the process of students' meaning making of NOS. We then describe the contexts and methodology of this study and present the findings to illustrate students' multimodal connections of NOS. Finally, we discuss how science teachers can harness multimodal connections in planning explicit-reflective instruction of NOS for secondary students.

10.2 Family Resemblance Approach to NOS: From List in Pieces to Coherent Wholes

The origins of the FRA in science education stem from Irzik and Nola's (2011) work where the authors adopted the notion of family resemblance to formulate NOS for inclusion in science education. The notion of family resemblance itself originated from Wittgenstein (1958) and suggests that there is not only a single definition but a cluster of interrelated components. For example, the term "game" can include ball games, card games or mind games (Wittgenstein, 1958). There are differences among different types of games, but these games share partly overlapping nets of similarities and differences with one another. The definition of games does not necessarily have a set of rigid conditions that determine their meaning (Wittgenstein, 1958).

Similarly, there are some characteristics that are shared among some science disciplines, but not all (Irzik & Nola, 2011). For example, human observations are required in some disciplines of science such as chemistry or biology, but not in astronomy. In astronomy, the satellites will capture the image of the Earth's surface, which does not require human eyes observation (Irzik & Nola, 2011). After examining a number of similarities and differences among various science disciplines, Irzik and Nola (2011) suggested four cognitive-epistemic categories that offer structural descriptions of nature of science, activities, aims and values, methods and methodological rules, and products. Irzik and Nola (2011b) replaced "activities" with the term "scientific process". They (2011b) further divided the social-related aspects into four finer categories: professional activities, scientific ethos, social certification and dissemination, and social values.

Based on the work of Irzik and Nola (2011b), Erduran and Dagher (2014) suggested that NOS comprise a cognitive-epistemic system and a social-institutional system by constructing a Family Resemblance Approach (FRA) wheel. In the FRA wheel, the cognitive-epistemic system of NOS comprises aims and values, scientific methods, scientific practices and scientific knowledge, while the social-institutional system encompasses social certification and dissemination, scientific ethos, social values, professional activities, financial systems, political power structures and social organisations and interactions. The FRA wheel comprises eleven revised broad categories, with emphasis on connections among them (Erduran & Dagher, 2014), which are inclusive of diverse understandings of NOS.

In this book chapter, we focus on students' meaning-making of two NOS categories as they emerged from the data:

- The *Methods and Methodological Rules* category addresses the diversity of scientific methods and a set of rules for scientific activities. Methodological rules include, for example, consideration of variables, setting up a control experiments and choosing suitable research objects;
- The *Practices* category comprises a set of cognitive and epistemic activities that lead to generation of scientific knowledge. The commonly addressed practices are those listed in 'A framework for K-12 science education: Practices, crosscutting concepts, and core ideas' (NRC, 2012), which includes asking a question, developing and using models, planning and carrying out investigation, analyzing and interpreting data, using mathematical thinking, constructing explanations, engaging in argumentation, and communicating information.

10.3 Multimodal Connections of NOS

The concept of multimodal connections originally refers to how students link different semiotic modalities of the same concept. Students need to critically constellate multiple modes of the same concept to fluently access disciplinary ways of knowing (Airey & Linder, 2009). Modes, which refer to a way of representing and communicating meaning in relation to socio-cultural contexts (Kress, 2010), can include the visual, linguistic, spatial, aural and gestural modes (New London Group, 2000). To name an example in science education, that of meaning making of work-energy relationships, students need to translate visual-spatial relationships in moving cars into their written mathematical formula (Tang et al., 2010). As there are different modalities to represent the same concept, it is important for students and teacher to pay attention to the subtle variations among different modalities of the identical concepts (Tang et al., 2010). The visual-spatial relationships between moving cars can show the physical distance, while the mathematical formula of work-energy relationships can be used to show how changes in one parameter affect the changes in another parameter. In a practical work that investigates the work-energy relationships of moving cars, it is important for students to make multimodal connections between the visual-spatial relationships among moving cars and the mathematical formula.

In the context of NOS, multimodal connections are more than using different modalities to represent identical ideas. Since NOS, particularly as conceptualised in the FRA framework, is an interconnected network of ideas, students can harness different modes to coherently represent such connections among different NOS categories. Young children in fact are able connect different ideas using a range of semiotic resources, such as colour (Taylor & Villanueva, 2014) or photographs (Littleton et al., 2010). Previous studies only investigated how a single mode can represent a single tenet of NOS. For example, in the context of third-grade American science classrooms, when students were asked to draw a scientist, they drew themselves engaging in doing experiments (Walls, 2012). Also, in Akerson et al. (2010)'s study, elementary students can use drawing to visualize creative NOS. However, as NOS ideas are intrinsically linked to each other, students, especially secondary school students, are capable of harnessing different semiotic modes to make multimodal connections across NOS categories.

10.4 Data Collection and Research Design

The following research question guided the present study:

How do students use different modes of representation to make connections across NOS categories?

The case chosen in this chapter is a focal group of three girls who were studying in a grade seven science classroom situated in a public girls' school in Hong Kong.

The three girls are randomly selected from a pool of students who consented to take part in this study. Cindy, who is the teacher of this class, has more than 27 years of teaching experience. In this research project, the first author co-planned a four-lesson instructional sequence which encompasses explicit-reflective instruction of NOS. Towards the end of each lesson, students were given opportunities to reflect on their NOS understanding in relation to the contexts of the lesson. The findings discussed in this chapter are drawn from the final lesson of the co-planned instructional sequence. During the lesson, a camera which was supported by a tripod captured students' drawing and writing, while another camera captured who was talking.

The fourth lesson was situated in the subtopic "Biodiversity". According to the local curriculum, students should be able to give examples of human activities that cause decrease in biodiversity. Cindy problematized the lesson by highlighting that the population of a native turtle species, the Chinese pond turtle, was decreasing in HK. At the start of this lesson, Cindy revisited four different types of scientific methods: manipulative hypothesis-testing, non-manipulative hypothesis-testing, manipulative parameter measurement and non-manipulative parameter measurement (Brandon, 1994). She then played a video showing that the red-eared slider turtle was more aggressive than the Chinese pond turtle in competing for food and shelter. Afterwards, students needed to draw an explanatory model to predict and account for what happens if a foreign species, the red-eared turtle, was introduced to the local habitat, based on their observations in the video. Cindy then questioned students on whether they had carried out a manipulative experiment or formulated a priori hypothesis. Students were then asked to reflect on the nature of scientific methods used by scientists to study ecological relationships between the foreign species and the native species. In the multimodal reflection task administered at the end of each lesson, students were given worksheets with guiding questions and a support box with all scientific practices, as Cindy perceived this as a useful way to scaffold her students in composing NOS multimodal representations.

The video recording was transcribed according to the conventions set out in Cheung et al. (2025). The video recording was segmented according to each utterance. Each utterance is denoted by the start and the end of an individual's speech. In the transcript, functions of the visual, verbal and written modes were analysed, which helped make sense of how different modes contribute to meaning making of NOS. Themes were generated after coding the functions of the visual, verbal and written modes. Three of the authors read and checked the themes.

10.5 Findings

10.5.1 Context of the Lesson

The last lesson of NOS instruction was grounded in the debates on whether people should buy turtles and release them to local pond due to religious reasons. Cindy explained that some individuals in Hong Kong had bought a more invasive and aggressive foreign species, the Red-eared Slider Turtle, and released them to local ponds. This threatens the life of a local species, Chinese Pond Turtle. Students watched a video on fighting between the two species of turtles, the Chinese Pond Turtle and Red-eared Slider Turtle in an aquarium. Based on their observations in the video, they drew a model to account for the relationship between the two species of turtle. Cindy explicitly explained that not all scientific methods require a hypothesis and an experiment, which is consistent to the ideas in Brandon (1994)'s matrix. After drawing a model to account for the relationships between the two species of turtles, the students reflected on the scientific method they had used in constructing an ecological model (Fig. 10.1).

10.5.2 The Focal Group's Discussion

We illustrate the focal group's discussion through an episode that began with Student 7C03 writing the first step of scientific method, "Having experiment or hypothesis" (turn 1). This was then checked by 7C05, who asked if they had to write these steps of scientific method (turn 2): "係要寫呢d咩?" (Is that we have to write this?). After that, student 7C03 suggested that they had to explain each step of scientific method, by saying "佢要解釋呀麻。" (They have to explain.) (turn 3). In turn 4, student 7C05 questioned, "唔係要畫咩? (.) 唔係要畫d烏龜咩?" (Do we

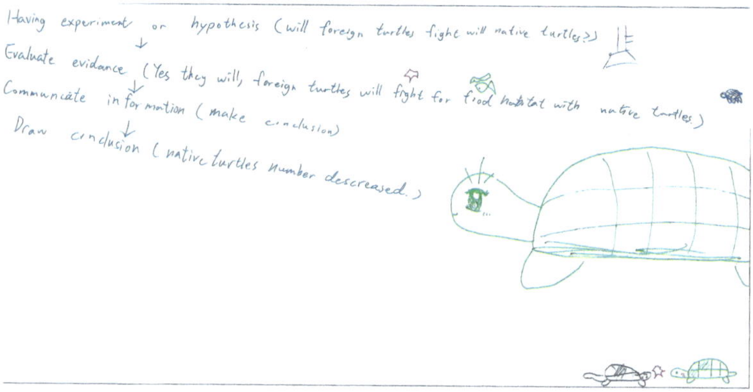

Fig. 10.1 Students' multimodal representations of nature of scientific methods

have to draw? (.) Do we need to draw the turtle?), as they were unsure if they needed to use drawings to visualize the turtles. Student 7C03 (turn 5) then agreed to use drawing to visualize the context of learning NOS. These verbal utterances on checking, suggesting and questioning ideas were mainly in Cantonese.

Starting from turn 4, student 7C03 drew an arrow pointing downwards under "Having experiment or hypothesis", in order to concretise the sequential relation between scientific practices and to show the steps of scientific methods. Afterwards, 7C03 wrote "Evaluate Evidence" (turn 6) and expanded the context of scientific method by writing "will foreign turtles fight with native species?" next to "Having experiment or hypothesis" (turn 7). Followed by that, student 7C03 drew another arrow pointing downward under "Evaluate evidence" (turn 8), to signify the next step of scientific method. The context of the written step of "evaluate evidence" was then expanded by 7C03 by writing "(Yes, they will, foreign turtles will fight for food habitat with native turtle)" (turn 9). To facilitate transduction from written steps of scientific method to visualization of context of scientific method, 7C05 used a mixed of English and Cantonese to suggest "Draw a Bunsen Burner to burn the turtle") (turn 10). Drawings of the scientific context for NOS meaning-making commenced from turn 11, with the use of a red star-shaped symbol to illustrate the intense fight between the two turtles. 7C03 then wrote down one of the steps of scientific methods, "Communicate information", under the arrow (turn 12).

From turn 13, the focal group students' discourse oriented from focusing on the nature of scientific method to scientific content. 7C05 asked in Cantonese and English, "food 佢地食咩架?" (What kind of food do they eat?). 7C03 expressed difficulty in drawing the turtle, by saying that "龜好難畫。" (So difficult to draw the turtle.)). The question asked by 7C05 was then answered by herself, saying in Cantonese, "養龜更係食龜糧。食大自然的蟲蟲。" (If you keep a turtle, they must eat turtle food. Eat worms in nature). She visualized her verbal discourse by drawing a fish and a worm above the word "food" in the second step of scientific method, illustrating the materiality of the scientific context. Afterwards, in turn 16, 7C03 drew an arrow pointing downwards below "communicate information", concretising the sequential relation between steps of scientific methods.

The focal group students used Cantonese to verbally express metacognition while they were constructing their multimodal representations of NOS. The verbal discourse in Cantonese between turns 2 and 5 allowed students to communicate knowledge of task variables. Knowledge of task variables includes that about the nature of the task and the demands of the tasks on individuals (Livingston, 2003). For example, in turn 3, 7C03 suggested in verbal mode that not only were they required to put down steps of scientific methods, but they would also have to provide explanation to elaborate each step. Moreover, 7C05 also questioned in turn 4 if the use of modalities other than the written mode, such as drawing, was necessary in illustrating the scientific contexts. Metacognition was empirically found in other studies to be related to students' nature of science understanding (Goren & Kaya, 2022; Peters & Kitsantas, 2010). Such verbal discourse in a more comfortable language, Cantonese, facilitated them to become metacognitive of the nature of multimodal tasks and expanded their visual and linguistic meaning of the nature of scientific methods. As seen in turns 7 and 9, each step of the scientific method was further expanded and contextualised in written mode. In turns 10, 11 and 15, drawings then extended the meaning of the students' written explanation of the steps of scientific method in relation to the context.

10.5.3 Multimodal Connections Across NOS Categories

From this episode, the combination of the verbal, written and visual modes prompts the rearrangement of discursive scientific practices into a coherent flow of scientific method. Regarding the verbal mode, the focal group checked (Turn 2), suggested (Turn 3) and questioned (Turn 4) if individual scientific practices or step of scientific method need to be included in their multimodal representations. The written mode layers down individual scientific practices, namely "Having an experiment or hypothesis" (Turn 1), "Evaluate evidence" (Turn 6) and "Communicate information" (Turn 12). These scientific practices were linked in a sequential order to students' drawing of downward arrows. Such multimodal representation of the scientific method engaged by the students (Fig. 10.1(c)) is different from that typically represented in science textbooks (Fig. 10.1(d)) which follows the sequence of asking a question, making a hypothesis, doing an experiment, recording the results, and drawing a conclusion. In their multimodal representations, the focal group students articulated the important stages of scientific methods including communicating information and evaluating evidence. More importantly, as indicated in the phrase 'having an experiment or hypothesis", students internalized that not necessarily every scientific investigation involves a hypothesis or an experiment. Brandon (1994)'s matrix indicates that scientific investigation can be about not testing a hypothesis or conducting an experiment. In other words, testing a hypothesis or manipulating variables in a controlled experiment are not exclusive criteria for a scientific method (Table 10.1).

10.6 Discussion and Conclusion

The empirical study reported in the chapter has potential implications for pedagogy and research in secondary science education. Different semiotic modes helped students to extend their understanding of a NOS category to their understanding of another NOS category. As seen in the episode, students extended their understanding of scientific practices into their understanding of methods and methodological rules. In the findings presented in this book chapter, students utilised different modes to rearrange these individual scientific practices into communicating the diversity of scientific methods.

Multimodal meaning making of NOS can be afforded by using the FRA framework from Erduran and Dagher (2014) as illustrated by the empirical data. Previous

Table 10.1 Lesson transcript for Lesson 4 (studying ecological relationships between native and foreign species of turtles)

Turn	Discursive action
1	*((7C03 wrote "Having experiment or hypothesis"))* 
2	7C05: 係要寫呢d咩? *((tr. Is that we have to write this?))*
3	7C03: 佢要解釋呀麻。 *((tr. They have to explain.))*
4	7C05: 唔係要畫咩? (.) 唔係要畫d烏龜咩? *((tr. Do we have to draw? (.) Do we need to draw turtle?))* *((7C03 drew an arrow pointing downwards, below the written phrase "(Having experiment or hypothesis)"))*
5	7C03: 好啦。 *((tr. Ok.))*
6	*((7C03 wrote "Evaluate Evidence"))* 
7	*((7C03 wrote "will foreign turtles fight with native species?" next to "Having experiment or hypothesis"))* 
8	*((7C03 drew an arrow pointing downwards, below the written phrase "Evaluate Evidence"))* 
9	*((7C03 wrote "(Yes, they will, foreign turtles will fight for food habitat with native turtle)"))* 
10	7C05: 畫個Bunsen Burner 燒左隻龜佢。 *((tr. Draw a Bunsen Burner to burn the turtle))* ((7C05 drew a Bunsen Burner next to "(*will foreign turtles fight with native species?)"))* 
11	7C05: 哼! *((tr. Hum!)) ((7C05 drew a star-shaped symbol using a red pen above the word "fight"))* 
12	*((7C03 wrote "Communicate information")* 
13	7C05: food 佢地食咩架? *((tr. What kind of food do they eat?))*
14	7C03: 龜好難畫。 *((tr. So difficult to draw turtle.))*
15	7C05: 養龜更係食龜糧。食大自然的蟲蟲 *((tr. If you keep a turtle, they eat turtle food. Eat worms in nature.))* *((7C05 used her green pen to draw fishes above the word "food))* 
16	*((7C03 drew an arrow pointing downwards, below the written phrase "Communicate information"))* 

work (Barak et al., 2023; Cheung et al., 2024; Erduran & Kaya, 2018; Erduran & Kaya, 2019) have demonstrated the affordances of drawing in pre-service teachers' meaning making of NOS. For example, pre-service science teachers in Ireland visualized and named four different branches extending from the centre to indicate four types of scientific methods. This book chapter further extends this body of literature by demonstrating through in-depth, micro-data analysis that multimodality can serve as a tool for meaning making across NOS categories. More specifically, we theorize, using the concept of multimodal connections from Tang et al. (2010), that secondary students can connect different NOS categories using the visual, verbal and written modes.

10.6.1 Harnessing Multimodal Connections in Designing Explicit-Reflective NOS Instruction

In designing explicit-reflective instruction of NOS, teachers can consider providing opportunities for students to reflect on multiple NOS categories using various modes of representation. Explicit-reflective NOS instruction in the previous research is often dominated by either verbal or written tasks (Khishfe, 2013; Park et al., 2020). Although our study did not plan for students' reflection of multiple NOS categories at the same time, students fluidly drew multimodal connections among NOS categories by themselves. As NOS ideas are intrinsically linked to each other (Cheung, 2020; Cheung & Erduran, 2022), teachers can provide explicit scaffolds on how students can utilise different modes to link various NOS categories together. For example, as revealed by our findings, teachers can encourage students to use arrows to foreground meaning making of scientific methods based on the category of scientific practices. More importantly, teachers can also encourage students to harness various modes to reflect on multiple NOS categories at the same time, instead of setting a task that only allows students to reflect on a single NOS category only. Although some NOS categories might not be explicitly planned in instruction, students' multimodal reflection of a single NOS category can be related to another NOS category.

10.6.2 Sequential Instruction of NOS Categories to Facilitate Students' Multimodal Connections

The findings of the study from which the data presented here were extracted also give insights on whether the introduction of NOS categories into the curriculum needs to be carefully planned in sequence. In the episode presented in this book chapter, students first approached the multimodal task by layering down individual scientific practices, followed by drawing of arrows to rearrange these scientific practices into a coherent representation of scientific methods. Hence, it could be the case that students extend their understanding of a NOS category to another owing to the multimodal connections among NOS categories. From students' perspective, understanding of scientific methods might be more complex than that of scientific practices, as scientific methods involve different stages of scientific practices and sequential relations among these scientific practices. Furthermore, scientific methods as defined in Erduran and Dagher's (2014) framework involve meta perspectives on methodology and not only particular methods or practices. Hence, they are likely to be more difficult for students to comprehend. To facilitate students' multimodal connections across NOS categories, teachers can begin to explain the meaning of individual scientific practices and their relationship to the inquiry context. After students successfully grasp these scientific practices, science teachers can begin to gradually introduce different types of scientific methods.

10.6.3 Limitations

A arises from the fact that the ways in which students connect individual scientific practices into scientific methods are dependent on curriculum design and classroom context. Hence, the findings might differ in other curriculum contexts. Findings generated in this study can be further validated by quantitative methods such as questionnaire surveys that take into account students' own reflections on their representations. More importantly, students' multimodal connections across NOS categories can be dependent on the time given to the multimodal tasks. In the current study, the teacher gave five minutes to students to draw and write their ideas of NOS. There would be more types of students' multimodal connections in evidence if longer engagement was provided. Despite such limitations, the empirical study provides useful insight into how students' multimodal meaning-making of NOS can be interrogated and what meanings they assign to scientific practices and methods.

Acknowledgement The first author was funded by Hong Kong Jockey Club Oxford Graduate Scholarships and supported by Clarendon Scholarship. The authors would like to thank the teacher and students who participated in this research study.

References

Airey, J., & Linder, C. (2009). A disciplinary discourse perspective on university science learning: Achieving fluency in a critical constellation of modes. *Journal of Research in Science Teaching, 46*(1), 27–49. https://doi.org/10.1002/tea.20265

Akerson, V. L., Cullen, T. A., & Hanson, D. L. (2010). Experienced teachers' strategies for assessing nature of science conceptions in the elementary classroom. *Journal of Science Teacher Education, 21*(6), 723–745. https://doi.org/10.1007/s10972-010-9208-x

Barak, M., Yachin, T., & Erduran, S. (2023). Tracing preservice teachers' understanding of nature of science through their drawings and writing. *Research in Science Education, 53*, 507–523. https://doi.org/10.1007/s11165-022-10069-3

Berland, L. K., Schwarz, C. V., Krist, C., Kenyon, L., Lo, A. S., & Reiser, B. J. (2016). Epistemologies in practice: Making scientific practices meaningful for students. *Journal of Research in Science Teaching, 53*(7), 1082–1112.

Brandon, R. (1994). Theory and experiment in evolutionary biology. *Synthese, 99*, 59–73.

Chan, H.-Y., Cheung, K., & Erduran, S. (2023). Science communication in the media and human mobility during the COVID-19 pandemic: A time series and content analysis. *Public Health, 218*, 106–113.

Cheung, K. (2020). Exploring the inclusion of nature of science in biology curriculum and high-stakes assessments in Hong Kong epistemic network analysis. *Science & Education, 29*(3), 491–512. https://doi.org/10.1007/s11191-020-00113-x

Cheung, K., Chan, H.-Y., & Erduran, S. (2023). Communicating science in the COVID-19 news in the UK during Omicron waves: Exploring representations of nature of science with epistemic network analysis. *Humanities and Social Sciences Communications, 10*(1), 1–14.

Cheung, K., & Erduran, S. (2022). A systematic review of research on family resemblance approach to nature of science in science education. *Science Education (Dordr)*, 1–37. https://doi.org/10.1007/s11191-022-00379-3

Cheung, K., Oancea, A., & Erduran, S. (2024). Students' meaning-making of nature of science: Interaction between visual, verbal, and written modes of representation. *Science Education*. https://doi.org/10.1002/sce.21916

Cheung, K. K. C., Erduran, S., & Oancea, A. (2025). 'Swirling' around translanguaging spaces of nature of science in multilingual classrooms. *International Journal of Science Education*, 1–25. https://doi.org/10.1080/09500693.2024.2444439

Curriculum Development Council [CDC]. (2017). *Science education key learning area curriculum guide (Primary 1 – Secondary 6)*. The Curriculum Development Council.

Department of Education. [DOE]. (2013). *Science programmes of study: Key stage 3 National curriculum in England*. Department of Education.

Erduran, S., & Dagher, Z. R. (2014). *Reconceptualizing nature of science for science education*. Springer.

Erduran, S., & Kaya, E. (2018). Drawing nature of science in pre-service science teacher education: Epistemic insight through visual representations. *Research in Science Education, 48*(6), 1133–1149.

Erduran, S., & Kaya, E. (2019). *Transforming teacher education through the epistemic core of chemistry*. Springer.

Gandolfi, H. E. (2020). "It's a lot of people in different places working on many ideas": Possibilities from global history of science to learning about nature of science. *Journal of Research in Science Teaching, 58*(4), 551–588.

Goren, D., & Kaya, E. (2022). How is students' understanding of nature of science related with their metacognitive awareness? *Science & Education (Dordr)*, 1–26. https://doi.org/10.1007/s11191-022-00381-9

Irzik, G., & Nola, R. (2011). A family resemblance approach to the nature of science for science education. *Science & Education, 20*(7–8), 591–607. https://doi.org/10.1007/s11191-010-9293-4. doi:10.1007/s11191-010-9293-4

Irzik, G., & Nola, R. (2011b). *A family resemblance approach*. Paper presented at the International History, Philosophy and Science Teaching Conference, Thessaloniki, Greece.

Khishfe, R. (2013). Transfer of nature of science understandings into similar contexts: Promises and possibilities of an explicit reflective approach. *International Journal of Science Education, 35*(17), 2928–2953.

Kress, G. (2010). *Multimodality: A social semiotic approach to contemporary communication*. Routledge.

Littleton, K., Twiner, A., & Gillen, J. (2010). Instruction as orchestration: Multimodal connection building with the interactive whiteboard. *Pedagogies: An International Journal, 5*(2), 130–141.
Livingston, J. A. (2003). Metacognition: An overview.
National Research Council (NRC). (2012). *A framework for K-12 science education: Practices, crosscutting concepts, and core ideas*. The National Academies Press.
New London Group. (2000). A pedagogy of multiliteracies. Designing social futures. In B. Cope & B. M. Kalantzis (Eds.), *Multiliteracies: Literacy learning and the design of social futures*. Macmillan.
Next Generation Science Standards (NGSS). (2013). *Next generation science standards: For states, by states*. The National Academies Press.
Osborne, J. (2014). Teaching scientific practices: Meeting the challenge of change. *Journal of Science Teacher Education, 25*(2), 177–196.
Park, W., Yang, S., & Song, J. (2020). Eliciting students' understanding of nature of science with text-based tasks: Insights from new Korean high school textbooks. *International Journal of Science Education, 42*(3), 426–450. https://doi.org/10.1080/09500693.2020.1714094
Peters, E., & Kitsantas, A. (2010). The effect of nature of science metacognitive prompts on science students' content and nature of science knowledge, metacognition, and self-regulatory efficacy. *School Science and Mathematics, 110*(8), 382–396.
Tang, K.-S., Tan, S. C., & Yeo, J. (2010). Students' multimodal construction of the work–energy concept. *International Journal of Science Education, 33*(13), 1775–1804.
Taylor, J., & Villanueva, M. G. (2014). The power of multimodal representations. *Science and Children, 51*(5), 60.
Walls, L. (2012). Third grade African American students' views of the nature of science. *Journal of Research in Science Teaching, 49*(1), 1–37. https://doi.org/10.1002/tea.20450
Wittgenstein, L. (1958). *Philosophical investigations*. Blackwell.

Kason Cheung is an Assistant Professor at the Department of Curriculum and Instruction, The Education University of Hong Kong. His research interests are multimodality, nature of science and more recently GenAI in science education.

Alis Oancea is Professor of Philosophy of Education and Research Policy at the University of Oxford. She specialises in philosophy of research, research on research and innovation, and metascience.

Sibel Erduran is a Professor of Science Education at the Department of Education, University of Oxford.

Chapter 11
The S'ence of Science Activities in Pre-school Science Education

Morten Rask Petersen, Linda Ahrenkiel, and Helle Hovgaard Jørgensen

11.1 Introduction

Research in science in early years is receiving increasing attention. Such research often focusses on professional development of staff (e.g. Barenthien et al., 2019; Vujičić & Čamber Tambolaš, 2017) or on the creation of science learning environments (e.g. Bonawitz et al., 2011; Fleer, 2019) rather than a focus on science as it is practiced in pre-school. However, some work has been conducted on the nature of early childhood science with different viewpoints on early years science (see e.g. Andersson & Gullberg, 2014; Broström, 2015; Johnston, 2013).

Children experience various natural phenomena in their everyday life. To convert such experiences into science activities, we need to conceptualise what is happening in the phenomenon experienced. Following Eshach and Fried (2005, p. 320): '… the way science ultimately allows us to see the world is by providing us with concepts with which we can frame its phenomena'. This means that science activities are present in the intertwining of phenomena and concepts (Fleer, 2019).

Science activities are often introduced as science instructions see e.g. (Bonawitz et al., 2011; Bulunuz, 2013; Gropen et al., 2017) However, science activities cover a broader span than instructions. Where science instructions are driven by the adult, science activities cover both adult and child driven science. Early childhood science education focusses on both conceptual and procedural learning, but often with focus only on one of these aspects (Russell & McGuigan, 2016). Procedural knowledge on science is often present in children's play but these procedures need to be identified and spoken aloud to address the science aspect in e.g. children's inquiries

M. R. Petersen (✉) · L. Ahrenkiel · H. H. Jørgensen
UCL University College, Odense, Denmark
e-mail: mrpe@ucl.dk; liah@ucl.dk; hhjo1@ucl.dk

G. Cakmakci, M. F. Tasar (eds.), *Connecting Science Education with Cultural Heritage*, Contributions from Science Education Research 15,
https://doi.org/10.1007/978-3-031-90944-3_11

(Andersson & Gullberg, 2014; Fleer, 2019). Hence there is great potential for expanding the view on science activities as situations initiated by both adults and children. The timespan of such science activities may vary. All these situations are here mentioned as science activities and the chapter will present a thematic analysis of different kinds of science activities in ECEC settings.

In the literature we often find approaches of addressing a specific phenomenon or doing spectacular experiments (e.g. Bonawitz et al., 2011; Bulunuz, 2013; Fleer, 2022; Vartiainen & Kumpulainen, 2020), and often science is done in a perspective of adjusting compulsory school science to an ECEC setting (Hansson et al., 2021). Larimore (2020) argues that a more holistic approach to science activities is needed in ECEC settings. We argue that science activities in ECEC needs to combine nature and natural phenomena with children's play to ensure their physical and sensory participation. Fleer (2022) argues that the approach to science activities starts with either a science phenomenon, e.g. a rainbow or dew on the grass, or the abstract science concepts such as the refraction of light or state forms of water. In this chapter, we direct our attention to 'how do we do science activities?' This means that we take the perspective of what is already happening in ECEC practice. What do practitioners see as science and how do practitioners conduct science in a pre-school setting? We there by combine the practitioner's view on science with the researcher's view. Together this forms our research questions:

How do science appear in practice in a Danish ECEC setting?

To address this question, we need to both specify our perspective on early years science and to clarify how ECEC staff sees and conducts science in practice.

11.2 What Characterizes a Science Activity

Science activities can be described in many ways. Neuman (1972) used the term "sciencing" to characterize science related activities for young children. He divided sciencing into three categories: formal sciencing, informal sciencing, and incidental sciencing. Formal sciencing are activities that are prepared and presented by adults for the children. Informal sciencing is characterized by the adults setting up a learning environment—hence the materials are selected by the adults and available to the children, who is interested in using the materials. Incidental sciencing is not planned or presented but is the result of an occurrence that is interesting to one or more children and is elaborated and expanded by the adult. However, these categories have been raised from a school setting and therefore within a school tradition of how to do science. Likewise, Haug et al. (2021) introduce a view on science from the perspective of science practices. These science practices also present science from a school setting. As mentioned earlier, science in an ECEC setting—emergent science—is not the same as school science (Johnston, 2013; Sundberg et al., 2018). Therefore, a framework specifically addressing to characterize and talk about science in an ECEC-setting is needed (Due et al., 2018).

According to Larimore (2020) science activities among other things could be about content knowledge and development of skills of practice. We argue, in line with Larimore (2020) that the emphasis should not be on a universal standard of specific bits of knowledge, but on content knowledge relevant to children's experiences with phenomena in their everyday world. Hence, science activities should be frequent, play based and flexible to focus on natural phenomena there are a part of children's daily lives to support them figuring out rather than knowing about science.

Therefore, the science activities must also support the development skills of scientific practice (Ahrenkiel et al., 2023; Haug et al., 2021; Larimore, 2020). The concept of science practices can help to clarify which actions children in ECEC settings carry out in situations with science, and characterises an investigative approach, thus giving ECEC staff opportunities to spot and support children's investigative actions and support the development of skills. This is an approach to emergent science, where we support both conceptual learning of science and a curious approach from the children towards natural phenomena instead of seeing it as two different perspectives on science in ECEC (Andersson & Gullberg, 2014).

When we direct our attention to what children do, it becomes concrete and possible to observe, analyse and develop science activities in ECEC contexts. In another study, we have identified 11 science practices specific for ECEC contexts within three dimensions: A dimension for exploration, a dimension with (body) language and a dimension on early mathematical awareness (Ahrenkiel et al. 2023).

The dimension for exploration is characterised by four distinct science practices: The tester who tests is seen when children are testing (something) based on 'what if'. The practice is a widespread science practice. The senser who senses takes place when children experience science phenomena and concepts with all their senses: feeling, hearing, seeing/observing or tasting differences. The word differences are important, as it is an attention hereto that makes the practice a part of a dimension for exploration. The planner who plans is about involving the children in planning science activities. The documentarian who documents takes place when involving children in documenting science phenomena or concepts, such as taking pictures.

The dimension of (body) language is about children expressing themselves to a greater extent with their body (non-verbal) than with verbal language: The questioner who asks, where the child asks questions using body-language, e.g. when pointing, shaping the mouth as an 'o', and/or directing wide-open eyes at a phenomenon. The narrator who narrates is about children narrating and relates to something that happens or has happened in relation to science phenomena or science concepts. The interpreter who interprets takes place when children either make their own interpretations of what is taking place or are supported in interpreting. This practice is characterised by 'if, then' realisations. The arguer who argues is characterised by 'because', for instance when the children relate to a previous experience from another context with a science concept or phenomena.

The dimension on early mathematical awareness has three science practices: The measurer who measures is about spotting and supporting situations where measurements are made. The measurement can take place in all units: hands, feet, blocks, centimetres, etc. The counter who counts is about counting. The sorter who sorts is about objects being grouped, sorted and classified according to different criteria (preferably some that the children come up with).

Together the 11 science practices constitute a framework for science activities. It is our experience that by focussing on science practices in science activities it is possible to learn about phenomena and concepts in everyday life in a frequent and playful way.

11.3 Learning Through Science Activities

Early years science education is often addressing how children can get experiences with science. Many of the phenomena and concepts that the children meet are new. So how do we serve the children best to get educative experiences to help the growth of further experience instead of precluding mis-educative experiences (Dewey, 1938/1953)?

The first step in getting new experiences is to recognize that something new is actually at stake. Chinn & Brewer (1993) introduces a psychological model on how we as humans has different approaches to anomalous data. When approaching something new and anomalous you can either accept or reject that this is something new. When rejected, the anomalous data is not seen and thereby do not become a new experience. When accepted, the anomalous data needs to be interpretated. This can happen in a range of ways in a span from retaining previous conceptual understanding by excluding data that does not fit into the explanation towards a fully change of conceptual understanding.

A challenge in early years science education is that the children have none or few experiences with the phenomena they meet and often even fewer conceptual understandings of these phenomena. A theoretical approach to how children learns science and the interplay between phenomena and concepts can be inspired by Kolb's experiental learning cycle (Kolb, 2014). This learning cycle describes how physical experience, and cognitive explanation can interact in learning.

A first step in the cycle can be to do active experimentation. This means manipulating with a phenomenon. Through such manipulation you meet the next step in the cycle, concrete experience. You get some kind of experiences with the manipulation and this experiences either confirms your understanding of the phenomenon or introduces anomalous data, that needs to be interpreted. Either way, the concrete experience can lead to a reflective observation, where you are doing reflections on the experience. Such reflections can lead to a conclusion and explanation of the observed, and thereby become an abstract conceptualization, which again can be tested through active experimentation.

Children are magnificent in doing active experimentation and getting concrete experiences. However, to place such experiences within science and to have these experiences to become educative, the ECEC staff needs to support the children in their reflective observation and abstract conceptualization. This changes the learning approach in part of the circle from a cognitive constructive approach towards a socio-constructive approach (Vygotsky, 1986). Vygotsky (1986) describes conceptual understanding as either a deductive approach with exemplifying scientific concepts with specific phenomena or an inductive approach in conceptualizing phenomena and thereby go towards a more general conceptual understanding. In combining Vygotskys conceptual approaches with the experiential learning of Kolb, we construct a theoretical framework, that allows the combination of trying out, sensing and doing with the conceptual understanding of both the specific phenomena and science concepts in general. It becomes the task for ECEC staff to challenge the children in their reflective observations and to help the children in conceptualizing these experiences and observations into science concepts. In this process the ECEC staff needs to be aware of the children's zone of proximal development. Our approach to emergent science is therefore that it is activities involving both children and ECEC staff in joint exploration and play.

11.4 The Five S's of ECEC Activities Addressing Science

To gain a more clarified picture of the characteristics of science activities in an ECEC context, we conducted an investigation. This was done in a project on 'Science & Play' with the purpose of combining a theoretical and practical perspective on early years science and play as enhancing each other. The project included 17 kindergartens from three different municipalities. During a three-year project period we developed a continuing in-service training program with a combination of workshops and supportive visits in practice.

11.4.1 Methods and Analysis

A two-step approach was used to identify and analyse science activities. First step was a mapping of science activities in practice. Data was retrieved through a survey on which science activities ECEC staff used. The informants to this survey were both participators from the three participating municipalities (n = 81) and participators in a FaceBook group on early years science (n = 28). This mapping was supplemented by continuous observations (Gulløv & Skreland, 2016) in the 17 kindergartens participating in the project. Second step was a thematic content analysis (Braun & Clarke, 2006) of the combination of survey data and observational records. In the thematic analysis, phase 4 'Reviewing themes', the themes were tested in ECEC practices by 15 ECEC staff to ensure that the evolved themes would be applicable to practice and not just stand as theoretical constructs. The thematic analysis revealed that science activities could be organized through different lenses. One perspective is to look at science activities as who is the driving force in the activities. Science activities look different depending on whether the ECEC staff or the child is the driving force in initiating the activity. Another useful lens to put on science activities was duration of the activities. Science activities also appear different depending on whether the activity lasts minutes, hours or even weeks. From the thematic content analysis two themes emerged: planned science and spontaneous science. The theme of planned science was further bifurcated in two significantly different themes: show science and slow science. Besides the science activity categories two additional non-science activities categories was found namely silent science and skewed science. In having all categories stating with s, we call the combination of these categories for the S'ence of science activities in pre-school science education.

11.4.2 The Three S's of Pre-school Science Activities

In the category of planned science activities, we found a need to make a distinction between the time needed for an activity to reach the end. We found that science appears in different forms and with different advantages and challenges depending on the form. In the following we introduce each type of science activity with a case from our field observations in the project Science and Play.

Case: The Colors of Nature
A group of 8 children aged 4–5 years are going on a trip to a nearby park. The ECEC staffs bring a roll of plastic bags. When arriving at the park the children are asked to find things from nature with different colors and sort them according to the colors in different bags. The children start exploring the area and bring back many different things like leaves, berries, gras, sticks etc. After a while they all gather around the bags and start counting how many things of each color they have found. It turns out that there are many green things, a few brown, red and yellow but no blue. After counting all the collected things, the ECEC staff write down the numbers and save these for later to compare them with other findings at another time of the year.

11.4.3 Show Science

Show science is a short-term planned activities initiated by the ECEC staff. Show science activities requires planning and often the purpose of such activities is to introduce the children to a scientific phenomenon (Fleer, 2022). Most of the responses in our survey had an approach of addressing a specific phenomenon or doing spectacular experiments in an ECEC setting, which was also found by Hansson et al. (2021). Examples of show science found in the data included 'cola and mentos' or 'we made volcanos'. However, show science activities can also be more of an introduction to nature in itself as seen in the case of show science above.

The science activities in show science are often found to be instructive and there is not much for children to do in the concrete science activities. However, show science holds the potential to involve children more in the activity. Having children doing investigations like: 'How much water can a diaper absorb?' involves the children in the show science activity along with withholding the opportunity to show children a specific phenomenon and articulate the science concepts involved.

The ECEC staff needs to become aware, that showing science also must include doing science, where doing science relates to the science practices (Ahrenkiel et al. 2023). In Table 11.1 the science practices involved in show science are highlighted. These include testing, sensing, questioning and interpreting. However, this highlighting is dependent on show science seen as both showing and doing science. If show science only appears as showing a phenomenon to the children, many of the highlighted science practices will not be present, and the science activity as a potential for learning science might lose its power.

Table 11.1 An overview of the properties of the five S's according to science practices and learning science

	Activity				
	Science activity			Non-science activity	
	Show science	Slow science	Spontaneous science	Silent science	Skewed science
Mostly initiated by children			x	x	
Mostly initiated by ECEC staff	x	x			x
Science practices present					
The tester who tests	x		x	x	
The senser who senses	x		x	x	
The planner who plans	(x)		x	x	
The documentarian who documents	(x)	x			
The questioner who asks	(x)	x	x		
The narrator who narrates	(x)	x	x		
The interpreter who interprets	(x)	(x)	x		
The arguer who argues			x		
The measurer who measures		x	(x)		
The counter who counts		x	(x)		
The sorter who sorts			x		
Learning element					
Active experimentation	x		x	x	
Concrete experience	x	x	x	x	x
Reflective observation	x	x	(x)		
Abstract conceptualization	x	x	(x)		

When involving the children in doing experiments with the shown phenomenon, show science holds the potential to address the full learning cycle from Kolb (2014). The children do active experimentation and get concrete experiences. Due to the planning, the ECEC staff are aware in which directions they can lead the children's reflective observations, and which abstract concepts can be introduced to the shown phenomenon. Planning show science activities are therefore essential to be prepared for the full learning cycle. In the case with the colours of nature, focus was on having the children to see the differences in nature and looking more into details. The case also connects to slow science in that the ECEC staff collects the numbers from each bag and during different seasons do the same activity with the same children. This allows a broader perspective on development as a key part of slow science.

Case: The Dead Cage

At the playground in an institution there is a small cage. Whenever a bird flies into a window and dies or the children find a small dead animal, they put these dead animals into the cage. When playing at the playground, the children often look into the cage and see how the dead animals inside decompose. On the day of the visit, one of the ECEC staff takes out a skeleton of a bird and a mouse from the cage. A group of 6 children aged 3–5 are all wearing plastic gloves. They are sitting at an outside table and start to examine the bones and feathers. While examining, the ECEC staff asks questions about how the children recall the decomposition of the animals. The children tell stories about when and how they found the animals and for how long they have observed them in the dead cage. After a while, one of the ECEC staff brings out a little bag with an owl fly. She tells how the owl eats animals and then vomits the leftovers. The children split the owl fly and compare it with the bones from the dead cage. One of the children says that it looks exactly like the bones from the mouse that was in the dead cage.

11.4.4 Slow Science

Slow science is characterized by continuous long-term science activities with reengagement, mostly planned by the ECEC staff. The science activities need an ECEC staff set-up, but the original idea can be child generated. Slow science is on and off activities meaning that it is an activity that children and ECEC staff engage and reengage in over a longer period. Between the reengagement the slow science activity just more or less present in the background with no need for attention. Slow science is thereby characterized as something you engage in and then leave, while doing something else, to then reengage with the activity. The time span in slow science activities can be quarters or hours like in raising a yeast dough or it can be seasonal like in growing vegetables or following the metamorphosis of butterflies from larva to flying insects.

Slow science thereby holds the potential to observe developmental processes. It gives the children the opportunity to experience for themselves the differences happening from last engagement to the present engagement. In Table 11.1 the most dominant science practices in slow science activities are the ones from the dimension of language. Slow science activities give excellent opportunities for narrating and documenting the developments and raising questions to these observations. This might lead to interpretations of the developments. In the case above, children continuously encounter the dead cage when playing at the playground. They make their own observations and experiences with the decomposition of dead animals, and they bring these experiences into play when examining the skeletons from the dead cage.

A challenge in many slow science activities is the lack of active experimentation. In most slow science activities, the development happens best without intrusion and since many of the slow science activities involves development of living organisms, an ethical issue can also be at stake in doing experimentation with the activity.

However, many of the observed development can be introduced in the setting of show science activities, where a specific phenomenon from the observations is demonstrated with other materials not interfering with the slow science activity. In the case of the dead cage, the ECEC staff let the children experiment with putting bones together and introduce the owl fly as a show science activity combining the children's experiences with a specific natural phenomenon.

This also gives the opportunity to show the children that concepts are generalizable while phenomena are contextual. The ECEC staff though needs to be aware of the children's age and not to make the gap between the observed phenomenon form the slow science activity and the explained concepts from the show science activity too wide.

Case: The Bomb

The children are playing in the playground. There are no prepared activities. The children are free to play whatever they want to. A boy aged 5 steps up to one of the ECEC staff. He is holding a ball of clay and dirt with a small stick pushed into the ball. 'Can we ignite this bomb?' he asks. The first thing that comes to the mind of the ECEC staff is NO. However, the ECEC staff instead says yes. This is the beginning of an examination of what burns and what does not burn. More children attend and together with the ECEC staff then try out different materials to see if they can burn or not. For each material they try out, the children make a guess and if able, they also argue for their guess. Then they try and afterwards follow up on the guesses and arguments.

11.4.5 Spontaneous Science

In contrast to the previous categories of science activities, the category of spontaneous science is unplanned activities caught in the moment and mostly initiated by a child. These activities are not prepared but arises and vanishes most often through children's play and curiosity (Ahrenkiel et al., 2023). Examples in the data are given when the respondents describe how science is initiated by the children as 'something that spontaneously arises because we are inspired by what the children ask about' or as 'something that occurs spontaneously as part of a play'. Spontaneous science combines nature and natural phenomena with children's play to elicit physical and sensory engagement. The activities can last for a minute, or they can go on for hours. In the case above, there was no intention from the ECEC staff to do investigations whether things can burn or not. However, by catching the moment introduced be the child a whole new play develops with offset in science.

The characteristics of spontaneous science clarifies, that since it is mostly child driven and not prepared, there is a risk of not connecting the activity to science and thereby instead having the non-science activity of silent science (see below). Children are not capable of capturing the science in their play and activities. Spontaneous science therefore happens when the science in children's activities and play is caught by the ECEC staff and emphasized together with the children.

As shown in Table 11.1, spontaneous science has the potential to involve most of the science practices, and often the ECEC staff recognize some of the science practices in the children's play and thereby capture this and introduces science into the play. The typical science practices are found in the dimensions of exploration and of (body) language. In spontaneous science the actions are performed by children. ECEC-staff engaged in spontaneous science activities have the potential to—at the right time—to include the science practices from the dimension of early mathematical awareness.

Spontaneous science thereby often requires a different position from the ECEC staff going from the already knowing teacher in show and slow science to becoming a co-investigator with the children in spontaneous science activities. This requires a gentle introduction of science and a sensory approach from the ECEC staff in order not to ruin the play in which the science became apparent.

When successfully catching the science in children's play, it gives the ECEC staff the opportunity to either start the learning cycle in entering with the children's concrete experiences or to introduce abstract concepts into the play, that can be investigated through active experimentation.

Spontaneous science activities thereby hold the potential to be part of the children's play and be relevant to the children due to the fact, that they themselves were doing it as a part of their play.

11.4.6 The Two S's of Pre-school Non-science Activities

Besides the three science activities of show, slow and spontaneous science, we also find activities that holds the potential to become science activities or is—with the best intentions- misunderstood as science activities. As a supplement to the science activities, these non-science activities show that not all science activities are recognized as such and that some science activities are not what they are intended to be.

Case: Skating in the Bathroom
Three children aged 5 have been playing a little too much with water in the bathroom. The floor is flooded. An ECEC staff asks the children to clean up the water and brings a stack of rags. He invites the children to step on a rag and clean up water from the floor while skating on the rags. This is a fun way to clean up the mess and after skating around for a while the children have cleaned the bathroom and carried on with other activities. In a follow-up talk with the ECEC staff, he says that he now sees that this was a situation containing so much science. He could have talked with the children about friction and fluidity etc. But this did not come to him in the situation. He would very much like to have science on his spontaneous repertoire so he can catch such situations and promote the science in these.

11.4.7 Silent Science

Silent science is an activity, mostly seen among children. It is an activity where there is a presence of science practices and content that entails science. However, there is no talking about the phenomenon and thereby no conceptualization of the science content within the phenomenon. A child might be caught up in investigating a natural phenomenon. But such an investigation is not science. To become science there needs to be a connection between the phenomenon and the science concepts. When such a conceptualization is absent the investigation of a phenomenon remains contextual and transfer to other contexts becomes harder.

In the above case, the children are experimenting with friction and fluidity in their play, but they are no more aware of this after the play than they were before they started.

The child might be approaching new and anomalous data (Chinn & Brewer, 1993). Without a supportive language and a conceptualization, it is not possible to help the child to understand the investigated phenomenon. It is not possible to see whether the child accept the anomalous data or not or how such data are interpreted.

Thereby, silent science is an activity that contains many opportunities for becoming science. However, it requires the engagement from the ECEC-staff to support the abstract parts of Kolb's learning cycle (Kolb, 2014). In practice, it is of course not always possible for the ECEC staff to recognize and underpin the silent science and since almost all contexts entail some sort of science, not all silent science should be underpinned. However, focusing on silent science may help raising the quality of science activities in early childhood science.

Silent science thereby holds an unredeemed potential to have more science activities without having to do a lot of preparation in advance. This perspective changes silent science to spontaneous science and the science practices related hereto (see Table 11.1). Moreover, the silenced science may work as inspiration in future science activities and thereby implicitly taking a starting point in children's initiatives.

As such, silent science can be seen as anomalous data. First you must accept or reject the presence. And if accepted, you can act on it and unfold the science potential of the activity. As hard as this can be in recognizing the science in children's activities it might be even harder as an adult to recognize the science in one's own activities. Our data shows that ECEC staff both initiates and engages in many activities that they do not ascribe as science. Hence, we argue on an awareness of activities with silent science who holds the potential to become a science activity according to science practices and learning science.

We find that learning about science practices helps the ECEC staff to better recognize activities with silent science. Recognizing these situations is the first step in transforming silent science activities into spontaneous science and thereby also underpinning the advantages from spontaneous science.

Case: Potting Daffodils

At a visit to a kindergarten the ECEC staff says that they are working with science today. The ECEC staff takes a group of 12 children aged 3–4 years outside and each of the children gets a small glass and a pot with a daffodil. The task is now for the children to take the daffodil from the pot and put it into the glass with their names pre-written on. During the task the ECEC staff talks with the children about how happy their parents will be for getting the daffodils as a present from the children. During the task there is no conversation about daffodils as plants. The science part of working with daffodils is absent.

11.4.8 Skewed Science

A second category of non-science activities is skewed science. This category is describes as arranged activities, where the ECEC staff have good intentions of doing science with the children, but where the activity does not contain science seen as the interplay between phenomena and concepts.

Skewed science activities are initiated by the ECEC staff, but the way in which they are practiced results in children not doing science practices. In the field study in the case above, we found that ECEC staff reported that they were doing science with the children in that they were taking daffodils out of pots and planting them in small glasses for the children to bring home to their parents. The activity has a content that could be articulated with science concepts like roots, stems, leaves and flowers, but it was not addressed. However, the ECEC staff in this case seemed to be convinced that the children were learning science by rearranging the flowers. Likewise, we saw an example of children collecting sticks in the woods. The ECEC staff saw this as science but there was no other activity than to collect sticks. Comparing the sticks and talking about different kinds of rind or leaves would turn this into a science activity.

Skewed science thereby becomes an activity with the potential to unfold science in that it contains one or more natural phenomena. However, these phenomena are not conceptualized or investigated by neither ECEC staff nor children. This leaves the science part of the activity to be dependent of children's own meaning making and conceptualization. According to our theoretical framework on how to learn science it thereby leaves the children with the opportunity to only have concrete experiences without connecting these experiences to any experimentation or conceptualization. The activity thereby becomes a non-science activity.

The potential in skewed science activities is that the ECEC staff are already aware that they have the intention of doing science with the children. Often, the activity always contains a phenomenon that can be conceptualized as science. With a few twists, skewed science may therefore in many situations be converted into show science. For this to happen the ECEC staff needs to both conceptualize the content and give space for the children to do an active experimentation. This can happen with a focus on science practices like e.g. testing, sensing and planning.

11.5 Discussion

There is a focus on how to support children's interest in and motivation for science (natural phenomena) in the early years in a holistic approach where play is fundamental. Play is children's primary practice form and the way they experience the world. Following Bogost (2016), play is deliberately working with the materials you encounter. Consequently, in play, children curiously investigate materials, and the potential of learning science is always present. Hence science is potentially in everything, and, taking the dialogic and open-ended nature of play (Jørgensen & Nielsen, 2023) into consideration, it becomes clear that a science approach that involves children's active experimentation and concrete experiences offers the potential to become playful.

In this study, the S'ence of science activities draws a picture, where science can become accessible to the children through play, especially when considering spontaneous science and silent science. Show, slow and skewed science also holds the potential to enrich children's play, but since these activities are driven by the ECEC staff it can be difficult to introduce these activities as play. However, by framing the environment for doing science practices this can lead to children's engaging in play (Petersen et al., 2025).

The five S's of science activities in ECEC context draws a powerful picture of different kinds of science activities with different kinds of potential. In introducing the five S's, we introduce a set of categories for emergent science and show how different types of emergent science have different learning qualities and different science practices involved seen in a learning perspective combining Kolb (2014) and Vygotsky (1986). In the discovering of the three different science categories—show, slow and spontaneous, it becomes clear that each of them addresses different aspects of early childhood science. Show science is the traditional approach to science activities with a focus on the cognitive and conceptual aspect of science (Fleer, 2022) whereas spontaneous science has a child centred and often affective approach to curiosity (Engel, 2011). Slow science offers the possibility of continuity (Dewey, 1938/1953) and curiosity (Engel, 2011). What also clarifies is that the three categories are not stand-alone categories. Each of the categories are intertwined with the others and a science activity from one category can lead to another activity from one of the other categories as also shown in the cases from show and slow science. The two non-science categories shows that there is a potential both for the ECEC staff to be more aware of what science in ECEC is and not is, and for helping the children with conceptualizing part of the science they do through play.

The five S's of ECEC science show that an instructional view on science miss opportunities to uncover science in the child-initiated activities e.g. play. Larimore (2020) argues that pre-school science education can be adjusted through science curricula from school settings. In such an approach, there is a risk that pre-school science turns in the direction of adult driven activities and that the potential from spontaneous child driven science might be missed. Emergent science needs to entail some opportunity for playing with science as a purpose in itself. With such a purpose, the S'ence of ECEC science introduces a new perspective on emergent science with value for ECEC instead of value for school.

A science situation can be characterized as show science, slow science or spontaneous science depending on who initiates and leads the activity and the duration of the activity. We argue that children must experience all types of science during their early years. Another conclusion of this chapter is that science phenomena and concept as well as processes is present in many situations. One way to find science is to grasp the science in children's play. To do so requires the ability of the ECEC staff to recognise science in play and to participate in the play, thereby introducing science concepts and phenomena as a natural part of the dialogue with children. Hence, recognising when children do science practices is a way to enriching science activities in ECEC settings.

References

Ahrenkiel, L., Petersen, M. R., & Jørgensen, H. H. (2023). Science practices as a tool for spotting and supporting children's investigative actions in Early Childhood Education and Care (ECEC). *Journal of Emergent Science, 25*(2), 5–11. https://www.ase.org.uk/system/files/JES%2025%20Ahrenkiel.pdf

Andersson, K., & Gullberg, A. (2014). What is science in preschool and what do teachers have to know to empower children? *Cultural Studies of Science Education, 9*, 275–296.

Barenthien, J., Oppermann, E., Steffensky, M., & Anders, Y. (2019). Early science education in preschools—The contribution of professional development and professional exchange in team meetings. *European Early Childhood Education Research Journal, 27*(5), 587–600.

Bogost, I. (2016). *Play anything: The pleasure of limits, the uses of boredom, and the secret of games*. Basic Books.

Bonawitz, E., Shafto, P., Gweon, H., Goodman, N. D., Spelke, E., & Schulz, L. (2011). The double-edged sword of pedagogy: Instruction limits spontaneous exploration and discovery. *Cognition, 120*(3), 322–330.

Braun, V., & Clarke, V. (2006). Using thematic analysis in psychology. *Qualitative Research in Psychology, 3*(2), 77–101. https://doi.org/10.1191/1478088706qp063oa

Broström, S. (2015). Science in early childhood education. *Journal of Education and Human Development, 4*(2 (1)), 107–124.

Bulunuz, M. (2013). Teaching science through play in kindergarten: Does integrated play and science instruction build understanding? *European Early Childhood Education Research Journal, 21*(2), 226249.

Chinn, C. A., & Brewer, W. F. (1993). The role of anomalous data in knowledge acquisition: A theoretical framework and implications for science instruction. *Review of Educational Research, 63*(1), 1–49.

Dewey, J. (1938/1953). *Experience and education*. The Macmillan Company.

Due, K., Tellgren, B., Areljung, S., Ottander, C., & Sundberg, B. (2018). Inte som i skolan-pedagoger positionerar naturvetenskap i förskolan: Preschool teachers talk about science–Positioning themselves and positioning science. *NorDiNa: Nordic Studies in Science Education, 14*(4), 411–426.

Engel, S. (2011). Children's need to know: Curiosity in schools. *Harvard Educational Review, 81*(4), 625–645.

Eshach, H., & Fried, M. N. (2005). Should science be taught in early childhood? *Journal of Science Education and Technology, 14*, 315–336.

Fleer, M. (2019). Scientific playworlds: A model of teaching science in play-based settings. *Research in Science Education, 49*(5), 1257–1278.

Fleer, M. (2022). Balancing play and science learning: Developing children's scientific learning in the classroom through imaginary play. *Journal of Emergent Science, 22*, 13–20.

Gropen, J., Kook, J. F., Hoisington, C., & Clark-Chiarelli, N. (2017). Foundations of science literacy: Efficacy of a preschool professional development program in science on classroom instruction, teachers' pedagogical content knowledge, and children's observations and predictions. *Early Education and Development, 28*(5), 607–631.

Gulløv, E., & Skreland, L. L. (2016). Ethnographic studies of young children. In J. Prior & Jo Van Herwegen (red.), *Practical research with children* (s. 127–144). Routledge.

Hansson, L., Leden, L., & Thulin, S. (2021). Nature of science in early years science teaching. *European Early Childhood Education Research Journal, 29*(5), 795–807.

Haug, B. S., Sørborg, Ø., Mork, S. M., & Frøyland, M. (2021). Naturvitenskapelige praksiser og tenkemåterpå vei mot et tolkningsfellesskap: [Scientific practices – towards a common understanding]. *Nordic Studies in Science Education, 17*(3), 293–310.

Johnston, J. (2013). *Emergent science: Teaching science from birth to 8*. Routledge.

Jørgensen, H. H., & Nielsen, M. P. (2023). Fortællekraft og legekraft: – Materialer, aktørskab og lytning i LegeKunst-forløb [Narrative power and play power: Materials, agency and listening in a play-art project]. *BUKS - Tidsskrift for Børne- & Ungdomskultur, 39*(67), 16. https://doi.org/10.7146/buks.v39i67.137350

Kolb, D. A. (2014). *Experiential learning: Experience as the source of learning and development*. FT Press.

Larimore, R. A. (2020). Preschool science education: A vision for the future. *Early Childhood Education Journal, 48*(6), 703–714.

Neuman, D. (1972). Sciencing for young children. In K. R. Baker (Ed.), *Ideas that work with young children*. National Association for the Education of Young Children.

Petersen, M. R., Ahrenkiel, L. & Jørgensen, H. H. (2025). The S'ence of early childhood science - Learning science from different approaches. In Fitzgerald, Earle, Preston & Georgiou (Eds.) *Primary science learning for children, teachers, and communities - Stories of practice and possibility for science educators*. Springer.

Russell,T. & McGuigan, L. (2016) Exploring science with young children. London: Sage.

Sundberg, B., Areljung, S., Due, K., Ekström, K., Ottander, C., & Tellgren, B. (2018). Opportunities for and obstacles to science in preschools: Views from a community perspective. *International Journal of Science Education, 40*(17), 2061–2077.

Vartiainen, J., & Kumpulainen, K. (2020). Playing with science: Manifestation of scientific play in early science inquiry. *European Early Childhood Education Research Journal, 28*(4), 490–503.

Vujičić, L., & Čamber Tambolaš, A. (2017). Professional development of preschool teachers and changing the culture of the institution of early education. *Early Child Development and Care, 187*(10), 1583–1595.

Vygotsky, L. S. (1986). *Thought and language*. Translated and edited by Alex Kozulin Cambridge, MIT Press.

Morten Rask Petersen is docent in science education at UCL University College, Denmark and holds a PhD in Science and Mathematics Education and Communication (SMEC). His primary research interests include interest development in and motivation for science, conceptual science learning and qualitative research methodology.

Linda Ahrenkiel is associate professor at the social education at UCL University College Denmark and holds a PhD in Science and Mathematics Education and Communication (SMEC). Her primary research interests include science in early childhood and playful learning.

Helle Hovgaard Jørgensen is docent in play and child culture at UCL University College, Denmark and holds a PhD in Media Literacy. Her primary research interests are play as culture, playful approaches to pedagogical practice and development of research methods for capturing children's experiences of play and playful approaches.

Chapter 12
Learning About Sound in Initial Teacher Training: Evaluation and Redesign of a Teaching-Learning Sequence

Aritz Ruiz-González, Arantza Rico, and Jenaro Guisasola

12.1 Introduction

We live surrounded by sounds, which we perceive unconsciously without knowing the complexity of a physical phenomenon produced by the "simple" vibration of an object (Rossing et al., 2001). If we consider that scientists themselves have historically faced serious conceptual difficulties in studying acoustic phenomena, it is not surprising that the proper understanding of the scientific model of sound (SMS) is a subject that involves learning difficulties for students and teachers, despite its continuous presence in everyday life. In fact, research in the teaching of physics has detected persistent difficulties of understanding at the phenomenological and interpretative level in students of all stages, which in the long run makes it difficult for them to conceptualise sound using the wave explanatory models and the corpuscular nature of matter (Faletič, 2023; Rico et al., 2021). In this sense, many alternative conceptions detected in elementary school students are still present in the first years of university (e.g. attribution of material properties, difficulty in identifying the vibrational source, confusion between tone and intensity, difficulty in relating the intrinsic properties of sound to the waveform magnitudes), which suggests that there is not an adequate learning progression of the SMS (Rico et al., 2021). In fact, effective teaching of this topic should involve not only a mastering of the subject

A. Ruiz-González (✉) · A. Rico
Faculty of Education and Sport, Department of Mathematics, Experimental and Social Sciences Education, University of the Basque Country, UPV/EHU, Vitoria-Gasteiz, Spain
e-mail: aritz.ruiz@ehu.eus; arantza.rico@ehu.eus

J. Guisasola
College of Dual Engineering, Institute of Machine Tools Campus, IMH, Elgoibar, Spain
e-mail: Jenaro.guisasola@imh.eus

G. Cakmakci, M. F. Tasar (eds.), *Connecting Science Education with Cultural Heritage*, Contributions from Science Education Research 15,
https://doi.org/10.1007/978-3-031-90944-3_12

matter but also content-specific science teaching core practices structured around evidence-based teaching interventions or products (Grossman, 2021).

In this chapter, we propose the use of Design Based Research (DBR) as a methodology to empirically fine tune a teaching learning sequence (TLS) about sound and to provide classroom intervention theories that takes into account the characteristics of the intervention context (Guisasola et al., 2023). In this framework, we define short-term TLS (ca. 12 h long), which includes well-researched and empirically adapted teaching and learning activities tailored to student reasoning, in an interventionist DBR process (Zuza et al., 2020). The TLS presented here integrates two fundamental aspects of prospective primary teacher (PPT) education that are interrelated: (a) the learning of scientific content, that is, to reach an understanding of the nature of sound and its propagation by developing a SMS and (b) the learning of content-specific teaching core practices, such as the use of guided-inquiry (Rivero et al., 2011). We argue that PPTs must use scientific practices in their learning process if they are to be able to guide their future pupils in inquiry-based learning.

12.2 Design Based Research for Developing Teaching Learning Sequences

There is a significant body of research produced over the last three decades focused on the design and evaluation of medium-scale curriculum products covering the teaching and learning of a scientific topic. These works include sequences of teaching activities with the aim of improving the students' learning of specific topics at a small-scale level (for instance, a few teaching sessions) or at a medium-scale level (a whole sequence of lessons on a particular topic) but typically not addressing the large-scale level of a whole program (of one or several academic terms). The literature refers to such teaching activities as TLSs (McDermott & Shaffer, 1992; Meheut & Psillos, 2004). A distinctive characteristic of the TLS is its dual character that implies both research and development, pointing at a strong link between the teaching and the learning of a particular topic. A TLS can be broadly defined as "both an interventional research activity and a product, like a traditional curriculum unit package, which includes well-researched teaching-learning activities empirically adapted to student reasoning" (Meheut & Psillos, 2004). Thus, TLS design reflects the interlink between the development of learning tools and environments and the development of theory. This interlinkage is a complex, cyclical process in which general education principles are applied to the teaching of specific topics in specific contexts (Lijnse & Klaassen, 2004; Juuti & Lavonen, 2006). In each phase, there are opportunities for testing conjectures about student learning and for refining those conjectures on the basis of experience, as well as redesigning the TLS proposal accordingly.

An emerging framework in science education that allows the research of methods for design, implementation and assessment of TLSs is DBR (Guisasola et al.,

2023). This approach takes into account the characteristics of local contexts to design feasible TLSs. In that sense, it is a flexible methodology that draws from diverse educational theories to generate useful products and humble intervention theories, but which should always include several cycles of design, implementation, assessment and redesign in an interactive process. Therefore, the use of DBR can help in solving some methodological issues, such as deepening the theoretical considerations that inform the TLS design, specifying the teaching strategies to overcome students' learning difficulties and clarifying the relation between the criteria of design and the evaluation and re-design process (Guisasola et al., 2017). Below, we describe in more detail the phases of the DBR methodology.

12.2.1 Design Phase

The first two elements that are considered in this phase are the educational context and the specific content of the topic to be taught. Besides, the general theories that support this phase include the social-constructivist cognitive psychology, the epistemology of science curriculum for the context of teaching and the results of the Science Education Research on teaching and learning (Zuza et al., 2020).

Another crucial step in this phase is the definition of learning goals based on the epistemological analysis and the identification of the gaps between these objectives and students' alternative ideas, which are usually defined as learning demands. The idea is that designers should avoid the definition of the learning goals based on particular idiosyncrasies or non-explicit traditional curricular elections. It is also crucial to review for known difficulties in learning in the target group and existing teaching solutions to overcome them. Guisasola et al. (2023) suggest the application of specific Design Tools that systematise the justification of the TLS and the learning demands required from the students. In the following section, we will give specific examples of the tools implemented in the TLS about sound.

Ultimately, a design solution will be drafted, which will include teaching materials, assessment guidelines and guidelines for teaching practice. The TLS product is presented as the solution for the relationship between epistemology, learning and teaching of the particular topic and context.

12.2.2 Implementation: Teaching Experiments

Classroom implementation of the TLS aims to investigate the hypothesis that the design will lead to better student learning based on the evidence provided by the defined assessment strategies and the learning environment. In DBR methodology, since the implementation is considered a teaching experiment, adjustments made to the TLS during the implementation to favour optimum conditions are considered part of the body of data.

12.2.3 Evaluation and Redesign

This phase involves a retrospective analysis to assess the efficiency of the design solution in relation to the stated objectives. The tools for this evaluation should be chosen by the researchers in accordance with the aspects evaluated and the academic context. Nevertheless, it is accepted that TLS efficacy is tested regarding two dimensions: (i) Analysis of the clarity of the sequence, regarding whether the activities were understood, the time required to complete the sequence and problems derived by the innovation as opposed to a traditional sequence; (ii) Analysis of students' learning outcomes regarding understanding of concepts, models and theories and development of scientific skills.

Both qualitative and quantitative research tools are used in this evaluation. The "teacher's journal", the "student's workbook" and the "external evaluator's report" are examples of tools that can be used in the retrospective analysis to overcome difficulties encountered in the implementation of the TLS (Guisasola et al., 2023). Quantitative tools, such as open-ended or multiple choice questions are normally used to evaluate the learning dimension, which can be complemented with the analysis of students' interviews probing particular issues arisen in the questions. Eventually, the results from this bi-dimensional analysis may suggest rewriting the text, modification of analogies and representations and the general focus of the TLS. Ultimately, the TLS will be redesigned to be subsequently re-assessed in the final evaluation (Zuza et al., 2020). In the following sections, we present detailed information of the Design Tools used in our research program for the topic "What is sound? A scientific model" in initial PPT.

12.3 Design and Evaluation of a TLS for the Topic of Sound in the Context of Prospective Primary Teacher Education

In this section, we start by presenting the context for which the TLS is designed. Then, we discuss the epistemological and educational arguments concerning the chosen topic. Finally, the learning goals and the gap between these goals and the students' difficulties are outlined followed by a description of the resulting learning path and activities.

12.3.1 Design Phase

12.3.1.1 Educational Context

The TLS was designed to be carried out within the compulsory 9 ECTS course "Natural Sciences in the Primary Education Classroom II" of the third year of a Primary Education degree at the University of the Basque Country (UPV/EHU), with an average enrolment of 110 students. The traditional course format is 2 h per week of lectures and another 2 h per week of seminar or laboratory sessions. The course covers pedagogical content knowledge about physical and chemical sciences for primary education and an introduction of the epistemology of science for educators, including the visions and treatment of the Nature of Science and Scientific Inquiry in the classroom. Students starting the Primary Education Degree show a diverse background regarding their secondary science education and, in general, only one third of the enrolled students took physics or chemistry courses in the last 2 years of secondary education (16–18 year olds). The topic of sound is included in the course program as a bridge between the topics of matter and energy, so in this way, students have worked on the main principles of the model of matter, including its conservation and its corpuscular nature. The topic usually takes 6 sessions of 2 h including laboratory sessions.

12.3.1.2 Epistemological Insights

Acoustic phenomena are closely related to our daily lives and therefore the understanding of the SMS is part of the science curriculum from elementary school to university (Rossing et al., 2001). Also, the sound—understood as a wave phenomenon—can contribute to the understanding of general concepts of current physics. However, the transition from the understanding of the SMS from a phenomenological and interpretative point of view to the understanding of sound as a wave phenomenon, does not seem trivial, as the students of all stages appear to have difficulties in describing and conceptualising sound (Rico et al., 2021).

There is a consensus in introductory-level physics to define sound as a wave generated by vibrating objects and propagated through an elastic medium, which gradually transmits a state of compression and rarefaction, without transport of matter. However, to describe sound using the 'vibration' model implies a difficult abstraction depending on the context where sound is being analysed. This "vibration" model leads to considering other characteristics, like vibratory motion and sound wave, its transmission, speed, volume and intensity and isolation (Rossing et al., 2001). In addition, the science curriculum in Primary and Secondary Education include intrinsic properties of sound, such as intensity, tone and timbre, while making links with other contexts such as music or the environment (Rico et al., 2021). Despite its daily use, tone and timbre are complex and difficult concepts to understand from an initial SMS. In this sense, the SMS explicitly distinguishes between

(i) amplitude and volume, (ii) frequency and tone, and indicates that (iii) when assessing acoustic energy it is necessary to take into account both properties (amplitude and frequency) and (iv) that intensity, tone and timbre are interdependent properties (Awad & Barak, 2018). A fundamental aspect of the SMS is to understand that what propagates through an elastic medium is the vibration and that sound waves are disturbances spreading in the air in a longitudinal way. In this way, the air molecules move back and forth in the direction of sound propagation but remain in equilibrium without moving with the wave (Rossing et al., 2001).

The previous epistemological content analysis allowed us to define the key ideas of the SMS:

Nature and properties of sound. Sound is defined as the event produced by the vibration of an object mechanically disturbed by a source of energy. We can use the three properties of vibrations to characterize sound: intensity, tone and timbre.

Sound propagation. Vibrations spread through an elastic medium without transport of matter. This is an emerging process of energy transfer where the vibration's behaviour is very different from its constituent parts, because the particles of the surrounding medium oscillate back and forth from their equilibrium position.

The wave model. The propagation of vibration is represented by the concept of pressure wave.

In accordance with DBR methodology, we consider the topic and contents defined in the official curriculum for the TLS about sound (context) and use contributions from the epistemology of science to justify and define the learning objectives. For brevity, we present the first three learning objectives that are related with the first key idea: *1. Nature and properties of sound.* A full description of the learning objectives for the full TLS was published previously (Rico et al., 2019).

12.3.1.3 Student Difficulties and Learning Demands

The second element in the design phase of our TLS, is the analysis of the gap between the defined learning goals and the students' learning difficulties, as shown in the literature. These learning demands are defined through an analysis of the ontological and epistemic differences between students' ideas and the scientific concept to be taught. These differences will guide the teaching layout of our TLS, by defining the type and level of difficulty that we expect the students to encounter. In consequence, the identification of learning demands is a useful guide in the design of activities, because a big learning demand should require a specific treatment in number of activities and structure, compared to a learning demand that is small.

In this section, we describe the learning demands for the Learning Objectives 1, 2 and 3, whose activities and evaluation are further discussed in the following sections. Table 12.2 shows the link between the learning goals and the learning demands. Regarding the first learning objective (LO1, Tables 12.1 and 12.2), the literature on students' difficulties shows that the understanding that sound is

Table 12.2 Alignment of Learning Difficulties (D) with Learning Objectives (LO) and identification of Learning Demands (LD) for a TLS about *sound* in initial Primary Teacher Training

Learning difficulties (D)	Learning objective (LO)	Learning demand (LD)
D1. An explanation that links sound with vibration is context-dependent (clear in a guitar string, confusing when hitting stones). Some students may not understand what "vibration" means. (Küçüközer, 2009).	LO.1 (*vibration*)	Low
D2. Difficulties on differentiating tone from intensity or relating the sound's tone with its frequency and sound's intensity with its amplitude (Awad & Barak, 2018).	LO.2, LO.3 (*properties*)	Medium

Table 12.1 Learning objectives (LO) that define the SMS (*Sound and its properties)* for a TLS in initial Primary Teacher Training

Elements from the epistemology of physics (E)	Learning objectives (LO)
Sound and its properties	
E1. The sound is generated when a typically audible vibration propagates through a medium and that the source of sound comes from a vibrating object. **E2.** The two main properties of vibrations can be used to characterise sound: the amplitude (intensity) and frequency (tone).	**LO.1.** Understanding that the vibration of an object produces a sound, although not all sounds are perceptible to the human ear. **LO.2.** Objects of different shapes and materials generate different sounds. **LO.3.** The sound vibrations can be differentiated by their intensity (high/strong and low/soft) and tone (low and high).

originated from the vibration of an object is context-dependent in primary students or Primary Teacher students (Küçüközer, 2009). We consider that this objective poses a low learning demand because providing different contexts for sound production and linking sound with everyday situations can help understanding sound phenomena. Regarding the second and third learning objectives (LO2 and LO3), there is abundant literature showing that students of all ages have difficulties differentiating the sound's intrinsic properties and relating them with their physical magnitudes (Awad & Barak, 2018). We propose a medium learning demand for these objectives because the literature has also shown that active and contextual learning situations improves students' understanding of sound as "vibration" and identifying its properties.

12.3.1.4 Designing TLS Materials

Taking into account the process described so far, we designed a series of tasks (questions and problems) that should, in principle, help students achieve the learning objectives. To organize the activities we followed a process that includes two phases that are integrated iteratively: (a) Sequencing the content; (b) specifying strategies and activities to help learning.

Our TLS about sound in PPT provides an interactive learning context that has four characteristics: (1) students are organized in groups of four, and they are asked to record all their observations and work in pairs in the tasks that involved predicting outcomes, designing and carrying out experimental activities; (2) the TLS is structured around several driving questions about the key ideas about sound; (3) in each driving question, there is a number of activities that guide students to make explanations for each key idea (Guisasola et al., 2014). This learning path is (4) organised around a scenario that enquires about how sound is produced in a music box. The discussion of this scenario leads to the definition of three driving problems (DP), which are stated as follows:

DP1. What is sound? How is it produced?
DP2. What are the properties of sound? Are there different types of sound?
DP3. How does sound propagate? How is it represented?

In this paper we will briefly present the structure of activities for DP1 and DP2. These driving problems structured an initial version of the TLS (henceforth TLS1) during the 2017–2018 and 2018–19 academic years with the following order of content presentation: (I) Activities to build the explanatory model of sound at a macroscopic level: *the music box; the metallic ruler, bands and pendulum; the human vocal cords*; (II) Activities to define and apply the vibratory model: *rules of different length, flutes and pan flutes; application with pan flute and vocal chords; use of tone and sound meters.* Further details of the activities of TLS1 are extensively described in Ruiz-González et al. (2021).

12.3.2 Evaluation and Refinement of the TLS

This section presents and discusses the assessment and consequent refinement of the TLS. First, we present the preliminary evaluation whose empirical evidence guided refinement of TLS1. Second, we present the results from implementing the refined TLS in relation to the learning achieved. There are two dimensions of the TLS evaluation: (a) analysis of the sequence quality and (b) measurement of the learning achieved by students. Our evaluation includes qualitative and quantitative tools in order to obtain complementary information on how to redesign the TLS. As Zuza et al. (2020) argue, this mixed methods evaluation requires the analysis of the students' learning results and the evaluation of the TLS implementation. In this section, we will first focus on the qualitative assessment and its impact on redesign and, secondly, on the quantitative analysis regarding students' learning. It is important to bear in mind that while the assessment shows which parts of the TLS and its implementation should be improved, the specific changes in TLS1 to obtain TLS2 come from the combination of professional content knowledge and domain-specific research evidence (see Table 12.3).

Table 12.3 Evaluation and refinement of TLS1 to TLS2[a]

Driving problems	Learning objectives and demands	Strategies to foster learning	Implementation and re-design TLS1. Activities and comments	 TLS2. Activities and comments
			Activities to build the explanatory model of sound at a macroscopic level	
DP1. What is sound? How is it produced?	LO.1 Low	A-Make explicit student's prior knowledge	A1.2. The music box scenario	No change in activities but more opportunities for reflection: (a) An introductory activity about guided problem solving; (b) Explicit pedagogic strategies: to evoke learnings from previous sessions and metacognition activities at the end of sessions
	LO.2 Medium	B-Make predictions and hypothesis C-Construct initial model of vibration	A1.3. Oscillations and vibrations of metallic ruler, rubber band and pendulum	
		D-Apply this initial model to human voice	A1.4. The human vocal chords	
			Activities to define and apply the vibratory model	
DP2. What are the properties of sound? Are there different types of sound?	LO.3 Medium	E-To predict, to design and carry out experiments	A2.1–2.4.. Vibration of short and long metallic rulers, flute and pan flute A2.5–2.7. Drawing intensity with metallic ruler	A2.1–2.4. Inclusion of straw-flute activity A2.5–2.7. Drawing intensity with metallic ruler A2.8- Application with slide whistle and mbira A2.9. Application with vocal chords and guitar A2.10. Tone detector and sound meter with different tuning forks and oscillometer (classroom and home activity)
		F-Complete and apply the vibratory model	A2.8. Application with slide whistle and music box A2.9. Application with vocal chords	
		G-Test and apply concepts with devices and sensors	A2.10. Tone detector and sound meter (home activity)	

[a]The fourth and fifth columns list the activities and how they relate to the skills to be worked on for TLS1 and TLS2. Each row presents the learning objectives (LO) and teaching strategies connected to each driving problem (DP) as well as the activities (A) proposed to address them in the TLSs. LO.1 to LO.3 refer to the learning objectives defined in Table 12.1

12.3.2.1 Analysis of the Quality of TLS1 and Subsequent Refinement

TLS1 was carried out during two academic years, taught to a total of 116 students and with a control group of 79 students. In both groups the topic of sound was developed during 6 sessions of 2 hours including laboratory sessions.

According to the social perspective of learning, the teaching strategy involves posing situations to students, with directions that guide them to find solutions, which could be considered an “oriented research activity” (Guisasola et al., 2014). The solution must be achieved in accordance with scientific practices that we want students to become familiar with. Students must have opportunities to make argued decisions in the small group and different ways of solving the task are discussed, guided by the teacher, and a classroom summary is formulated. By carrying out these activities, students engage in scientific practices because we expect that students (i) start assessing the question using a qualitative approach, avoiding straight mathematical operations; (ii) design an experiment and predict the outcome before carrying it out; (iii) collect data and explain the results; (iv) discuss the results.

Analysis of the information included by the professor in his “teacher’s journal” and the students’ workbooks clearly shows that students had difficulties designing experiments and that the number of exercises to experiment with tone and sound did not seem enough to properly explore the relationship between the objects and the properties of the sound. The teacher also noted in his journal that “we have not given them enough opportunities to illustrate and revise their mental models about the vibratory nature of sound and its propagation”. Accordingly, the sequence of activities was reformulated and more time was allotted for experimental design, more activities about tone were included by presenting more objects and musical instruments (guitars, straws and Mbira) and tune forks were provided to “visualise” tone when using the sound sensors (see Fig. 12.1 and Table 12.3).

On the basis of the data obtained from the teacher’s journal, students’ workbooks, the feedback received by the classroom observers and the results of the post-test -including the comparison between an experimental and a control group- we refined the sequence of the contents in TLS1. The changes are mainly made in reformulating activities to adapt them so that the students understand their aim (metacognitive difficulty) and to stimulate production of hypotheses and arguments

DP2-Which are the properties of sound? Are there different types of sound?
Activities to define and apply the vibratory model: A2.1 to A2.10

TLS1	TLS2
A2.1-2.4. Vibration of short and long metallic rulers, flute and pan flute, A2.5-2.7. Drawing intensity with metallic ruler. A2.8. Application with slide whistle and music box A2.9. Application with vocal chords A2.10. Tone detector and sound meter (home activity)	A2.1-2.4. Vibration of short and long metallic rulers, flute and pan flute. **Inclusion of straw-flute activity** A2.5-2.7. Drawing intensity with metallic ruler. A2.8- **Application with slide whistle and embira** A2.9. Application with vocal chords and guitar A2.10. Tone detector and sound meter with ≠ **tuning forks and oscillometer** (**classroom** and home activity) 

Fig. 12.1 Depiction of the activities regarding Driving Problem 2 (DP2) for TLS1 and TLS2. The changes made and new activities for TLS2 - related to the students’ difficulties in test, apply and complete the vibratory model of sound (LO.3) - are highlighted.

for the conclusions. The second version of the TLS applied in the spring semester of 2020–2021 academic year presents the rewritten activities.

12.3.2.2 Results Regarding Students' Understanding

To see how much students had improved their understanding in relation to the learning objectives of Table 12.1, we used a pre- and post-test design. The post-test was given to students from experimental groups and control group students in exam conditions and the result was included as a part of the final mark for the subject unit. The scores for each group were compared. The questionnaires consisted of six questions (Q1–Q6) that covered all the learning objectives defined for the TLS. Table 12.4 presents questions Q1 to Q3, which aim to evaluate student's understanding of the vibratory nature of sound and its properties.

The results were analysed by a phenomenographic approach to investigate the different ways students experience, conceptualise, perceive and understand different sound phenomena. Regarding the establishment of categories, after defining the final categories, a 90% of agreement among the pairs of researchers was reached, with a reliability coefficient of Cohen's kappa of 0.95 in average. Here we present the results obtained from questions Q1 to Q3. As shown in Table 12.4, question Q1 aimed to evaluate learning about the vibration model and understand if students integrate the corpuscular model of matter in order to explain how sound is produced and propagated and how we hear it. Question Q2 aimed to evaluate whether students understand how the size of an object affects its vibration frequency and thus, its tone, while amplitude is kept as a constant variable by using bands of different thickness. Question Q3 aimed to evaluate whether students understand how the strength applied to the vibrating object affects the intensity of the sound, which

Table 12.4 Evaluation questions Q1 to Q3 and alignment with the Learning Objectives (LO.1-LO.3)

Learning objective	Post-test question
LO.1	Q1. Why do we hear a sound when we beat a drum? Reason your answer.
LO.2 LO.3	Q2. Look at the image. In this activity, students are analysing the characteristics of sound. This box has three elastic bands of different widths. If they vibrate the three bands the same way (i.e. each rubber will move the same distance): How will the characteristics of the three produced sounds differ? Reason your answer. 
LO.3.	Q3. We can produce sounds at different volumes, for example, screams and whispers. Why? Reason your answer.

relates to the amplitude of the vibration. A final system of five hierarchic categories of growing complexity was defined for each question.

In Table 12.5 we provide a comparison of students' learning gains focusing on the frequency of correct answers to questions Q1 to Q3. During the three years that the experience lasted, the percentages of correct answers in the pretest did not vary significantly, so we have presented the average of the percentages in the first column. We subsequently compared the data between the Pre-test and Control and Experimental groups post-implementation. A chi-square test of independence showed that there was a significant association between the frequencies of correct answers and the type of group for the three questions with a significance level of alpha = 0.05 (Q1: $Chi^2 = 56.22$, $p < .000001$; Q2: $Chi^2 = 80.5$, $p < .000001$; Q3: $Chi^2 = 69.91$ $p < .000001$; Table 12.5). Regarding the learning gains in TLS1 and TLS2, we can see that there are between 15 and 20 points of difference in the frequencies of correct answers between both TLSs for the three questions. However, the Chi square test of independence did not show a significant association between the TLS version and the learning gain for these questions (see Table 12.5; Q1: $Chi^2 = 2.193$, $p = 0.144$; Q2: $Chi^2 = 2.42$, $p = 0.12$; Q3: $Chi^2 = 3$ $p = 0.083$). This suggests that after TLS2, it is probable that the cycles of redesign have reached its ceiling for the intended learning gain (Juuti & Lavonen, 2006).

Regarding students' learning, we also examined how the frequency with which students experience difficulties changed in relation to defined objectives. In particular, we are going to refer to two difficulties that were located in categories B and C in the analysis for questions Q1 and Q2. The difficulty D1 (understanding sound as movement or vibration without considering the need of a medium and a receptor) appeared in question Q1 (>80% pre-implementation). In question Q2 we identified difficulty D2, which is associated with the distinction of the two key characteristics of sound. Figure 12.2 shows the evolution of these two difficulties post-implementation in the control and experimental groups. Regarding D1, although the difficulty persists in the experimental group, it does with less frequency (around 20% in TLS1 and 9% in TLS2). Regarding difficulty D2, Fig. 12.2 shows the persistence of such difficulty in TLS1 and a diminution in the experimental group in TLS2 (from 19% to 11%). These results suggest that the changes introduced in the TLS1 have contributed to a higher learning in the TLS2 regarding the analysed difficulties.

Table 12.5 Percentages of the correct answer (most advanced category) for all questions in the Pre-Test (Control-CONT and Experimental-EXP groups merged) and in the Post-Test for the Control groups (merged) and the TLS1 and TLS2 experimental groups

	Pre-test	Post-test		
	ALL courses	CONT (17/18, 18/19, 20/21)	EXP (17–18, 18–19) TLS1	EXP (20–21) TLS2
Questions	$N = 293$	$N = 133$	$N = 116$	$N = 56$
Q1	0.68	6.77	39.66	58.93
Q2	0.34	30.83	55.56	83.93
Q3	0	3	38.5	53.6

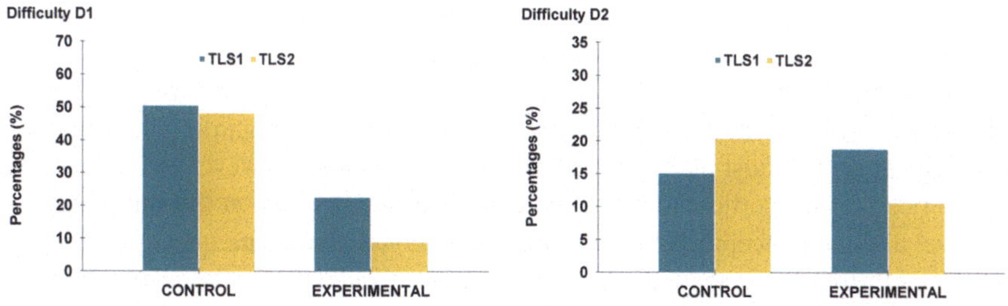

Fig. 12.2 Evolution of learning difficulties D1 and D2 in control and experimental groups in the post-test

12.4 Conclusions

The purpose of this study was to apply the DBR methodology to make explicit the choices in material design and in the systematic refinement of the TLS based on assessment tools. Our application to the topic of sound in Primary Teacher Training indicated that the DBR methodology allowed us to build and refine a teaching strategy for our chosen topic with a higher probability of obtaining a viable instructional approach, and helped us to build design principles and didactical tools to design TLSs. DBR methodology emphasises the importance of justifying the decisions made in the design and how and why to use the activities and the evaluation tools. This is evident in the way that we use it for designing the learning objectives explicitly based on epistemological and psychological arguments, rather than on research group idiosyncrasy. Later in the evaluation, following DBR, we introduce some tools that involve the evaluation of the design itself and the learning obtained by students. The epistemological analysis and learning demands have been used as design tools to define learning activities located in the appropriate development zone for our students (Zuza et al., 2020). The evaluation tools include the students' progression when overcoming learning difficulties. Another novel aspect, in relation to the chosen design principles, is the need of confronting the design with its classroom implementation and analysing the coherence between the TLS activities aims and the obtained students' results. To that end, we have presented data obtained with the evaluation tools on the quality of the TLS. This study contributes to the body of research applying the DBR methodology in a specific topic for PPT students.

The analysis of the qualitative and quantitative data of the progression in the learning of PPT has allowed us to demonstrate the usefulness of the didactic proposal in the understanding of the SMS because our data shows that the experimental group students achieve a stronger conceptual understanding. Since the current study's sample is limited to participants from only one university (UPV-EHU; University of the Basque Country), we cannot generalise about benefits of our TLS in terms of learning in general, even though some of the teaching and learning difficulties encountered are similar to those detected in other international studies. As we have mentioned elsewhere (Guisasola et al., 2023), this type of studies allow us to generalise a model of design from the findings of a particular case of DBR. In this sense, we are confident that the evidence presented here regarding the students' learning gains was most likely caused by our intervention. However, some learning difficulties prevail, particularly the conceptualization of sound at the submicroscopic level and the conceptualization of the wave model. These are questions that are examined further in the second part of our sequence and that, for the sake of brevity, have not been the focus of this article but will be discussed in the future.

References

Awad, N., & Barak, M. (2018). Pre-service science teachers learn a science, technology, engineering and mathematics (STEM)-oriented program: The case of sound, waves and communication systems. *Eurasia Journal of Mathematics, Science and Technology Education, 14*(4), 1431–1451. https://doi.org/10.29333/ejmste/83680

Faletič, S. (2023). Sound and waves. In: M. F. Taşar and P. R. L. Heron (eds.) *The international handbook of physics education research: Learning physics* (p. 26: 1–34). AIP Publishing. doi:https://doi.org/10.1063/9780735425477_001.

Grossman, P. (Ed.). (2021). *Teaching core practices in teacher education*. Harvard Education Press.

Guisasola, J., Ceberio, M. & Zuza, K. (2014). Problem solving by developing guided research in introductory university physics courses. In: D. W. Sunal et al. (eds.) *Research-based undergraduate science teaching*. (p. 9:379–402). Information Age Publishing.

Guisasola, J., Zuza, K., Ametller, J., & Gutierrez-Berraondo, J. (2017). Evaluating and redesigning teaching learning sequences at the introductory physics level. *Physical Review Physics Education Research, 13*(2), 020139. https://doi.org/10.1103/PhysRevPhysEducRes.13.020139

Guisasola, J., Zuza, K., Sarriugarte, P., & Ametller, J. (2023). Research based teaching-learning sequences in physics education: A rising line of research. In M. F. Taşar & P. R. L. Heron (Eds.), *The international handbook of physics education research: Special topics* (pp. 1–26). AIP Publishing LLC. https://doi.org/10.1063/9780735425514_026

Juuti, K., & Lavonen, J. (2006). Design-based research in science education: One step towards methodology. *Nordic Studies in Science Education, 2*, 54. https://doi.org/10.5617/nordina.424

Küçüközer, A. (2009). Prospective science teachers' understanding of sound. *Procedia-Social and Behavioral Sciences, 1*(1), 1889–1894. https://doi.org/10.1016/j.sbspro.2009.01.332

Lijnse, P. L., & Klaassen, C. W. J. M. (2004). Didactical structures as an outcome of research on teaching-learning sequences? *International Journal of Science Education, 26*, 537. https://doi.org/10.1080/09500690310001614753

McDermott, L. C., & Shaffer, P. S. (1992). Research as a guide for curriculum development, Part II: Design of instructional strategies. *American Journal of Physics, 60*, 994. https://doi.org/10.1119/1.16979

Meheut, M., & Psillos, D. (2004). Teaching-learning sequences: Aims and tools for science education research. *International Journal of Science Education, 26*, 515. https://doi.org/10.1080/09500690310001614762

Rico, A., Ruiz-González, A., & Guisasola, J. (2019). Designing a teaching-learning sequence about sound for pre-service primary teachers. In O. Levrini & G. Tasquier (Eds.), *Electronic proceedings of the ESERA 2019 conference*. University of Bologna.

Rico, A., Ruiz-González, A., Azula, O., & Guisasola, J. (2021). Dificultades de aprendizaje del modelo de sonido: una revisión de la literatura *(Learning difficulties about the sound model: A review of the literature). Enseñanza de las Ciencias, 39*(2), 5–23. https://doi.org/10.5565/rev/ensciencias.3217

Rivero, A., Azcárate, P., Porlán, R., Martín-del-Pozo, R., & Harres, J. (2011). The progression of prospective primary teachers' conceptions of the methodology of teaching. *Research in Science Education, 41*, 739–769. https://doi.org/10.1007/s11165-010-9188-z

Rossing, T. D., Moore, R. F., & Wheeler, P. (2001). *The science of sound* (3rd ed.). Addison-Wesley.

Ruiz-González, A., Rico, A., & Guisasola, J. (2021). Tocar la melodía conociendo la partitura: Una secuencia sobre el sonido en el Grado de Primaria. *(Play the tune knowing the score: A sequence about sound in the Primary Degree). Alambique: Didáctica de las ciencias experimentales, 105*, 16–23.

Zuza, K., Sarriugarte, P., Ametller, J., Heron, P. R., & Guisasola, J. (2020). Towards a research program in designing and evaluating teaching materials: An example from dc resistive circuits in introductory physics. *Physical Review Physics Education Research, 16*(2), 020149. https://doi.org/10.1103/PhysRevPhysEducRes.16.020149

Aritz Ruiz-Gonzalez is an Associate Professor in Science Education in the Department of Didactics of Mathematics and Experimental and Social Sciences at the University of the Basque Country (UPV/EHU). His primary research interests are STEM Education and Design-Based Research in primary education and teacher training. He is a member of the consolidated research group funded by the Basque Government "*IKASGARAIA. Education, Culture and Sustainable Development*" a multidisciplinary group of educational researchers with a specific sub-group focus on science education.

Arantza Rico is an Associate Professor in Science Education in the Department of Didactics of Mathematics and Experimental and Social Sciences at the University of the Basque Country (UPV/EHU). She specializes in Design-Based research and STEM and sustainability education. She is a member of the consolidated research group funded by the Basque Government "*IKASGARAIA. Education, Culture and Sustainable Development*", a multidisciplinary group of educational researchers with a specific sub-group focus on science education.

Jenaro Guisasola is an Associate Professor of Physics in the Department of Applied Physics at the University of the Basque Country (UPV/EHU). Since 2023, he has been serving as a visiting researcher at the Dual Engineering School at IMH Campus. His main research focus is Design-Based Research, particularly in the teaching of Physics at both university and high school levels. He is also interested in exploring the relationship between the Nature of Science and Science Education.

Chapter 13
Acceptance of Evolution by Israeli Students from Diverse Religious Groups

Netta Dagan, Masha Tsaushu, Tali Tal, Rachel S. A. Pear, Nigmeh Abu Toameh Kadan, and Hanan A. Alexander

13.1 Introduction

13.1.1 Challenges in Evolution Education Stem from Religious Views

The theory of evolution is considered one of the most challenging topics in biology teaching. There are various reasons for these difficulties, including inadequate content knowledge and a limited understanding of the nature of science. In addition, research indicates that some of the challenges are due to the assumption that the theory of evolution contradicts religious beliefs (Smith, 1994). From a science education perspective, it is commonplace to make a clear distinction between science

N. Dagan (✉) · M. Tsaushu
Faculty of Education in Science and Technology, Technion - Israel Institute of Technology, Haifa, Israel
e-mail: netta.dagan@campus.technion.ac.il

T. Tal
Faculty of Education in Science and Technology, Technion - Israel Institute of Technology, Haifa, Israel

Samuel Neaman Institute, Technion - Israel Institute of Technology, Haifa, Israel
e-mail: rtal@technion.ac.il

R. S. A. Pear · N. Abu Toameh Kadan
Faculty of Education, University of Haifa, Haifa, Israel

H. A. Alexander
Faculty of Education, University of Haifa, Haifa, Israel

Center for Jewish Studies, University of California, Berkeley, CA, USA
e-mail: hanana@edu.haifa.ac.il

G. Cakmakci, M. F. Tasar (eds.), *Connecting Science Education with Cultural Heritage*, Contributions from Science Education Research 15,
https://doi.org/10.1007/978-3-031-90944-3_13

and religious beliefs (Reiss, 2010). Yet learning does not occur in a vacuum. It is multidimensional, encompassing internal and external factors (Huleihil & Huleihil, 2016), and linked to previous knowledge, including worldviews derived from culture and community (Bransford et al., 2000). This suggests that students' religious and social identity influences their motivation to study culturally controversial issues such as evolution (BouJaoude, 2018), as they see it as an attempt to change their belief system (Barnes & Brownell, 2016).

Numerous studies, worldwide, have investigated the acceptance of evolution (e.g., Kuschmierz et al., 2021a; Miller et al., 2006). Despite the differences in instruments and student age, many studies indicate that misconceptions about evolution are common and that sociocultural context determines the acceptance or rejection of evolution (Kuschmierz et al., 2021b). In Israel, attitudes concerning the relationship between science and religion, particularly evolution and faith, have only been surveyed in adults (Pew Research Center, 2016), and little is known about high school students' attitudes.

13.1.2 The Israeli Education System and Science Curriculum

Israeli society and its education system are diverse regarding ethnicity, religion, and religious observance (Central Bureau of Statistics [CBS], 2018). Israel's education system has four educational streams aligned with the religious and cultural structure of society (Tal, 2021):

(a) State schools for Hebrew speakers—serve secular and traditional Jewish students.
(b) State religious schools for Hebrew speakers—serve orthodox and traditional Jewish students.
(c) State schools for Arabic speakers—serve Sunni Muslim, Christian and Druze students.
(d) Autonomous ultra-orthodox schools—serve ultra-orthodox Jewish students.

In state schools, humanities such as literature, history, and religious studies have modified curricula to serve different student populations. Jewish students study the Bible as a mandatory subject, whereas others study a parallel topic aligned with their religion (e.g., Islam, Christianity). By contrast, there is one national science curriculum for all students where the only difference is the language of instruction, i.e., Hebrew or Arabic. Even though the science curriculum is constant across all communities, science-education researchers indicate differences among cultural groups in Israel regarding science capital (Diamond, 2020), perceptions about scientists (Koren & Bar, 2009), and views on the purpose of science education (Tal, 2021).

13.1.3 Evolution and Faith in Biology Education

In Israel, evolution is part of the middle school science and technology curriculum, and since 2015 has been a mandatory topic for high school biology majors (Ministry of Education, 2015, 2016). However, student proficiency scores for the evolution questions on the biology matriculation exam leave much to be desired (Ministry of Education, 2019). Previous studies have suggested that denial of the theory of evolution based on religious beliefs may account for this low performance (Siani & Yarden, 2020; Stahi-Hitin & Yarden, 2022a). Israeli curriculum writers have expressed concerns about the faith issues students raise when studying evolution (Siani & Yarden, 2021), since this will likely influence learning.

Different approaches have been suggested to cope with the conflict between evolution and faith in biology education. One is 'The pedagogy of difference' which encourages explicit discourse about students' beliefs and scientific evidence (Owens et al., 2018). This approach allows students to engage with the theory of evolution from three perspectives: 'in,' 'from,' and 'about.' The 'in' perspective acknowledges their commitment to self-identity. The 'from' perspective introduces them to alternative worldviews, encouraging them to adopt additional values. The 'about' perspective focuses on acquiring knowledge and understanding without changing one's values (Alexander, 2017). Sensitive management of discourse, using culturally responsive methods, can help reduce tensions during the teaching of evolution, even when participants do not share the same worldview (Reiss, 2019).

An alternative strategy for addressing religious challenges in teaching evolution involves integrating aspects of the Nature of Science (NOS). NOS encompasses how science is defined, how it is conducted, and how it is influenced by society (McComas & Clough, 2020). Explicit instruction of NOS is important for developing a deeper understanding of science (Abd-El-Khalick & Lederman, 2000) and the distinction between science and other ways of knowing (Reiss, 2010). It also may improve the acceptance of the scientific validity of the theory of evolution (Nelson et al., 2019).

13.1.4 Goals

This study aimed to probe how students' religious beliefs in the different parts of the Israeli school system impact their acceptance of the theory of evolution. Understanding the perspectives of different student populations, including their views and beliefs about various aspects of evolution that they may find difficult to accept, is required to find the most suitable approach to cope with the conflict between evolution and faith in Israeli biology classes.

Awareness of students' views will hopefully contribute to the development of culturally responsive learning materials that can improve their understanding and acceptance of evolution in Israel and elsewhere.

13.2 Methodology

13.2.1 Participants

The participants were Jewish, Muslim, Druze, and Christian secondary school students with religious and non-religious orientations (N = 1028; Table 13.1). They represented almost every religious subgroup in Israel except Ultra-Orthodox Jewish students, who hardly study science. Another reason to exclude this group is its reluctance to take part in educational studies related to topics they see as harmful.

In addition to identifying the students by their school stream (e.g., Jewish religious, Jewish secular, Arab), students were asked to indicate their faith and level of observance. In Israel, a person who identifies as 'religious' is committed to observing the commandments of their faith. 'Traditional' refers to people with religious sentiments who believe in God but are not necessarily observant. The term 'secular' is used only by Jews and refers to those who do not believe in God and do not observe religious commandments. Muslims, Druze, and Christians prefer the term 'not religious'.

The research team included Muslim and Jewish, religious, non-religious, secular, and native speakers of Hebrew, Arabic and English.

13.2.2 Data Collection

The students completed the MATE (Measure of the Acceptance of the Theory of Evolution) questionnaire. This questionnaire is often used in evolution education studies (e.g., Metzger et al., 2018). It was developed to examine the level of self-reported acceptance of the theory of evolution in different populations (Rutledge & Warden, 1999). The original English MATE questionnaire consists of 20 items (for example, "Living organisms existing today are the result of evolutionary processes that have occurred over millions of years"). We added seven items that explicitly address possible contradictions between evolution and faith (e.g., "Evolution and religious belief do not contradict each other"). Based on the results of a pilot study,

Table 13.1 Participant demographics (N = 1028)[a]

Educational stream	397 from State schools for Hebrew speakers 265 from State religious schools for Hebrew speakers 351 from State schools for Arabic speakers
Self-reported religious observance	311 Religious 366 Traditional 308 Secular/not religious
School level	425 Middle school 549 High school

[a]The numerical difference stems from respondents who only provided incomplete information

which showed that students hesitated as to what the expected "right answer" was, we added the phrase 'In my opinion' to the beginning of each item. This was done to make it clear that we were looking for the students' opinions rather than their knowledge.

The original English MATE questionnaire was translated into Hebrew and Arabic employing multi-step validation. Piloting led to minor cultural adaptations before the final versions were administered. Respondents were asked to rate their agreement with the statements in the questionnaire on a Likert scale ranging from 1 to 5. Statistical analysis of the responses was conducted using SPSS. The findings below relate to 25 of the 27 items of the questionnaire. Two items were removed due to low internal consistency ($\alpha = 0.33$).

13.3 Findings

The goal of this study was to assess the acceptance of the theory of evolution in middle and secondary school students across religious groups in Israel. Factor analysis of the responses to the MATE questionnaire yielded four categories (1) Acceptance of the theory of evolution *(acceptance)*; (2) Rejection of the theory of evolution *(rejection)*; (3) (there is) No contradiction between evolution and faith *(no-con)*; and (4) Acceptance of the scientifically acknowledged age of the Earth *(Earth-age)*. We found significant differences between school streams, levels of religiosity and students' grade level, in most categories. Due to space limitations, we only present the main patterns.

13.3.1 Educational Stream

As presented in Fig. 13.1, there were significant differences in *acceptance* among students from different educational streams. Hebrew-speaking students from non-religious schools exhibited the highest levels of acceptance of evolution, compared to students in the other educational streams. Hebrew-speaking students from religious schools, had lower levels of acceptance of evolution than students from non-religious schools. Arabic-speaking students had the lowest level of acceptance of evolution. This was reinforced by the mirror image findings on the *rejection* of evolution, as presented in Fig. 13.2. The highest level of rejection was observed among Arabic-speaking students, with lower levels of rejection across all other groups: there was a lower rejection of evolution among Hebrew-speaking students from religious schools, and the lowest level of rejection was found for Hebrew-speaking students from non-religious schools.

No-con views and views on the *age of the Earth* (faith-related beliefs) are presented in Figs. 13.3 and 13.4. The highest levels of agreement with items stating that evolution and faith are non-contradictory were found among Hebrew-speaking

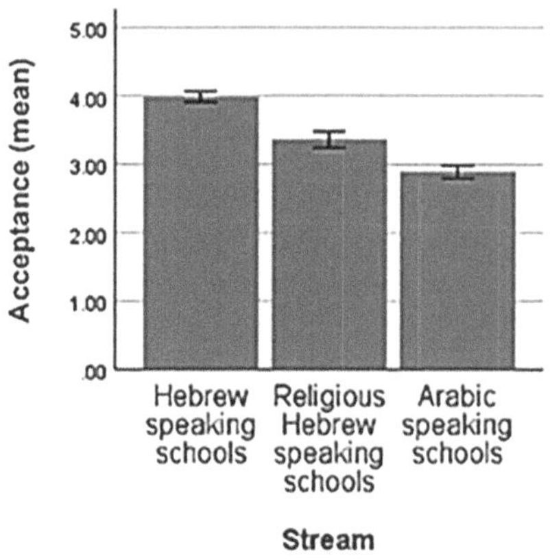

Fig. 13.1 Acceptance of evolution by the educational streams

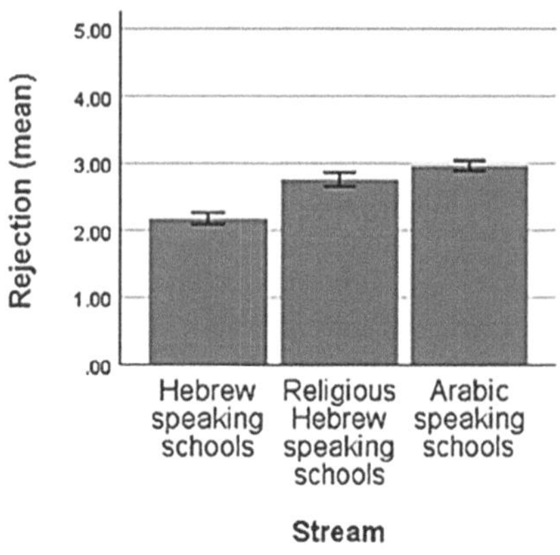

Fig. 13.2 Rejection of evolution by the educational streams

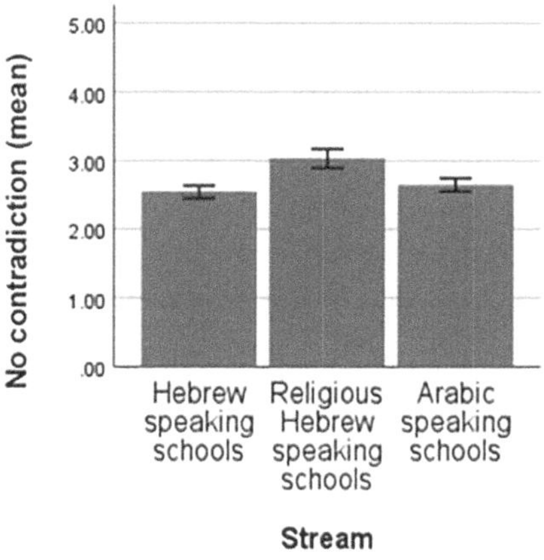

Fig. 13.3 Non-contradictory views of evolution by the educational streams

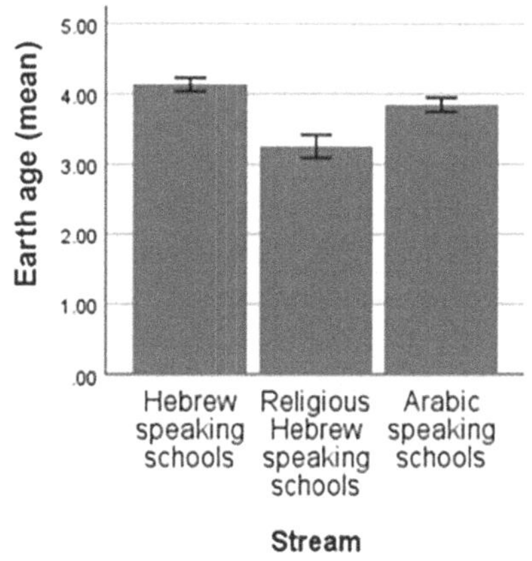

Fig. 13.4 Scientifically accepted Earth-age views by the educational streams

students from religious schools, despite their low acceptance of the scientific age of Earth. By contrast, the lowest level of support for the non-contradictory items and the highest level of support for the (scientific) Earth-age items were found among Hebrew-speaking students from non-religious schools. Arabic-speaking students fell between the other two groups.

13.3.2 Self-Reported Level of Religious Observance

To examine the relationships between religious observance and the acceptance of evolution, the participants were asked to identify themselves as religious, traditional, non-religious or secular. In the Hebrew-speaking non-religious schools, 20%

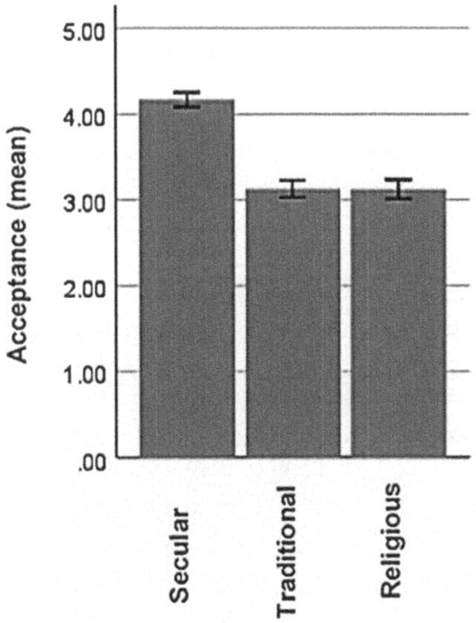

Fig. 13.5 Acceptance of evolution by self-reported religious identity

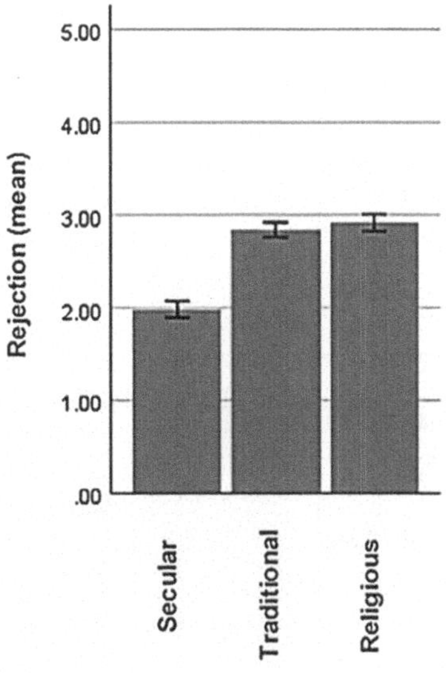

Fig. 13.6 Rejection of evolution by self-reported religious identity

of the students defined themselves as traditional. In the Arab-speaking schools, approximately 75% of the students defined themselves as traditional. As presented in Fig. 13.5, there were significant differences in acceptance of evolution between students who identified themselves as secular and all other students. Secular students showed the highest level of acceptance. Traditional and religious students showed similar levels of acceptance, which were significantly lower than secular students. This was reinforced by the findings on rejection, where we found the mirror image, as presented in Fig. 13.6. Secular students showed low levels of rejection of evolution, while traditional and religious students showed similar levels of rejection, which were significantly higher than secular students.

13.3.3 School Level and Acceptance of Evolution

Given the differences in science curricula for middle school and high school, the participants were asked to indicate their grade level. There was a positive relationship between students' school level and the degree of acceptance of evolution. As presented in Figs. 13.7 and 13.8, middle school students (ninth grade; aged 14–15) were more uncertain about accepting the theory of evolution and more likely to reject it than high school students (10th–12th grades; aged 15–18). No differences were found between middle school and high school students in terms of non-contradictory views, as presented in Fig. 13.9. Middle school students showed low levels of acceptance of the scientifically acknowledged age of the Earth compared to high school students, as presented in Fig. 13.10.

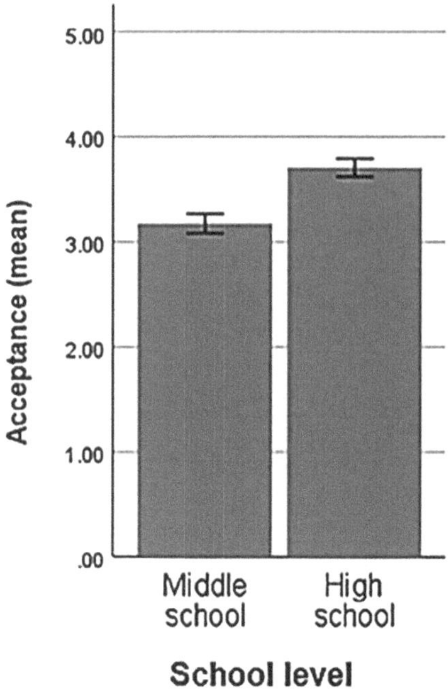

Fig. 13.7 Acceptance of evolution by students' school level

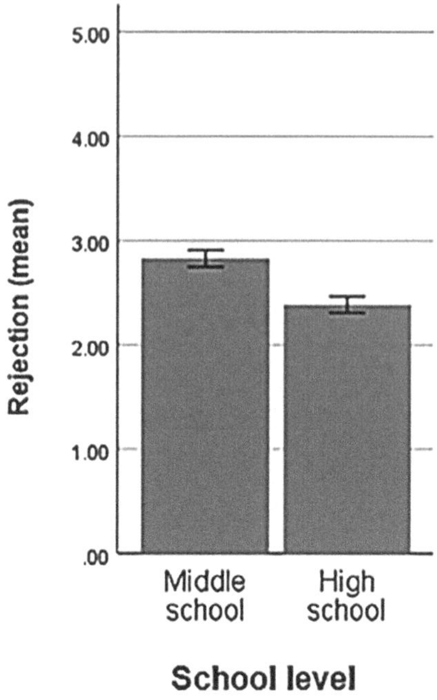

Fig. 13.8 Rejection of evolution by students' school level

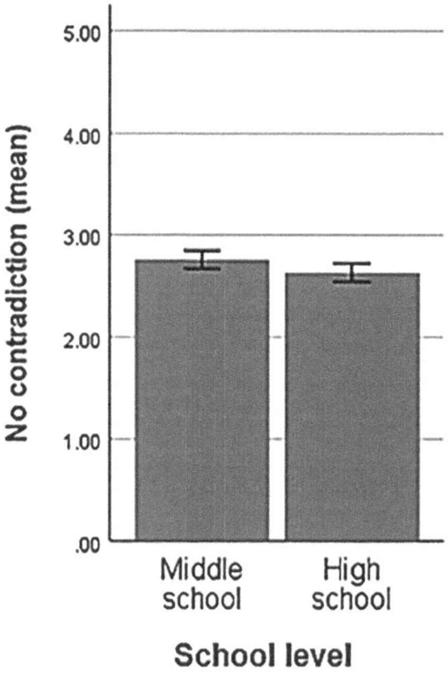

Fig. 13.9 Non-contradictory views of evolution and faith by students' school level

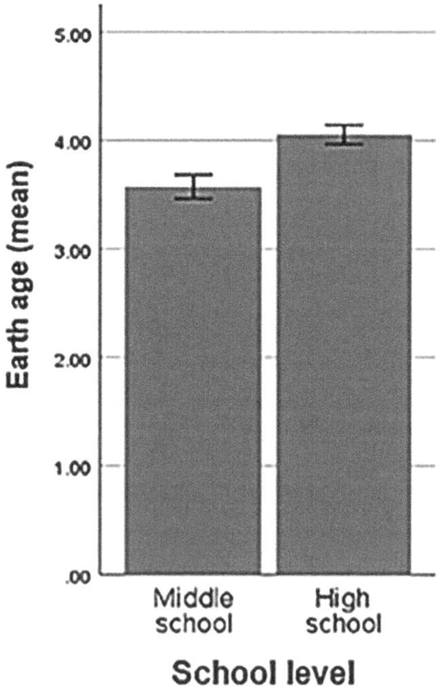

Fig. 13.10 Scientifically accepted Earth-age views by students' school level

13.4 Discussion

This study explored views on the theory of evolution in a large sample of middle and high school students representing different religions and levels of religious observance.

The findings indicated a low level of acceptance of evolution, particularly among religious and traditional students, irrespective of their religion (Jewish, Christian, Muslim, or Druze). Previous studies have shown that teachers encounter student religious-based resistance when teaching evolution (Stahi-Hitin & Yarden, 2022a). Calls have been made to address students' religious views to create an educational climate that is more conducive to learning (Barnes & Brownell, 2016).

The differences between formal education streams (schools for non-religious Hebrew speakers /religious Hebrew speakers/Arabic speakers) do not capture the entire picture. Although students identifying themselves as "traditional" typically study in non-religious schools, they were found to be more hesitant about evolution than their counterparts identifying themselves as secular. As a result, even in schools associated with the State's non-religious stream, faith-related challenges in evolution education remain common. These challenges suggest that professional development programs should provide teachers with approaches that engage with evolution and faith (Stahi-Hitin & Yarden, 2022b), which could provide them with strategies to deal with issues that are likely to arise in class.

Interestingly, despite their high support of the non-contradictory items, students from religious Hebrew-speaking schools strongly rejected the scientifically acknowledged age of the Earth. The view that there is no contradiction between evolution and faith reflects attempts to bridge between modern ways of life and religious observance in religious Hebrew-speaking schools. Students are taught that faith and modern life are not contradictory, but this is not sufficient to affect their views about evolution in general and the age of the Earth in particular. In Israel, the Hebrew calendar is omnipresent in public life, in addition to the Gregorian calendar, and all Jewish holidays are celebrated according to the Hebrew calendar date. In religious schools and for religious families, it determines many religious practices. According to the Hebrew calendar that reflects the biblical age of the Earth in Judaism, 2024 is considered year 5784. This is an apparent contradiction for many religious students. As suggested by Reiss (2010, 2019), a religious worldview shapes how and what students learn. Scientific facts rarely change people's worldviews, even when well taught. However, a respectful attitude toward students' views could help them cope with this conflict and promote more openness to learning science. In addition, a better understanding of the nature of science (NOS) may foster a deeper appreciation for the scientific method and could help students understand how the age of the Earth is determined. It also may help students to distinguish between science and religion as two perspectives on the world, addressing different questions (Reiss, 2010), and, as a result, to refer to the scriptures interpretively rather than literally.

The opposite picture was found among students from non-religious Hebrew-speaking schools who acknowledged the contradiction between faith and science. In secular schools, a distinction is made between myth (the biblical creation) and science. This educational approach is aligned with recommendations by Smith (1994), who suggested avoiding non-scientific discussions in science classes and issues that involve a literal interpretation of the Bible, in particular, such as the age of the Earth. For many students, this appears to imply that there is indeed a conflict between evolution and faith.

Student school level was also found to be a predictor of acceptance or rejection of evolution, in that middle school students showed significantly higher hesitation to accept evolution compared to high school students. Previous studies have indicated that curricula appear to play a significant role: when evolution is not included in the curriculum, this tends to correlate with higher levels of rejection (Kuschmierz et al., 2021b). In Israel, despite the inclusion of evolution in the middle school science curriculum (Ministry of Education, 2016), many teachers opt not to teach evolution when running out of class time to cover the already overloaded curriculum. In addition, the absence of proper learning materials and the complexity of the topic affect teachers' priorities and the decision to teach it. This suggests that middle school students' perception of evolution may derive from insufficient knowledge and be primarily influenced by sociocultural backgrounds that are significant to their identity (BouJaoude, 2018). To augment middle school students' acceptance of evolution, we suggest that middle school teachers should be encouraged to teach evolution and be provided with suitable learning materials for this age bracket. Otherwise, evolution will remain another example of an official topic that is not taught (Glatthorn, 1999).

In Israeli high schools, introductory science is compulsory in tenth grade. In the 11th–12th grades, science classes are elective, and biology majors study evolution as a mandatory topic (Ministry of Education, 2015). Additionally, at this level, students usually expand their horizons and are more exposed to information from various sources. Thus, it makes sense that high school students' attitudes towards evolution are grounded in more comprehensive knowledge. At least some of them accept evolution even if they experience a conflict related to their religious background. This finding is consistent with Smith and Siegel (2016), who distinguished between knowledge about evolution, understanding the theory of evolution, belief in evolution, and acceptance of evolution. This distinction helps show that, rather than belief, acceptance depends on a choice, is related to debating ideas, and has nothing to do with sentiment.

This study points to the importance of understanding students' views and resistance to evolution. It innovates by comparing different religious groups and levels of observance in a diverse education system that serves different religious and ethnic groups who all study the same science curriculum. The findings can thus contribute to the development and implementation of more sensitive teaching (Reiss, 2019), as well as to the development of culturally responsive learning materials (Gay, 2002) which can reduce tensions in evolution and science classes, and contribute to students' understanding and achievements as it addresses the unique characteristics of the learner (Huleihil & Huleihil, 2016).

13.4.1 Limitations

The demographics of the students who have responded to the questionnaire so far do not accurately represent the different religious groups in Israel; students from religious Hebrew-speaking schools were over-represented, and students from non-religious Hebrew-speaking schools were underrepresented due to the low cooperation of teachers in these schools. We continued the data collection, distributing questionnaires to students from non-religious Hebrew-speaking schools through social networks, and will add these data in future publications.

References

Abd-El-Khalick, F., & Lederman, N. G. (2000). Improving science teachers' conceptions of nature of science: A critical review of the literature. *International Journal of Science Education, 22*(7), 665–701. https://doi.org/10.1080/095006900050044044

Alexander, H. A. (2017). Public theology and liberal education. *International Journal of Public Theology, 11*(3), 313–327. https://doi.org/10.1163/15697320-12341495

Barnes, M. E., & Brownell, S. E. (2016). Practices and perspectives of college instructors on addressing religious beliefs when teaching evolution. *CBE Life Sciences Education, 15*(2), 1–19. https://doi.org/10.1187/cbe.15-11-0243

BouJaoude, S. (2018). Evolution education in the Arab States: Context, history, stakeholders' positions and future prospects. In *Evolution education around the globe* (pp. 297–312). Springer International Publishing.

Bransford, J. D., Brown, A. L., & Cocking, R. R. (2000). *How people learn* (Vol. 11). National Academy Press.

Central Bureau of Statistics [CBS]. (2018). *The Israeli society - Religion and self-definition of the degree of religiosity*, p. 21 (HEBREW). https://www.cbs.gov.il/he/publications/DocLib/2018/rep_10/h_print.pdf

Diamond, A. H. (2020). The social reproduction of science education outcomes for high school students in Israel. *British Journal of Sociology of Education, 41*(7), 1029–1046.

Gay, G. (2002). Preparing for culturally responsive teaching. *Journal of Teacher Education, 53*(2), 106–116. http://jte.sagepub.com

Glatthorn. (1999). Curriculum alignment revisited. *Journal of Curriculum and Supervision, 15*(1).

Huleihil, M., & Huleihil, H. (2016). Status of science and technology education at Arab high schools in Israel. *Athens Journal of Education, 3*(4), 365–386.

Koren, P., & Bar, V. (2009). Pupils' Image of 'the Scientist' among Two Communities in Israel: A comparative study. *International Journal of Science Education, 31*(18), 2485–2509. https://doi.org/10.1080/09500690802449375

Kuschmierz, P., Beniermann, A., Bergmann, A., Pinxten, R., Aivelo, T., Berniak-Woźny, J., Bohlin, G., Bugallo-Rodriguez, A., Cardia, P., Cavadas, B. F. B. P., Cebesoy, U. B., Cvetković, D. D., Demarsy, E., Đorđević, M. S., Drobniak, S. M., Dubchak, L., Dvořáková, R. M., Fančovičová, J., Fortin, C., & Futo, M. (2021a). European first-year university students accept evolution but lack substantial knowledge about it: A standardized European cross-country assessment. *Evolution, 14*(1). https://doi.org/10.1186/s12052-021-00158-8

Kuschmierz, P., Meneganzin, A., Pinxten, R., Pievani, T., Cvetković, D., Mavrikaki, E., Graf, D., & Beniermann, A. (2021b). Towards common ground in measuring acceptance of evolution and knowledge about evolution across Europe: A systematic review of the state of research. *Evolution, 13*(1). https://doi.org/10.1186/s12052-020-00132-w

McComas, W. F., & Clough, M. P. (2020). Nature of science in science instruction: Meaning, advocacy, rationales, and recommendations. In W. F. McComas (Ed.), *Nature of science in science instruction. science: Philosophy, history and education*. Springer. https://doi.org/10.1007/978-3-030-57239-6_1

Metzger, K. J., Montplaisir, D., Haines, D., & Nickodem, K. (2018). Investigating undergraduate health sciences students' acceptance of evolution using MATE and GAENE. *Evolution, 11*(1). https://doi.org/10.1186/s12052-018-0084-8

Miller, J. D., Scott, E. C., & Okamoto, S. (2006). Public acceptance of evolution. *Science, 313*(5788), 765–766. https://doi.org/10.1126/science.1126746

Ministry of Education. (2015). *Biology curriculum for high school interns*. (HEBREW) https://pop.edu cation.gov.il/tchumey_daat/biologya/chativa-elyona/biology-pedagogy/curriculum-specialize/

Ministry of Education. (2016). *Science and technology curriculum for middle school*. (HEBREW) https://pop.education.gov.il/tchumey_daat/mada-tehnologia/chativat-beynayim/mada-technologia- pedagogia/curriculum/#pos_248323

Ministry of Education. (2019). *Biology matriculation exams: Questionnaires, indicators, sample answers and analysis*. (HEBREW) Link to The National Center for Biology Teachers.

Nelson, C., Scharmann, L., Beard, J., & Flammer, L. (2019). The nature of science as a foundation for fostering a better understanding of evolution. *Evolution: Education and Outreach, 12*, 1–16. https://doi.org/10.1186/s12052-019-0100-7

Owens, D. C., Pear, R. S. A., Alexander, H. A., Reiss, M. J., & Tal, T. (2018). Scientific and religious perspectives on evolution in the curriculum: An approach based on pedagogy of difference. *Research in Science Education, 48*(6), 1171–1186. https://doi.org/10.1007/s11165-018-9774-z

Pew Research Center. (2016). *Israel's religiously divided society*. https://www.pewforum.org/2016/03/ 08/education-values-and-science/. Accessed 18 Jul 2023.

Reiss, M. J. (2010). Science and religion: Implications for science educators. *Cultural Studies of Science Education, 5*(1), 91–101. https://doi.org/10.1007/s11422-009-9211-8

Reiss, M. J. (2019). Evolution education: Treating evolution as a sensitive rather than a controversial issue. *Ethics and Education, 14*(3), 351–366. https://doi.org/10.1080/17449642.2019.1617391

Rutledge, & Warden, M. A. (1999). The development and validation of the measure of acceptance of the theory of evolution instrument. *School Science and Mathematics, 99*(1), 13–18.

Siani, M., & Yarden, A. (2020). "Evolution? I don't believe in it": Theological tensions surrounding the implementation of evolution in the Israeli curricula. *Science and Education, 29*(2), 411–441. https://doi.org/10.1007/s11191-020-00109-7

Siani, M., & Yarden, A. (2021). "I think that teachers do not teach evolution because it is complicated": Difficulties in teaching and learning evolution in Israel. *International Journal of Science and Mathematics Education, 20*(3), 481–501. https://doi.org/10.1007/s10763-021-10179-w

Smith, M. U. (1994). Counterpoint: Belief, understanding, and the teaching of evolution. *Journal of Research in Science Teaching, 31*(5), 591–597.

Smith, M. U., & Siegel, H. (2016). On the relationship between belief and acceptance of evolution as goals of evolution education: Twelve years later. *Science and Education, 25*(5–6), 473–496. https://doi.org/10.1007/s11191-016-9836-4

Stahi-Hitin, R., & Yarden, A. (2022a). Should high-school biology teachers relate to students' religious faith when teaching evolution? The perspective of Jewish teachers. *International Journal of Science Education*, 1–22. https://doi.org/10.1080/09500693.2022.2070936.

Stahi-Hitin, R., & Yarden, A. (2022b). Evolution education in light of religious science teachers' and scientists' conceptions of evolution and religion. *Evolution: Education and Outreach, 15*(1), 1–16. https://doi.org/10.1186/s12052-022-00167-1

Tal, T. (2021). Inequity and cultural differences in out-of-school science education: The case of the Arab Minority in Israel. *Asia-Pacific Science Education, 6*(2), 427–455. https://doi.org/10.1163/23641177-BJA10011

Netta Dagan is a PhD student at the Faculty of Education in Science and Technology of the Technion – Israel Institute of Technology. Her research focuses on the relationship between evolution and faith in biology education. She is an experienced biology teacher in high school and a teacher educator

Dr. Masha Tsaushu is an adjunct lecturer at the Faculty of Science and Technology Education at the Technion – Israel Institute of Technology. She was a biology teacher in high school for many years, a teacher mentor and a teacher educator.

Tali Tal the the Ruben and Tallu Academic Chair of Science, is a professor of science and environmental education at the Faculty of Education in Science and Technology, Technion – Israel Institute of Technology, and a Senior Research Fellow at the Samuel Neaman Institute for National Policy Research. She served as president of NARST (2019–2022) and as the Faculty Dean (2020–2024). Currently, she chairs the Technion's Pre-University Education Center. Her research focuses on out-of-school science education, inquiry-based learning, and environmental education.

Dr. Rachel S. A. Pear is a research fellow at the University of Haifa's Center for Jewish and Democratic Education. She received a PhD from Bar-Ilan University within the Graduate Program on Science, Technology and Society. Her dissertation examined changes in Jewish American engagement with evolution over the 20th century. She co-teaches a continuing education course for teachers on Judaism and evolution at Herzog College.

Nigmeh Abu Toameh Kadan is a PhD student at the Faculty of Education at the University of Haifa. She was an instructor of the Open University's "Introduction to Academia" course for high school students, through which she taught courses on evolution in Arab Israeli schools. The experience of exposing students to evolution was the basis of her MA thesis. Her current research takes this subject further, exploring historical and cultural elements of the reception of evolution in the local context.

Hanan A. Alexander is a professor of Philosophy of Education at the University of Haifa, where he served as Dean of the Faculty of Education, Dean of Students, Head of the International School, and Head of the Center for Jewish Education. President of the Religious Education Association, he has taught at American Jewish University, UCLA, UC Berkeley, Graduate Theological Union, Jewish Theological Seminary, and Bar-Ilan University. He was a Visiting Fellow at the University of Cambridge. He is a past editor of *Religious Education* and has published more than 130 essays and 7 books.

Chapter 14
Assessing Science Competencies in Danish Primary and Lower Secondary Education

Jørgen Løye Christiansen

14.1 Introduction

The integration of competence-based approaches into the Danish educational system occurred relatively late. In 2003, the so-called FNU reports outlined science competencies for Danish science education (Andersen et al., 2003; Busch et al., 2003), and in 2014, the Danish Ministry of Education introduced *Simplified Common Goals* to enhance the relevance and practicality of lower secondary school examinations. This initiative included a mandatory oral-practical examination in physics/chemistry, biology, and geography for ninth grade students, aimed at testing students' competencies in inquiry, modelling, perspective-taking, and communication. However, despite these efforts, current general assessment methodologies primarily focus on testing knowledge and skills rather than competencies (e.g., Dolin, 2016; Nielsen & Dolin, 2016). Consequently, assessing students' science competencies in primary and lower secondary schools remains challenging.

Centred on the research question: What are the characteristics of competence-oriented teaching, and which assessment tools would be beneficial in assessing students' level of competence achievement?, this study seeks to address this challenge. In collaboration with primary and lower secondary school science teachers, through studies of the concept of competence and the characteristics of competence-oriented teaching, new formative assessment designs, also known as assessment for learning, were developed and tested. These designs are expected to provide a more effective means of assessing students' inquiry, modelling, and perspective-taking

J. L. Christiansen (✉)
Department of Science Education, University of Copenhagen, Copenhagen, Denmark

Department of Teacher Education, University College Absalon, Roskilde, Denmark
e-mail: jlc@ind.ku.dk

G. Cakmakci, M. F. Tasar (eds.), *Connecting Science Education with Cultural Heritage*, Contributions from Science Education Research 15,
https://doi.org/10.1007/978-3-031-90944-3_14

competencies, along with their science communication skills, within the daily ongoing science education.

14.2 Theoretical Background

Central to this work is the concept of competence. However, there is no consensus in the literature on how the concept of competence should be defined. Competence often means different things in different contexts and at different times. Blömeke et al. (2015) have called competence a "messy construct," and Hodges (2012) argues that competence is a word that is so frequently used, and in so many different ways, that it risks having no meaning at all.

In more recent interpretations of competence, it is often mentioned that competencies are more than just knowledge and skills (see Rychen & Salganik, 2000; Wagenaar & Ferreras, 2008; Drisko, 2014). The Tuning project, for example, defined competence as *"… a dynamic combination of cognitive and metacognitive skills, knowledge and understanding, interpersonal, intellectual and practical skills, and ethical values"* (Wagenaar & Ferreras, 2008, p. 147). *"Competence thus encapsulates a complex amalgamation of knowledge, skills or practices, attitudes, values and behaviors that are transversal and interdisciplinary, yet also domain-specific"* (Oyao et al., 2015, p. 2239).

For a general science competence definition, we applied Weinert's broader definition of competence to the realm of science:

> Science competencies are those, for individuals, available or learnable cognitive skills and abilities that, coupled with motivation, willpower, social readiness, and social skills, produce successful and responsible problem solving in variable science situations (Weinert, 2001, translated after Wiesner & Schreiner, 2020, added "science").

Evidently, this definition underscores the multifaceted nature of competence beyond the mere aggregation of knowledge and skills.

It is clear that such a definition, although generally sensible, are not particularly action-orientated in relation to teaching and learning in science. Therefore, many countries, including Denmark, operate with several subordinate science competencies, sub-competencies or whatever we want to call them (see e.g. Waddington et al. 2007 and Mathelitsch, 2013).

In Denmark, primary and lower secondary school science objectives is organised into four science competencies: inquiry competence, modelling competence, perspective-taking competence and communication competence (e.g. Ministry of Children and Education, 2019). Therefore, these four science competence areas form the context categories of this project.

The preferred assessment approach in the project was formative assessment, more specifically formal formative assessment. The formative assessment approach has emerged from constructivist learning theory (Fosnot, 1989), where the purpose is to promote effective and meaningful learning. It builds on what is already known and allows new ideas to develop as we adapt and change old ones (Fosnot, 1989). Effective formative assessment prerequisites (1) knowledge and skills regarding the entire assessment process, (2) psychological factors (e.g., positive attitude, ownership, social pressure), and (3) social factors (e.g., collaboration with colleagues and students) (see also, Schildkamp et al., 2020).

Formal formative assessment is characterised by the use of various techniques and tools, such as forms, rubrics, logbooks and portfolios. The teacher will initially, based on questions and/or assessment tools, try to determine where the students currently place themselves within a goal category. Based on this, the teachers will subsequently observe and interpret the students' work and scaffold them towards a higher degree of goal fulfilment. The teacher will typically adjust their planning and teaching in a direction that allows students to achieve a sufficient level of goal fulfilment, and this assessment cycle continues thereafter.

Zhang et al. (2023) have shown that practicing formative assessment has a positive direct effect on students' learning behaviour in science and a positive indirect effect on students' motivation for science education.

14.3 Method

The methodology employed in this study is based on the well-proven Design-Based Research (DBR) approach, wherein a specific intervention undergoes continuous iteration of design, enactment, analysis, and redesign (Brown, 1992). The fundamental premise of DBR, that only by intervening with new designs can we develop better theories of practice at the same time as we seek to improve practice, is embraced in this study.

This study is based on activities in the 3-year Assessment of Science Competence project named NATKOM. Throughout a span of 13 months, 28 science teachers from 18 Danish schools engaged in an initial one-day course aimed at imparting the latest insights into the concept of competence, particularly within the domain of science. Given that many of the participating teachers initially struggled to view competence as something beyond knowledge + skills, we addressed this obstacle by applying Weinert's broader definition of competence to science (see definition above).

Upon establishing a shared comprehension of the concept of competence, we proceeded to delineate and mutually agree upon the defining characteristics of competence-oriented science teaching, paving the way for investigating and testing different formative assessment designs capable of assessing science competencies.

Three workshops (WS1–3) were conducted, each focusing on distinct aspects of science education: inquiry (WS1), modelling (WS2), and perspective-taking competencies (WS3), respectively. The communication competence was embedded in all workshops, as it here was considered a fundamental aspect of the other three competencies and contributes to the students' ability on a scientific basis to argue, discuss, etc. (see Christiansen, 2022).

Between workshops, teachers carried out teaching and assessment of the science competencies in question in the following order: inquiry competence in combination with communication (after WS1), modelling competence in combination with communication (after WS2), and perspective-taking competence in combination with communication (after WS3). Other or overlapping competence areas were 'allowed' to co-exist (and they mostly do), but they had no focus in planning, implementation and assessment. The participating teachers collaborated with teacher educators to develop new assessment designs aimed at assessing students' science competencies more effectively than existing practices. Teachers implemented these assessment designs, while teacher educators conducted observations, recorded student dialogues, and engaged in interviews with teachers about their perceptions of competence-based teaching, learning, and assessment. Through the three workshops and three sessions of competence-oriented teaching and assessment, the study employed three iterations with assessment designs predominantly focused on formative assessment, as the focus was on assessment for learning rather than assessment of learning.

The teachers' involvement ended with a validation seminar, which included a written questionnaire about difficulties in competence-orientated science teaching.

14.4 Preliminary Findings

A total of 27 of the participating teachers took part in an initial interview to uncover their preconceptions regarding competence-oriented science teaching and assessment of competencies. When asked, *"what do you think is characteristic of competence-oriented science teaching?"*, 6 out of 27 teachers, through various expressions, articulated that it is characteristic of competence-oriented teaching that the students are the ones who are the active part (some of them adding that students must have the opportunity for independent thinking). Additionally, two teachers explicitly noted that it is characteristic of competence-oriented teaching that there are real degrees of freedom/choices for the students.

Furthermore, three teachers associated the concept of curiosity with competence-oriented teaching. However, overall, the majority of the participating teachers (14 out of 27) find it difficult to characterise competence-oriented science teaching, ranging from direct statements such as *"I don't really know"* (Teacher 24), via statements of a more general nature, such as *"you can acquire knowledge and communicate it yourself"* (Teacher 14), to statements that to a greater extent represent an absence *"competencies .... come at the expense of knowledge"* (Teacher 10).

When asked *"How do you think competence-orientated science education can be assessed?"*, 9 out of 27 teachers expressed, in various ways, that it is difficult to assess competencies. For example, one teacher remarked, *"I actually think that's one of the things that's difficult. That's why we're here. That's what we want to get better at"* (Teacher 7). Another teacher echoed a similar sentiment, stating, *"that's exactly why I signed up, because I think it's really hard, I really think it's difficult"*

(Teacher 8). A third teacher has the same general opinion, albeit expressed more bluntly: *"Oh my gosh, that's why I've taken this course, isn't it? How can it be assessed? Christ, that's what's so damn difficult ..... I think that was a really difficult question, I don't know what to answer"* (Teacher 3).

In five instances, the time-intensive nature of assessments was cited as an obstacle to assessing science competencies.

A few teachers also acknowledged the discrepancy between current official conventional assessment practices and competence-oriented assessment methodologies. One teacher expresses it as follows: *"I find it hard to see how you can really test students' competencies in a written test. I think the best way is to somehow collect some data through dialogue with the students"* (Teacher 9).

Conversational approaches emerged as a favoured means of assessing competencies, with 12 out of the 27 teachers specifically mentioning conversations as a mean of assessing competencies.

However, it was clear that most of the participating teachers described competence-centred assessment in such general terms that it could be difficult to separate the approach from more traditional assessment. For example, one teacher says *"observing what they do and what they say"* (Teacher 2). However, a few teachers express what competency-based assessment is not about, namely multiple-choice assessment forms and Kahoot assessments.

In order to get a common understanding of what competence-oriented teaching is, we started, in collaboration with The Natural Sciences Evaluation and Development Centre (NEUC), to investigate criteria for competence-oriented teaching. The work was initially based on a network analysis of 320 keywords related to the concept of competence, in which a researcher from NEUC had been involved. The analysis identified eight central concepts, which NEUC and educators from University College Absalon synthesized into the following six characteristics of competence-oriented science education adapted by the project.

1. Teaching is based on a concrete (science) problem, which is evident to the students and in which the majority of students are engaged.
2. Teaching has a clear and explicit science competence focus.
3. Teaching contains collaboration and dialogue clearly scaffolded by the teacher.
4. Teaching provides a significant degree of freedom, while catering to the different needs of students for degrees of freedom.
5. Teaching provides enough time and opportunity for students to reflect and includes scaffolding processes that support student reflection.
6. Teaching is based on core science knowledge and skills.

Observations of teachers' practices in the classroom reveal that, in several instances, it has been a challenge to ensure that the implemented teaching programmes were competence-oriented according to the six criteria mentioned above. Specifically, three criteria posed significant challenges for the teachers (criteria 1, 4, and 5):

- It was often not clear to the students that the teaching was based on a concrete problem.

- The teaching did frequently not provide a sufficient degree of freedom for the students in order to demonstrate competence.
- The teaching did frequently not provide time and opportunity for students to reflect on the process/product.

During a subsequent validation seminar, teachers highlighted in a questionnaire that it was difficult to teach competence-orientated due to general time pressure and because they experienced a dilemma between teaching competence-orientated and ensuring students' knowledge and skills through more traditional teaching. At the seminar, teachers confirmed that two of the criteria for competence-orientated science teaching were particularly difficult to work with. These were the criteria relating to students' degrees of freedom and time to think. Notably, these criteria were among the lowest-rated in the observations as well.

14.5 Formative Assessment Findings

Through the three DBR iterations, assessment designs underwent development and testing. Recognizing the challenges teachers may face in observing the activities of all student groups, we discovered that continuous self-assessments, along with peer assessments, within student groups proved to be a valuable formative assessment tool for assessing competencies. Such an approach enabled both teachers and students to track progress in the science competence in question as the work progressed.

In the two first examples outlined below, students are provided with a set of questions to address. In the first case, which illustrates aspects of the inquiry competence, students offer responses to various inquiry questions as they progress through their group's inquiry design. Here (Fig. 14.1), students write their answers to the questions. It can be seen in the example that they manage to complete three answers to three questions in the first lesson. In the figure, answers are supported on a background of shades of blue. If students revise or refine their answers in subsequent lessons, these updates are indicated with a darker shade of blue. This approach ensures that previous answers are preserved, while new answers are added during the corresponding lesson where they were revised. In this example, the group answers all questions after the fourth lesson. Due to the large textual volume of especially answers to the later questions, Fig. 14.1 only shows a possible example of textual answer progression for the first three questions. Both students and teachers can here visually track changes to responses over time, fostering discussions regarding the rationale behind these modifications and the quality of the varied responses, thereby considering potential enhancements in addressing similar challenges in future teaching situation.

Some teachers implemented educational scenarios primarily structured around an assessment framework depicted in Fig. 14.1, but allowed students to develop their own criteria for what constitutes a good science inquiry. First, the students discussed among themselves what characterises a scientific inquiry in general and

Create your own science investigation - a document for students

	First lesson	Second lesson	Third lesson	Fourth lesson
1) The science question to be investigated is as follows:	What is going on inside a windmill?	How does a wind turbine work?		
2) Why do you want to investigate it?	I don't know what goes on inside a windmill.	I don't know exactly how a wind turbine works.	Wind turbines play an important role in Danish energy production, so it is important to know about it.	Wind turbines play a significant role in Danish green energy production, so it is important to know how kinetic energy is transformed into electrical energy.
3) How and in what order will you get the information you need to answer your science question?	I will look it up on the Internet.	First find information in books and on the Internet about what's going on in the Nacelle, Then on how the electricity gets from the wind turbine to the consumer.		
4) How will you keep track of your information, i.e. record your collected data?		…………	…………	
5) How will you analyse your data?		…………	…………	…………
6) What was the result (conclusion) of your study?			…………	…………
7) How can you argue in favour of your conclusion based on the data collected?			…………	…………
8) How will you present your results?			…………	…………
9) If you had the opportunity to improve your study, what would you do?				…………

Fig. 14.1 Student Document—Create your own science investigation/inquiry. The document that students work in is in a word processing programme. This means that there is significantly more space available for the students' text answers than the figure shows. Examples of possible student answers are included in the first 3 categories of the form to give an insight into how students are expected to work their way through the form. Showing a complete and detailed filled out form would be too voluminous here

later more specifically a fieldwork inquiry. Then, in dialogue with the teacher, the students drew up a joint list of criteria for a good field inquiry. The criteria held greater relevance for students as they were articulated in their own language. These criteria served as a guideline for students, both when they were in the process of planning their study and along the way, when they carried out the inquiry and finally when they had to assess their results. The engagement of students in defining the criteria against which their inquiry was assessed fostered a motivating learning environment.

The second example (Fig. 14.2), taken from the work with the perspective-taking competence, is comparable to the previous example. However, in contrast to the previous example, in this case, students are instructed to write within the same designated area but utilizing different font colours across various lessons. The dialogue between students and teachers, as well as among peers, remains akin to the example described earlier.

Furthermore, structured forms containing relevant questions aimed at fostering scientific argumentation were employed in a "design a model" activity. Here students would have to consider; *what is the aim of your model?* (Answer and justify your answer), *what is the model for? What should it not be used for? Who is the target audience of the model? To what extent does the model reach the target audience?* etc. One of the participating schools devised a self-assessment form for lower secondary education based on these prompts (see Fig. 14.3). This form provided students with a clear outline of assessment objectives allowing them to gauge their

Perspective-taking – Student document

A little about putting things into perspective:

Perspective-taking is about putting something into a larger perspective, going beyond your study/results.
Perspective-taking is about putting your topic/results into other contexts, i.e. relating it to something else.

Fill out this form in your work with the perspective-taking competency:

Things you need to consider:	Your answers:	
Which study or results are the starting point for your perspective:	XXXXXXXXXXXX (LESSON ONE)	
How can you put that into perspective:	Your answer:	Reason for the answer (it is because....)
•in your everyday life? (Neighbourhood?):	XXXXXXXXXXXX (LESSON ONE) XXXXXXXXXXXX (LESSON TWO) XXXXXXXXXXXX (LESSON THREE)	XXXXXXXXXXXX (LESSON ONE) XXXXXXXXXXXX (LESSON TWO) XXXXXXXXXXXX (LESSON THREE
•in relation to other regions:		
•in relation to other science issues:		
•in relation to social issues:		
•in relation to conflicts of interest		
•Time-wise (the same 'subject area' but in a different time; past or future):		

Fig. 14.2 Student document—perspective-taking

Modelling: Designing models				
Grade level	Adequate (02-4)	Fairly good (4-7)	Very good (7-10)	Excellent (10-12)
Purpose	Can describe the purpose of own model with help	Can describe the purpose of own model with increasing complexity		Can explain the purpose of own model
Construction	Can reproduce a model with simple modifications	Can reproduce a model with increased level of detail	Can combine elements from different models into their own model	Can independently plan and design own model
Usage	Can identify elements from own model using simple relevant terminology	Can describe the coherence of own model using simple relevant terminology	Can explain correlations and details in their own model using relevant terminology	Can use own model to predict (put into perspective) specific issues using relevant terminology
Valuation	Can identify simple differences between own model and reality	Can compare own model with similar models	Can mention advantages and disadvantages of own model	Can make reasoned suggestions for improving own model

Fig. 14.3 Student document—self-assessment designing models

perceived level of attainment at any given point. However, the descriptions here were formulated by the teacher team and therefore did not have the same degree of freedom as the previous examples. Subsequently, these self-assessments served as the basis for group discussions and conversations with the teacher, prompting students to justify their positions and discern the requisite effort and understanding needed for progression asking what kind of work and insight was needed to be able to tick higher up the hierarchy, so to speak.

14.6 Discussion and Conclusion

Formative assessment enriches the process of learning by promoting dynamic dialogues and collaborative interactions between students and educators, thereby advancing the learning process (see Arnold, 2022). Teachers must observe how their students react to different forms of feedback to customize it to meet each student's specific needs. However, it is challenging the teacher to gain insight into students' way of thinking and to frame feedback that guides them towards specific learning goals (Black & Wiliam, 2009). Student self-assessment becomes a potent tool for enhancing student achievement, especially when students are trained in utilizing predefined sets of performance criteria like rubrics to assess their own work, or when they receive other direct instruction on self-assessment (Ross, 2006).

In the context of peer assessment, students take on roles akin to instructional resources for one another, similar to collaborative learning. The feedback exchanged among peers during this assessment process provides valuable insights into their comprehension levels (Black & Wiliam, 2009).

For students to embrace these roles effectively, it is crucial that they possess a clear understanding of both the learning objectives and the performance criteria. Some strategies to facilitate this include teachers and students collaboratively constructing assessment criteria to enhance the reliability of peer assessment (Topping, 2010), or teachers demonstrating the process to students to encourage their active participation (Black & Wiliam, 2009). Assessment of competence in education requires a framework of unknown and uncertainty (cf. definition of competence used in this study). Therefore, the teacher must organize teaching in a way that allows for a significant degree of freedom for students, with opportunities for collaboration and dialogue integrated into the teaching and clearly scaffolded. This forms the basis for assessing competencies, as learners are observed in specific and authentic situations. In this project, this was typically conducted within the context of problem-based learning and group work. For instance, students formulated their own hypotheses or research questions, and designed and explored their own self-selected paths from question to answer. Through this dialogue-based approach, teachers scaffolded students by posing questions rather than providing answers, thereby fostering scientific thinking and argumentation. However, teachers may not always be able to fully observe the work of all student groups. To address this, this study found that continuous self-assessments within student groups, primarily based on rubrics and forms, were an effective formative assessment tool for assessing competencies. This approach enabled both teachers and students to monitor progress in competence as the work progressed.

14.6.1 Implication of Future Practice

Some teachers are frustrated by the fact that even though the official Danish curriculum objectives is competence, with a mandatory oral-practical science competence examination for ninth grade students, the corresponding official written self-correcting electronic examination in the various science subjects (biology, geography, physics/chemistry) are in reality a test of knowledge and skills. This can lead teachers to focus their teaching almost solely on subject knowledge and skills. However, for competence-orientated teaching, knowledge and skills are also very important, but the hierarchy here is simply that knowledge and skills are the means and competence is the end.

Another attention point that the teachers in the project have addressed is time. Science teaching in Danish primary and lower secondary schools is often conducted in 45-min lessons, which makes it difficult to implement competence-oriented teaching with numerous degrees of freedom with time and opportunity for students to reflect, while the teacher, through formative assessment processes, often dialogue-based, scaffolds the groups' and individual students' competence development.

As mentioned, competence-oriented science teaching posed a challenge for many participating teachers in the project, despite their presumed specialized academic qualifications and perceived importance of competence-oriented teaching. The analysis identified some of the six criteria that were rated low, indicating insufficiently competence-oriented teaching. This also suggests that there is still a need for efforts that can support more teachers to be able to work in a competence-oriented way and that assessment practices support competence-oriented science teaching.

Acknowledgments I thank my co-educators at Department of Teacher Education (John Andersson, Dorrit Hansen, Mari-Ann Skovlund Jensen, Lars Bo Kinnerup & Jørgen Haagen Petersen) and the 28 science teachers who took part in the NATKOM project, as this article largely is based on results obtained in the NATKOM project. The NATKOM project was carried out with financial support of the Novo Nordisk Foundation grant no. NNF20OC0065745.

References

Andersen, N. O., Busch, H., Troelsen, R. & Horst, S. (2003). *Fremtidens naturfaglige uddannelser* [Future science education]. Uddannelsesstyrelsens temahæfteserie, nr. 7, 2003. Undervisningsministeriet.

Arnold, J. (2022). Prioritising students in Assessment for Learning: A scoping review of research on students' classroom experience. *Review of Education, 10*, e3366, 36 p.

Black, P., & Wiliam, D. (2009). Developing the theory of formative assessment. *Educational Assessment, Evaluation, and Accountability, 21*(1), 5–31.

Blömeke, S., Gustafsson, J.-E., & Shavelson, R. J. (2015). Beyond dichotomies: Competence viewed as a continuum. *Zeitschrift für Psychologie, 223*, 3–13.

Brown, A. L. (1992). Design experiments: Theoretical and methodological challenges in creating complex interventions in classroom settings. *Journal of the Learning Sciences, 2*(22), 141–178.

Busch, H., Horst, S., & Troelsen, R. (Eds.). (2003). *Inspiration til fremtidens naturfaglige uddannelser [Inspiration for future science education]*. København.

Christiansen, J. L. (2022). Naturfaglige kompetencer i grundskolen – hvilke bør der fokuseres på? [Science competencies in primary and lower secondary schools - which ones should be prioritised?]. *MONA, 2022*(3), 6–17.

Dolin, J. (2016): Idealer og realiteter i målorienteret undervisning [Ideals and realities in aim-oriented education]. In E. Krogh & S. E. Holgersen (Eds.). *Sammenlignende fagdidaktik 34. Cursiv 19*. pp. 67–87.

Drisko, J. W. (2014). Competencies and their assessment. *Journal of Social Work Education, 50*, 414–426.

Fosnot, C. T. (1989). *Enquiring teachers, enquiring learners: A constructivist approach for teaching*. Teachers College Press.

Hodges, B. D. (2012). The shifting discourses of competence. In B. D. Hodges & L. Lingard (Eds.), *The question of competence: Reconsidering medical education in the twenty-first century* (pp. 14–41). Cornell University Press, ILR Press.

Mathelitsch, L. (2013). Competencies in science teaching. *Center for Educational Policy Studies Journal, 3*(3), 49–64.

Ministry of Children and Education. (2019). *Læseplan for faget fysik/kemi*. [Curriculum for the subject physics/chemistry]. see https://emu.dk/sites/default/files/2020-09/Gsk_l%C3%A6seplan_fysikkemi.pdf

Nielsen, J. A., & Dolin, J. (2016). Evaluering mellem mestring og præstation [Assessment between mastery and performance]. *MONA, 2016*(1), 51–62.

Oyao, S. H., Holbrook, J., Rannikmäe, M., & Pagunsan, M. M. (2015). A competence-based science learning framework illustrated through the study of natural hazards and disaster risk reduction. *International Journal of Science Education, 37*(14), 2237–2263.

Ross, J. A. (2006). The reliability, validity, and utility of self-assessment. *Practical Assessment, Research, and Evaluation, 11*, Article 10.

Rychen, D. S. & Salganik, L. H. (2000). *Ines general assembly 2000. A contribution of the OECD program definition and selection of competencies: Theoretical and conceptual foundations definition and selection of key competencies* (pp. 1–15). , OECD.

Schildkamp, K., van der Kleij, F. M., Heitink, M. C., Kippers, W. B., Veldkamp, B. P. (2020). Formative assessment: A systematic review of critical teacher prerequisites for classroom practice *International Journal of Educational Research, 103*, 101602, ISSN 0883-0355.

Topping, K. J. (2010). Peers as a source of formative assessment. In H. L. Andrade & G. J. Cizek (Eds.), *Handbook of formative assessment* (pp. 61–74). Routledge.

Waddington, D., Nentwig, P., & Schanze, S. (2007). *Making it comparable. Standards in science education*. Waxmann.

Wagenaar, R. & Ferreras, J. M. G. (2008). *Robert tuning educational structures in Europe. Universities' contribution to the Bologna process. An introduction* (2nd Edn). Publicaciones de la Universidad de Deusto.

Wiesner, C., & Schreiner, C. (2020). Ein Modell für den kompetenzorientierten Unterricht und als Impuls für reflexive Unterrichtsentwicklung und –forschung [A model for competence-oriented teaching and as an impulse for reflective teaching development and research]. In U. Greiner, F. Hofmann, C. Schreiner, & C. Wiesner (Eds.), *Bildungsstandards: Kompetenzorientierung, Aufgabenkultur und Qualitätsentwicklung im Schulsystem* (pp. 319–352). Waxmann.

Zhang, Y., Zhan, Y., Hong Wan, Z., & Sun, D. (2023). What are the effects of formative assessment on students' science learning motivational beliefs and behaviours? Comparison between Western and East Asian learners. *International Journal of Science Education*.

Jørgen Løye Christiansen is an associate professor at University College Absalon, Department of Teacher Education and currently a PhD student at the Department of Science Education, University of Copenhagen in Denmark. His primary research interests concentrate on the concept of science competence, competence assessment, modelling in science and motivation for and in science.

Part III
Teacher Agency, Professional Development, and Leadership in Science Education

Chapter 15
Science Teachers' Agentic Practice in the Pedagogy of Scientific Argumentation

Zeynep Guler

15.1 Introduction

Science educators have expressed longstanding concerns about the teaching of science in schools, highlighting the need for a stronger connection between the process of science learning and the practical application of scientific knowledge in real-life contexts (Duschl et al., 2007). Over the past two decades, research on how science is taught and learned through scientific argumentation has grown significantly (Lin et al., 2019). This growth reflects a recognition that incorporating scientific argumentation into science education could be beneficial in addressing the need. Studies (e.g., Henderson et al., 2018; Osborne, 2010) have identified scientific argumentation as a critical and fundamental component of science teaching and learning in schools.

However, as Osborne (2010) indicates, teaching and learning science through scientific argumentation rarely emerge in science classrooms and remain challenging for both students and science teachers (Henderson et al., 2018). To address this challenge, previous studies designed professional development workshops (McNeill & Knight, 2013; Osborne et al., 2013; Simon et al., 2006) and online instructional materials, including technologically enhanced learning resources (Simon & Davies, 2019), to support science teachers in incorporating this approach into their practices. Despite participating in intensive professional development, it is observed that teachers often lack a comprehensive understanding of this approach (McNeill & Knight, 2013).

Z. Guler (✉)
IOE-UCL's Faculty of Education and Society, University College London, London, UK

Board of Education, Turkish Ministry of National Education, Ankara, Türkiye
e-mail: zeynep.guler.16@ucl.ac.uk; zeynep.guler@meb.gov.tr

G. Cakmakci, M. F. Tasar (eds.), *Connecting Science Education with Cultural Heritage*, Contributions from Science Education Research 15,
https://doi.org/10.1007/978-3-031-90944-3_15

Understanding a new approach requires teachers to share their experiences and values within a community, engage in practice, be willing to take risks (Loucks-Horsley et al., 2010), and reflect on patterns of change to expand their knowledge and experience with the approach (Hoban, 2002). Furthermore, the process of developing competency in effectively implementing a new approach demands significant amount of time and continuous professional development through reflection and collaboration (Clarke & Hollingsworth, 2002). Studies on teacher professional development have increasingly emphasised teacher agency as a crucial and emerging component of professional development (Jones & Charteris, 2017; Leijen et al., 2024). Therefore, it is essential to understand the concept of teacher agency and its implications for both theoretical and practical change, assisting teachers in adopting new pedagogies or instructional approaches in their practice (Biesta et al., 2015).

15.2 Theoretical Framework

15.2.1 Teacher Agency

Agency has become an essential concept in social discussions and is defined through various disciplinary perspectives (Archer, 2000; Biesta & Tedder, 2007; Emirbayer & Mische, 1998; Giddens, 1984). According to Giddens (1984), agency is described as the deliberate actions that shape and reshape educational institutions and systems. Giddens (1984) identifies intentionality and knowledgeability as key components of agency, emphasising that individuals are deliberate actors who engage in intentional actions with an awareness or belief that these actions will lead to specified outcomes. This is achieved through their understanding of their activities, referred to as competence. Another aspect of agency relates to the degree of influence individuals can exert within social organisations, which often requires both autonomy and collaboration. Furthermore, Archer (2000) argues that humans have agentic power due to their ability to reflect on and evaluate social circumstances, anticipate alternative outcomes, and collaborate with others to bring about change.

Emirbayer and Mische (1998) conceptualise agency in a way that addresses the limitations of theories focusing solely on either individuals' expertise or experience, motivations, or decision-making regarding their actions. They argue that developing agency requires integrating past implications, future orientations, and active engagement in the present. This perspective considers the dynamic interplay among these three elements and the way in which agency develops in response to the structural environment of action (Emirbayer & Mische, 1998). Building on this understanding, Priestley et al. (2015) situate teacher agency within an ecological framework, constructing a model to understand teacher agency in relation to classroom activities. Their model draws on Emirbayer and Mische's (1998) view of agency as a temporal and relational phenomenon, as well as Biesta and Tedder's (2007) ecological perspective on agency. According to this perspective, teacher agency develops through a specific combination of individual information, available resources, contextual and structural influences, and interactions between situations and individuals (Biesta & Tedder, 2007). This view illustrates how individuals enhance agency within particular contexts by acting in alignment with their beliefs, values, and attributes.

Further extending this understanding, Pantić (2015) provides a model that investigates teacher agency, emphasising its dynamic and relational nature in the context of educational reforms and teachers' roles as change agents, drawing on the influential theories by Archer (2000) and Giddens (1984). According to Pantić (2015), teacher agency is characterised as the intentional and purposeful actions of teachers, along with their ability to accomplish objectives while having some degree of autonomy. Pantić (2015) argues that these three elements foster reflection along with potential enhancements. Therefore, teachers' actions in a specific environment may be seen as the outcome of complex relationships between "their personal and professional beliefs and dispositions; degrees of autonomy and power; and interactions with other actors within the social contexts in which they work" (Pantić, 2015, p. 760). Similar to Pantić (2015), van der Heijden et al. (2015) conceptualise teacher agency as a context-sensitive and developmental process, driven by the interaction between individual and environmental factors. They identify key components of teacher agency, including a strong foundation in practice mastery, reflective practice, risk-taking, and collaboration. Teachers with agency are required to engage in consistent and systematic self-reflection throughout their careers, have an extensive understanding of their subject area and instructional methods, and effectively address challenges that arise in their work and the broader educational context (van der Heijden et al., 2015).

In this study, teacher agency is approached as a dynamic, relational, and context-dependent construct shaped by personal beliefs, professional knowledge, and systemic factors (van der Heijden et al., 2015; Pantić, 2015). This approach emphasises the role of professional development, and the empowerment of teachers as drivers of change aligns with our investigation into how teachers enhance their understanding of new pedagogies and shape their educational contexts. The frameworks introduced by van der Heijden et al. (2015) and Pantić (2015) inform the approach of our study to understanding teacher agency as a critical factor in educational reform, particularly in relation to teachers' ability to enhance the pedagogical and instructional practices. Teacher agency is defined here as comprising four elements: a teacher's sense of purpose (the belief that a specific practice is valuable for achieving a particular outcome), mastery (the understanding of how to effectively influence the desired outcome in practice), autonomy to act (the ability to take risks and have an impact within a given context), and reflexivity (the capacity to monitor, evaluate, and modify one's own practice). We believe that an examination of the four interconnected components of teacher agency simultaneously can provide valuable insights into both the processes and outcomes of professional learning. As highlighted by Pantić (2015), there remains a lack of understanding about the concept of teacher agency and how it manifests in classroom practices. Therefore, it is essential to understand teacher agency in teacher development and its contributions to pedagogical development.

15.2.2 Teacher Agency in Professional Development

The concept of teacher agency has gained growing interest in discussions on teacher professional development (Jones & Charteris, 2017; Leijen et al., 2024; Oolbekkink-Marchand et al., 2017). This concept is recognised as a critical element of professionalism, emphasising that agency is essential for understanding the decisions and practices of teachers (Arnold & Clarke, 2014) and for sustaining the (re)construction of the professional knowledge linked to these practices (Jones & Charteris, 2017). This perspective suggests that examining teacher agency in relation to teacher development could clarify the procedures and outcomes of professional development. Therefore, recent research has highlighted the significance of teacher agency in professional development, analysing teacher practice within educational contexts (e.g., van der Heijden et al., 2015; Kauppinen et al., 2020; Leijen et al., 2024; Lu et al., 2021; Pantić 2015).

Furthermore, an increasing emphasis on teacher agency has emerged, particularly in the context of science education. Various approaches have been utilised to analyse teacher agency, including studies on the implementation of science education reforms (Leander & Osborne, 2008), and the examination of teachers' science knowledge and their pedagogical development (King & Nomikou, 2017). The potential of using teacher agency as a lens for evaluating pedagogical development (such as the science capital pedagogical approach) was investigated by King and Nomikou (2017), who found that teachers actively shape their practices in ways that extend beyond changes to their pedagogy. Their findings also indicated that while most teachers demonstrated a degree of improvement in their agency as a consequence of professional development, some displayed less agency due to an unwillingness to change the way they teach and make their own decisions.

The extent to which teachers engage agency in their decision-making processes might significantly influence how they adopt and implement new teaching approaches. Despite its importance, little is known about the characteristics of teacher agency or how it really works in practice (Pantić, 2015). Similarly, Miller-Rushing and Hufnagel (2023) argue that the current conceptualisation of science teacher agency remains predominantly theoretical, with limited attention given to the profession and experiences of science teachers. This offers science education researchers a valuable opportunity to deepen their understanding of science teacher agency by further developing its theoretical foundations and practical application.

Despite recent studies above investigating the concept of teacher agency across different contexts, there remains a lack of understanding of how the development of agency in teacher development occurs while enhancing and implementing a new approach in practices over time. To address that, this research uses the concept of teacher agency as a lens to understand teachers' pedagogical development while supporting science teachers in enhancing their pedagogy of scientific argumentation with an emphasis on lesson planning, classroom practices, and reflective discussions on these practices. There is also a need to gain a deeper understanding of teacher agency in practice to effectively support teachers in adapting and developing new pedagogies (King & Nomikou, 2017). Therefore, the purpose of this research was to use the concept of teacher agency as a lens to gain an in-depth understanding of how (if at all) science teachers develop their agency and agentic practice in the pedagogy of scientific argumentation over time.

15.3 Research Methods

A comparative case study approach (Goodrick, 2020) was applied to examine how three science teachers develop their agency and agentic practice in teaching scientific argumentation over 2 years, with the aim of analysing and synthesising the similarities, differences, and patterns that are present in two or more instances. This research was conducted during the 2017–2018 and 2018–2019 academic years at a lower secondary school located in the south-eastern part of the UK. Three science teachers —one male and two females, referred to by the pseudonyms Ray, Mala, and Padma—volunteered to implement the pedagogy of scientific argumentation into their teaching practice. Ray had a total of 8 years of teaching experience, three of which were specifically in teaching science. Mala had completed 1 year of training in teaching science, while Padma had no prior experience in teaching science at the time of the research. In addition, all three teachers lacked previous experiences or specialised training in teaching science through scientific argumentation. After discussions with the teachers, Year 7 students in Key Stage 3 were identified as the most suitable group for this research, since they are not constrained by curriculum requirements or public examinations at this level.

Given that teachers are rarely provided with initial training sessions on integrating scientific argumentation into their practice (Murphy et al., 2018), the teachers participated in an initial training session at the beginning of the research to collaboratively enhance their theoretical understanding of argument and scientific argumentation, as well as develop their understanding of how scientific argumentation may be effectively applied in science classrooms. Furthermore, the Ideas, Evidence, and Argument in Science (IDEAS) (Osborne et al., 2004) and online resources from Continuous Professional Development Units (Simon & Davies, 2019) were used as primary resources to present the pedagogy of scientific argumentation while designing and planning activities and implementing classroom strategies. Argumentation activities, relevant to the science curriculum, were selected from the IDEAS resources or designed by the teachers by using resources. The teachers were allowed to adapt or change argumentation activities to improve their practice and, if needed, change their objectives or design their own activities.

Collaborative lesson planning meetings were conducted before each argumentation lesson to support the teachers in discussing how to integrate scientific argumentation approaches into their practice and designing the argumentation activities. The meetings lasted between 90 and 120 min, covered topics such as scaffolding argumentation lessons, introducing argument and scientific argumentation, organising student groups and group discussions, the topics they brought up from their practices, the lesson activities that required further elaboration, and promoting reflection on practice. The planning sessions provided teachers with opportunities to enhance their understanding and competence in scientific argumentation, exchange experiences with colleagues, and reflect on their practices using this approach. Each teacher implemented scientific argumentation based on their understanding of what was addressed in the initial training session and lesson plan meetings. It was essential to provide teachers with guidance on planning and strategies for adapting activities to meet a diverse range of teaching objectives (Simon & Richardson, 2009).

Each participant teacher agreed to regular classroom observations. The teachers were expected to implement six argumentation activities in their practices each year. However, minor adjustments were made to the frequency of lesson observations and discussions (ranging from once a month to once every 2 months) to accommodate the busy schedule of the teachers (such as attending department meetings and teaching multiple classes). Therefore, Ray implemented three argumentation activities while Padma carried out five activities each year, and Mala implemented five activities in the first year and six in the second year, showing variations in their implementation of the planned activities over 2 years. After each argumentation lesson, the teachers engaged in both individual and collaborative reflective discussions on their teaching practices.

Data sources included video recordings of teachers' practices, audio recordings of lesson planning meetings, and audio recordings of teachers' reflective interviews after each argumentation lesson. These data were used to generate a "profile" of the teachers and establish a baseline for comparing changes in their agentic practice. To identify instances of the four components of teacher agency, each lesson planning meeting, teachers' interview, and classroom practice was transcribed in its entirety and analysed thematically into 'instances of agency' inductively involving repeated reading of the data and discussion of codes to ensure validity and reliability (Flick, 2013). This approach examined and deduced instances of the interconnected elements of agency, as well as observed changes in teachers' agentic practice and the development of their agency.

15.4 Findings

The purpose of this research was to investigate how science teachers develop their agency and agentic practice over time while implementing the pedagogy of scientific argumentation in their practices. Drawing on the conceptualisation of teacher agency by Pantić (2015) and van der Heijden et al. (2015), four key components were identified: a sense of purpose, mastery, autonomy to act, and reflexivity.

15.4.1 A Sense of Purpose

This component of agency focuses on how science teachers engage in practices they believe would lead to specific outcomes. We identified instances where teachers' purpose, motivation, and commitment to using scientific argumentation supported both their own development, aligning with their beliefs about the values of science teaching through scientific argumentation.

The data from the classroom observations of various argumentation lessons over 2 years demonstrate significant differences in how the three science teachers implemented the argumentation approach in their practice. Despite collaborative lesson planning and access to the same resources, the observational data show that teachers incorporated various instructions and objectives of the argumentation approach differently into their teaching. These variations were observed in lesson content, scaffolding of scientific argumentation, timing, and student organisation for group discussions, reflecting their distinct purposes and values. These purposes often shaped how the teachers adapted their own pedagogy of argumentation approach to align with their own pedagogical purposes. The initial interviews revealed that all three emphasised the importance of utilising scientific argumentation as a pedagogical approach in science teaching, not only to provide discussions and justifications but also as a valuable means of improving students' understanding of scientific concepts and identifying any misunderstandings they may have regarding these concepts.

During lesson planning meetings and teaching practices, the teachers demonstrated individualised decision-making processes based on their students' needs and their current practices. The teachers' different choices in implementing scientific argumentation illustrated how this approach can be adapted in various ways in scientific contexts. The data from the classroom observations indicates that Ray prioritised resolving the scientific topics over engaging in argumentation processes (McNeill & Knight, 2013), while Padma spent significant time on discussions, risking off-task engagement. In contrast, Mala's instruction clearly emphasised scientific argumentation, incorporating a wide range of argumentation processes in her lessons.

The reflective interviews with Ray, Mala, and Padma also revealed their own approaches, each adapted to implementing scientific argumentation within scientific contexts. Ray stated:

> *I tried to help them (students) figure out the correct answers and then gave some explanations on the right answers. I was guiding them towards the correct answer by asking questions and encouraging them to draw from their knowledge to answer the questions, rather than simply saying no.*

In contrast, Mala stated:

> *The way I did it was they had to write their justification for each one (statement), and then they moved around to see if they agreed with other groups' works. If they were telling me, 'We think this because of this, and it was right.' My scientific questions were slightly different*

as I was not trying to tease out the correct answer. I was trying to get them to explain why they think that answer is right scientifically.

Similarly, Padma said:

I was trying to get them to justify what they were saying, so I just let them have a bit of discussion.

As seen in the observational data and the provided excerpts, Ray's objective was to teach established knowledge within a scientific context using a teacher-centred method. He emphasised his efforts to provide explanations aimed at resolving his students' conceptual misunderstandings. In contrast, Padma and Mala placed greater emphasis on fostering student discussions and encouraging justifications during student discussions. Moreover, Mala prioritised supporting her students in evaluating and critiquing one another's contributions during small group discussions.

Over time, the three teachers focused more on their students' needs while planning and implementing argumentation activities. They employed different instructional strategies and objectives, reflecting their individual sense of purpose in making pedagogical decisions to best meet their students' needs. In the second year, particularly Mala, they drew on their previous experiences, emphasising the importance of students' prior knowledge for scientific argumentation. Therefore, the teachers aimed to use the argumentation activities as a recap after teaching the topic to assess students' understandings and address any misconceptions they might have regarding scientific contexts.

Teachers like Mala, who prioritised scientific argumentation processes and goals, as shown in her practices and lesson activities, are more likely to enhance students' argumentation skills. Mala articulated an enhanced motivation to incorporate this approach into her instructional practices to assess students' conceptual understanding. This illustrates how Mala's objective for 'revision' evolved as a result of her developing agency, as seen by her practices and the following comment:

We did a formative assessment with Year 7. We gave them a scenario that they had to write down. I think this is a better way to do it because we can collect their sheets afterwards and look at what has been said. As teachers, we can go around and observe the students who have good discussions and the students who struggle and need support. So, it is a nice way of assessing them without telling them that you are assessing them. It's a good way of checking their understanding as well.

The observational and interview data reveal that the teachers' objectives were shaped by how they intended to implement scientific argumentation into their practices. The development of agency was derived from the pedagogical decisions and purposes teachers make in their practices. The teachers demonstrated developing agency in argumentation approaches by making pedagogical decisions about what and how to engage in, as well as adjusting the level of engagement to develop their professional approaches.

15.4.2 Mastery

Regarding the mastery of agency, we identified the teachers' competence and confidence in applying the pedagogy of scientific argumentation, including (a) their knowledge across various contexts, (b) confidence in adapting the approach, and (c) changes in their practice over time. Increased mastery was evident in how teachers' pedagogy evolved and their growing competence as a result of experience (van der Heijden et al., 2015).

The observational data reveals that the teachers enhanced their pedagogical knowledge of scientific argumentation, achieved a better degree of competence through their experiences in the planning and implementation of the argumentation approaches over time, and refined their approach each year, though their initial strategy remained fundamentally unchanged. Mala and Padma had more experiences in teaching science through scientific argumentation compared to Ray, since Ray used this approach less frequently in his practice. Therefore, they developed their mastery to various degrees. In the first year, they recognised the need for a more organised approach in the planning stage to improve decision-making while also requiring more flexibility in the implementation stage to facilitate the development of their practice. In the second year, both Mala and Padma expressed the need to incorporate scientific argumentation more frequently into their teaching, covering a variety of contexts. Mala expressed:

> *I need to consider my class and focus more on the teaching strategies that I use, not seeing an argumentation lesson as something separate or something different. Instead of planning argumentation as a stand-alone lesson, it should just fit; here is my scheme of work: I am going to fit argumentation in there.*

Padma said:

> *I think with the argumentation lessons it is so different, even though it fits in a range of topics. I think I will try to use this approach outside of the argumentation lessons; then it will not be so much of a surprise, and it will be helpful to students as well in their writing.*

As observed in the classroom practices, the teachers planned to integrate it into their lessons during the second year, rather than considering scientific argumentation as an independent instructional approach. It was also observed that the teachers' pre-existing teaching practices, values, and understanding of science teaching significantly influenced their implementation of scientific argumentation (Simon et al., 2006). The teachers implemented scientific argumentation in various ways over 2 years, with fewer differences in instructional focus, timing, scaffolding, and student organisation for group discussions. The observational data reveals that Padma emphasised discussions, justifying opinions, supporting arguments with evidence, and evaluating arguments through modelling and exemplification, with a stronger focus on these in year 2. Mala, on the other hand, focused on listening, explaining arguments, justifying positions, anticipating counter-arguments, and reflecting on the process. Both teachers found teaching the same lesson in different classrooms over 2 years beneficial, using their prior experiences to enhance their pedagogy of scientific argumentation.

During lesson planning meetings, the teachers considered their previous experiences and reflections when organising group discussions, demonstrating an improved understanding of groupwork. They were concerned about students' lack of engagement and participation in the argumentation activity. Padma and Mala were able to make constructive decisions regarding the organisation of group discussions, lesson structure, and the scaffolding of argumentation activities. These decisions were more closely linked to their developing competency. Across the 2 years, they implemented the argumentation approach within a variety of activities, drawing from their own understanding and knowledge of scientific argumentation.

In the final interview, as observed in the classroom practice, Mala described how her teaching approaches had evolved over time. Her account indicates that she had attained a high level of competence in her practice. She explained how the scientific argumentation approach assisted in the development of her own pedagogical knowledge and expertise, stating:

> It (scientific argumentation) has helped me think more about being a scientist, how to express our ideas, how to change opinions, and how to explain to students why something is wrong. It has given me another tool to use in planning, and it's made me think a lot more about what students think and how something might come across to them. It has made me want to find out students' ideas with questions. Before, it was probably just … I had an idea about misconceptions, but it was like, "I need to teach them this." Now, I realise I need to let them express their ideas as well.

The excerpt above and observational data illustrate how Mala enhanced her pedagogy by prioritising her students in the design and implementation of scientific argumentation, demonstrating her development in agency. As both Mala and Padma recognised the benefits of this approach, they committed to improving their understanding and increasing their agency. As van der Heijden et al. (2015) note, teachers who commit to an approach and recognise its benefits are more likely to develop mastery. Mala, Padma, and, to a lesser extent, Ray, were observed engaging with this practice in their classrooms and developing their mastery over time.

15.4.3 Autonomy to Act

To examine this component of agency, we analysed instances of decision-making and risk-taking to examine the teachers' autonomy, focusing on their willingness to make informed decisions, take risks, and implement changes in their instructional practices over time.

The analysis of the initial interviews reveals that the three teachers were unfamiliar with the pedagogy of scientific argumentation. Nevertheless, their willingness to participate in the research and implement a new approach in their practices reflects their autonomy to act. During the research, all three teachers established their autonomy by displaying their willingness to make decisions, take risks, and make changes in their instructional practices.

In the first year, all three teachers followed the suggested instructions that were provided in the IDEAS resource, indicating a limited level of autonomy in determining how to implement the argumentation approach. Over 2 years, as the three teachers gained confidence in using this approach in their practices, they each became more willing to make their own decisions and try out various strategies. Observational findings reveal that the teachers considered their students' dynamics when planning argumentation lessons, focusing on introducing argumentation, organising groups, and facilitating student engagement and discussions. The key difference between the teachers was their purpose for using the activity: Padma and Ray implemented argumentation activity in a scientific context as a recap to assess students' understanding, while Mala incorporated it as part of a sequence in the scheme of work. Despite planning together, they had distinct objectives and made individual decisions.

Padma and Mala tried out various strategies to incorporate scientific argumentation, assessing their effectiveness for their students. By year 2, they drew on prior experiences and students' needs to plan argumentation activities. For example, while they used the envoy strategy as a group strategy in year one, they adopted a different group strategy in year 2 to better align with students' abilities and needs. This demonstrates their developing autonomy in considering their own students when making decisions.

Observational findings also reveal that Mala was the only teacher who consistently applied a variety of group strategies in various argumentation lessons, while Padma and Ray were more willing to make decisions that they felt enabled them to manage their practice. Mala made the following comment:

> *I am thinking of using group work in nearly all my lessons. They (students) work with somebody else, which is basically with the person next to them. That helps teachers a lot as well because sometimes students can help each other. While you get around, you can learn more about students' understandings and misconceptions about the topics.*

Nevertheless, it is noteworthy that Ray expressed concerns about including group discussions in his instructional approach. He noted that group work was often ineffective for students' learning and made classroom management more difficult:

> *I don't do a lot of group work in my lessons. When it comes to doing group work, it tends to be more practical based on equipment rather than out of a desire for them to be working together. I've phased out a little bit because it's been more effective to get them to be working independently rather than with each other. It can be relatively inefficient and more challenging to monitor the classroom as a teacher in terms of behaviour as well.*

Similarly, Padma encountered behavioural challenges and felt forced to change her plans during implementing argumentation activities. Both Padma and Ray expressed a preference for using group discussions only with students they believed they could manage more effectively. The teachers' comments above reveal the reasons behind the limited and inconsistent use of different group work.

The observational and interview data indicate that not all three teachers utilised this approach to the same degree. Padma and Mala's sustained practice and dedication show that they were able to develop their pedagogy of teaching argumentation, make autonomous decisions, and take risks. That indicates the development of autonomy to act, as reflected in their practice over the 2-year period.

15.4.4 Reflexivity

Reflexivity was observed when teachers engaged in practice, reflected on their experiences, and implemented changes informed by those reflections. This research included collaborative lesson planning meetings where teachers discussed integrating scientific argumentation into their practice. Additionally, individual reflective interviews conducted after each lesson further encouraged their reflection, fostering reflexivity across all data sets. The data analysis shows that through collaboration and reflection on prior experiences, the teachers enhanced their understanding of scientific argumentation and gradually restructured their approach to implementing it. They also considered class dynamics more when planning argumentation lessons, dedicating additional time to organising student groups and ensuring engagement in the activity.

After gaining a better understanding of scientific argumentation, the teachers began by defining the purposes and procedures for implementing the activity, envisioning how the lesson would unfold, including student organisation for group discussions, encouraging them to participate in group discussions, addressing the science content through resources, and concluding the activity. The teachers recognised the complexity of the argumentation activities and the need to consider students' prior knowledge, experiences, and access to resources to ensure successful engagement in the activity as a result of reflecting on their practice. They all highlighted the value of sustained practice and the requirement of self-reflection in their practice following each argumentation lesson. Padma and Mala, in particular, emphasised the importance of engaging more deeply with this practice to develop their reflexivity. Mala's comment further exemplifies this:

> *I do think that I am thinking more about it than last year, like what is a good argument in science because we don't really think about it daily. We just teach it as a very narrow view. Sometimes, you don't think actually outside the box. It (scientific argumentation) gives you a wider view and makes you think; it gives students a chance to improve their skills and express themselves when we ask them why. I think it did prompt me to think about science as a bigger thing other than just a topic I'm teaching.*

The teachers' reflections provided valuable insights into their own assumptions, values, and previous practices, and how these influence their decisions, practice and pedagogical outcomes. The process of reflection enhanced their reflexivity and facilitated their learning and application of the pedagogy of scientific argumentation. The teachers' approaches to addressing challenges in teaching argumentation revealed a renewed practice that facilitated the development of their self-awareness.

While the teachers made consistent efforts to learn throughout the research, they were aware of the pedagogical difficulties they faced. The teachers demonstrated an increasing level of self-awareness and critical thinking in response to these challenges, as seen by their commitment to their own learning to improve their understanding and knowledge of scientific argumentation. This also revealed their developed agency in reflecting on their own practices.

15.5 Discussion

The purpose of this research was to investigate how (if at all) science teachers develop their agency and agentic practice in the pedagogy of scientific argumentation over time, demonstrating the potential of using the concept of teacher agency as a lens for examining their pedagogical development throughout four components.

One key finding in this research was that all three teachers were able to acquire some degree of agency, with some teachers developing a greater degree of agency in the pedagogy of scientific argumentation and agentic practice than others (Oolbekkink-Marchand et al., 2017; Priestley et al., 2012). Mala, as an agentic teacher, took advantage of opportunities for her students to learn in response to changing situations, which influenced both her approach and her own perspective on future practice.

Another key finding was that, despite shared resources and collaborative lesson planning, the teachers demonstrated distinct approaches to scientific argumentation. Ray's emphasis on providing knowledge, focusing on resolving scientific misconceptions, contrasts with Padma's and Mala's emphasis on student engagement and discussion, even at the risk of off-task engagement. This variation highlights the tension between teacher-centred and student-centred pedagogies in scientific argumentation, as discussed by McNeill and Knight (2013). The differences also indicate that teacher agency, as articulated by Biesta et al. (2015), extends autonomy to act and is closely connected to teachers' pedagogical approaches and the educational purposes.

Empirical research, such as that by Priestley et al. (2012), supports the claim that teacher agency evolves through the pedagogical choices teachers make and their subsequent implementation in practice. In the context of this research, the teachers developed their agency in the domain of scientific argumentation by making deliberate pedagogical decisions regarding their engagement with the approach, demonstrating varying degrees of commitment to shaping their professional practices. The teachers' sustained practices also enabled them to identify opportunities to implement and design 'rich and meaningful educational experiences for their students' (Priestley et al., 2012, p. 210), thereby exemplifying a notable degree of autonomy in their pedagogical decisions. The teachers proactively sought opportunities for professional learning, adapting to evolving contexts that influenced both their pedagogical strategies and their instructional objectives for argumentation lessons.

Teachers with a strong sense of agency, such as Mala, acknowledged their deficiencies in knowledge and understanding and actively sought to develop them. Their experiences with this approach, such as their belief on the value of teaching science through scientific argumentation, reinforced their commitment to continuous practice over time (Biesta et al., 2015; Priestley et al., 2012; Oolbekkink-Marchand et al., 2017). Furthermore, the findings clearly illustrate how teachers' previous experiences and achievements in practices influenced their ability to take risks, and improved their teaching approaches and agency (Lu et al., 2021).

The significance of fostering collaboration and reflective practices emerged as critical components for enhancing teachers' pedagogical learning (Osborne et al., 2013) and developing teacher agency. This finding aligns with previous research (Leijen et al., 2024), which emphasising that such elements play an essential role in advancing professional development and empowering teachers to refine their teaching approaches. Applying teacher agency as a lens allowed us to gain a greater understanding of how science teachers developed their agency and agentic practice in teaching science through scientific argumentation, considering the interconnected four components of teacher agency.

By examining teacher agency in practice, we were able to deepen our understanding of the relationship between teacher professional development and teacher agency, which could potentially influence our future efforts to support teachers in the development of pedagogical knowledge and understanding of new approaches. Therefore, we believe that further examination of teacher professional development from the perspective of teacher agency would be beneficial. Continuous practice, reflection, and collaboration with colleagues may help science steachers further develop their pedagogical knowledge and agency in their teaching practices. Therefore, science teachers should be provided with these opportunities to enhance their pedagogical knowledge and practice. From the results, it would be interesting to identify factors that either support or hinder the development of agency in teaching practice of scientific argumentation for future research.

Acknowledgments This manuscript is a part of Zeynep Guler's PhD thesis. I acknowledge the financial support provided by the Turkish Ministry of National Education (within the scope of Law No.1416) for my PhD study at the IOE-UCL's Faculty of Education and Society, University College London, U.K. I also express my gratitude to Prof Shirley Simon for her guidance, support, and contributions throughout my PhD research. Her passing is a great loss, and I will forever be grateful for the opportunity to have worked under her supervision.

Ethics Approval This study has been fully approved by the UCL Institute of Education/UK ethics committee.

References

Archer, M. S. (2000). *Being human: The problem of agency*. Cambridge University Press.

Arnold, J., & Clarke, D. J. (2014). What is 'agency'? Perspectives in science education research. *International Journal of Science Education, 36*(5), 735–754.

Biesta, G., & Tedder, M. (2007). Agency and learning in the life course: Towards an ecological perspective. *Studies in the Education of Adults, 39*(2), 132–149.
Biesta, G., Priestley, M., & Robinson, S. (2015). The role of beliefs in teacher agency. *Teachers and Teaching, 21*(6), 624–640.
Clarke, D., & Hollingsworth, H. (2002). Elaborating a model of teacher professional growth. *Teaching and Teacher Education, 18*(8), 947–967.
Duschl, R. A., Schweingruber, H. A., & Shouse, A. W. (2007). Taking science to school: Learning and teaching science in grades K-8. *Eurasia Journal of Mathematics, Science and Technology Education, 3*(2), 163–166.
Emirbayer, M., & Mische, A. (1998). What is agency? *American Journal of Sociology, 103*(4), 962–1023.
Flick, U. (Ed.). (2013). *The SAGE handbook of qualitative data analysis*. Sage.
Giddens, A. (1984). *The constitution of society: Outline of the theory of structuration*. University of California Press.
Goodrick, D. (2020). *Comparative case studies* (Vol. 9). Sage.
Henderson, J. B., McNeill, K. L., González-Howard, M., Close, K., & Evans, M. (2018). Key challenges and future directions for educational research on scientific argumentation. *Journal of Research in Science Teaching, 55*(1), 5–18.
Hoban, G. F. (2002). *Teacher learning for educational change: A systems thinking approach*. Open University Press.
Jones, M., & Charteris, J. (2017). Transformative professional learning: An ecological approach to agency through critical reflection. *Reflective Practice, 18*(4), 496–513.
Kauppinen, M., Kainulainen, J., Hökkä, P., & Vähäsantanen, K. (2020). Professional agency and its features in supporting teachers' learning during an in-service education programme. *European Journal of Teacher Education, 43*(3), 384–404.
King, H., & Nomikou, E. (2017). Fostering critical teacher agency: The impact of a science capital pedagogical approach. *Pedagogy, Culture and Society, 26*(1), 87–103.
Leander, K. M., & Osborne, M. D. (2008). Complex positioning: Teachers as agents of curricular and pedagogical reform. *Journal of Curriculum Studies, 40*(1), 23–46.
Leijen, Ä., Pedaste, M., & Lepp, L. (2024). Supporting teacher agency during a collaborative inquiry-based in-service teacher education course. *European Journal of Teacher Education, 47*(4), 769–786.
Lin, T.-J., Lin, T.-C., Potvin, P., & Tsai, C.-C. (2019). Research trends in science education from 2013 to 2017: A systematic content analysis of publications in selected journals. *International Journal of Science Education, 41*(3), 367–387.
Loucks-Horsley, S., Stiles, K. E., Mundry, S., Love, N., & Hewson, P. W. (2010). *Designing professional development for teachers of science and mathematics*. Corwin Press.
Lu, X., Leung, F. K. S., & Li, N. (2021). Teacher agency for integrating history into teaching mathematics in a performance-driven context: A case study of a beginning teacher in China. *Educational Studies in Mathematics, 106*(1), 25–44.
McNeill, K. L., & Knight, A. M. (2013). Teachers' pedagogical content knowledge of scientific argumentation: The impact of professional development on K-12 teachers. *Science Education, 97*(6), 936–972.
Miller-Rushing, A., & Hufnagel, E. (2023). Trends in K-12 teacher agency research: A review of science education research. *Journal of Science Teacher Education, 34*(2), 157–180.
Murphy, P. K., Greene, J. A., Allen, E., Baszczewski, S., Swearingen, A., Wei, L., & Butler, A. M. (2018). Fostering high school students' conceptual understanding and argumentation performance in science through Quality Talk discussions. *Science Education, 102*(6), 1239–1264.
Oolbekkink-Marchand, H. W., Hadar, L. L., Smith, K., Helleve, I., & Ulvik, M. (2017). Teachers' perceived professional space and their agency. *Teaching and Teacher Education, 62*, 37–46.
Osborne, J. (2010). Arguing to learn in science: The role of collaborative, critical discourse. *Science, 328*(5977), 463–466.

Osborne, J., Erduran, S., & Simon, S. (2004). *Ideas, evidence and argument in science. In-service training pack, resource pack and video*. Nuffield Foundation.

Osborne, J., Simon, S., Christodoulou, A., Howell-Richardson, C., & Richardson, K. (2013). Learning to argue: A study of four schools and their attempt to develop the use of argumentation as a common instructional practice and its impact on students. *Journal of Research in Science Teaching, 50*(3), 315–347.

Pantić, N. (2015). A model for study of teacher agency for social justice. *Teachers and Teaching, 21*(6), 759–778.

Priestley, M., Biesta, G., & Robinson, S. (2012). *Teachers as agents of change: An exploration of the concept of teacher agency*. University of Stirling Press.

Priestley, M., Biesta, G., & Robinson, S. (2015). *Teacher agency: An ecological approach*. Bloomsbury Publishing.

Simon, S., & Davies, P. (2019). Teachers as participatory designers of a professional development website. In *Material practice and materiality: Too long ignored in science education* (pp. 201–215). Springer International Publishing.

Simon, S., & Richardson, K. (2009). Argumentation in school science: Breaking the tradition of authoritative exposition through a pedagogy that promotes discussion and reasoning. *Argumentation, 23*(4), 469–493.

Simon, S., Erduran, S., & Osborne, J. (2006). Learning to teach argumentation: Research and development in the science classroom. *International Journal of Science Education, 28*(2-3), 235–260.

van der Heijden, H. R. M. A., Geldens, J. J. M., Beijaard, D., & Popeijus, H. L. (2015). Characteristics of teachers as change agents. *Teachers and Teaching, Theory and Practice, 21*(6), 681–699.

Zeynep Guler I am an education expert at the Ministry of National Education in Türkiye. I currently work as an editor for the Journal of *Milli Eğitim*. My academic background is in science education, and my research interests include curriculum development, teacher professional learning, and educational program evaluation. Within this scope, I lead research-based projects aimed at enhancing the quality and effectiveness of curriculum reform through data-driven insights.

Chapter 16
Stages of the Professional Development of Mediators of Non-Formal Science Education Settings

Natália Cândido Vendrasco and Ainoa Marzábal

16.1 Introduction

The contextual model of learning suggests that various factors related to the personal, physical, and sociocultural contexts of visitors and visits are crucial for reaching the full potential of non-formal education spaces such as museums, zoos, aquariums, and botanical gardens (Falk & Dierking, 2016). As highlighted in previous studies, adapting educational programmes to diverse audiences can be challenging, especially in settings that receive a wide range of visitors (Daza-Caicedo et al., 2020; Massarani et al., 2015).

Mediation, in this context, refers to the process carried out by mediators to facilitate all visitors' access to exhibition-related experiences and knowledge (Norberto Rocha & Marandino, 2020; Cândido Vendrasco et al., 2024; Weiler & Smith, 2009; Massarani, 2020; Falk & Dierking, 2012). It is essential for realizing the potential of visits in educational settings and influencing visitors' learning and experience (Dillon, 2016; Falk & Dierking, 2016; Massarani, 2020).

In non-formal science education contexts, mediators' educational practices can vary widely, affecting the effectiveness of visitors' learning and experience (Cândido Vendrasco et al., 2024; Marandino, 2008). Therefore, researching mediators' decision-making during mediation processes has become crucial (Busch et al., 2022). According to Busch et al. (2022), Berdnikoff and Silva (2020), and Tran and King (2011), decisions are primarily based on the mediators' knowledge, which is derived from their education and professional experience.

There are theoretical proposals regarding mediators' knowledge (Busch et al., 2022; Berdnikoff & Silva, 2020; Tran & King, 2011), which can be classified into

N. C. Vendrasco (✉) · A. Marzábal
Facultad de Educación, Pontificia Universidad Católica de, Santiago, Chile
e-mail: natalia.candido@uc.cl

G. Cakmakci, M. F. Tasar (eds.), *Connecting Science Education with Cultural Heritage*, Contributions from Science Education Research 15,
https://doi.org/10.1007/978-3-031-90944-3_16

four main types: content knowledge, didactic and facilitation knowledge, context knowledge, and knowledge about visitors and their learning (Cândido Vendrasco & Marzábal, 2023). However, characterising this knowledge and its evolution with experience and professional development is challenging due to the diverse professional profiles, contexts, and experiences of mediators, as well as the limited opportunities for specialised educational programs (Bevan & Xanthoudaki, 2008; Norberto Rocha & Marandino, 2020; Robinson, 2019). The purpose of this paper is to explore how mediators' perception about their knowledge develops throughout their professional careers and to identify the different stages of this trajectory.

16.2 Theoretical Background

16.2.1 Mediators: Definition and Background

The vision of museums, zoos, botanical gardens, and aquariums as places of education and learning makes the role of mediation explicitly relevant, due to its significant impact on visitor experiences and outcomes. Various terms such as mediators, informal educators, monitors, museum educators, facilitators, explainers, and interpreters are used in the literature to describe the professionals responsible for this process (Falk & Dierking, 2012; Specht & Loreit, 2021). These concepts reflect the diversity of perspectives on mediation and the importance of structuring knowledge in the field through the sharing of a common framework for research and practice, that can contribute to the recognition of the professional role and expertise of these actors (Daza-Caicedo et al., 2020; Norberto Rocha & Marandino, 2020).

In the English-speaking world, the terms museum educator and informal educator are often used, while in Spain and Latin America the terms mediator and mediation are more commonly used. The term mediator refers more to nontraditional educational practices, dialoguing with visitors during museum visits and leading different educational programs based on visitors' experiences (Peters, 2019). From a Latin American standpoint, we propose that the term "mediator" be employed to designate the professionals bearing responsibility for the reception and recognition of the audience, its ideas and prior knowledge, and for the promotion of constructive dialogues surrounding the exhibitions and their associated content. We will therefore focus on the concept of mediators in the remainder of this article.

These definitions are significant because they go far beyond the transmission of scientific knowledge, implying that the non-formal settings are understood as an environment that fosters dialogue, encounter, learning, and discussion and that the mediator would have an essential role in these processes because of their ability to personalize visits and facilitate connections between audiences and exhibits (Daza-Caicedo et al., 2020). The practice of mediators and their ability to tailor the visits to the audience requires applying knowledge from multiple domains to facilitate the interactions between the public and the exhibit (Geveke et al., 2017) and consequently allow achieving the expected learning for educational visits (Busch et al., 2022).

16.2.2 Mediators´ Professional Knowledge

There is a wide variety of possible mediation processes. Depending on how mediators orient these practices, they can be more or less effective in facilitating visitors' learning and may or may not be responsive to the characteristics of the visiting groups. As a concept, responsive mediation involves a scaffolding process that acknowledges and leverages the sociocultural and personal characteristics of diverse audiences to facilitate more effective and equitable interactions with exhibits and science experiences (Cândido Vendrasco et al., 2024; Hernandez et al., 2013).

Several authors argue that the way in which mediators conduct their guided tours and their ability to be responsive to the different groups respond primarily to the mediator's available knowledge, derived from their theoretical training and professional expertise (Busch et al., 2022; Berdnikoff & Silva, 2020; Tran & King, 2011).

In light of the above, Tran and King (2011) made a first approach to the professional knowledge of mediators, proposing categories that would make it possible to identify and organize different types of knowledge that make up this framework: (a) educators' orientation towards facilitation; (b) knowledge of the affordances of the objects; (c) knowledge of visitors' learning and understanding of science; (d) knowledge of facilitation strategies.

Extending this framework, Busch et al. (2022) propose a theoretical framework for educators' pedagogical knowledge of content in non-formal settings (ISE-PCK framework), articulating five types of pedagogical knowledge: (a) knowledge of the objectives of non-formal education spaces, (b) of the programs of the educational space, (c) of the audiences, (d) of participation strategies, and (e) of evaluation. It is emphasized that integration and coherence among these components contribute to the quality of ISE-PCK, since an effective educator would be one who is able to select appropriate objectives for non-formal education and non-formal education settings in general and for his or her own institution; and to choose appropriate engagement strategies to suit the context of his or her audience.

Finally, according to Cândido Vendrasco and Marzábal (2023), these frameworks in conjunction with Latin American perspectives (Berdnikoff & Silva, 2020), can be organized around four dimensions of professional knowledge of mediators: knowledge of content, didactic and facilitation knowledge, knowledge of context, and knowledge of visitors and their learning, characterized in their previous work (Fig. 16.1).

Content knowledge is composed of two areas: knowledge about the disciplinary contents associated with the space and collection in which the mediator works professionally, and the processes through which this knowledge has been historically constructed. This type of knowledge allows mediation practices to be disciplinary and epistemologically appropriate and requires structured and intentional instances

Fig. 16.1 Dimensions and subdimensions of mediator's knowledge associated with mediation traits (own production adapted and translated from Cândido Vendrasco & Marzábal, 2023)

of training to be developed (Cândido Vendrasco & Marzábal, 2023; Tran & Halversen, 2021).

Didactic or facilitation knowledge is composed of four areas: knowledge of facilitation strategies, focused on communicating content in an understandable way; knowledge of didactic strategies, focused on the role of the mediator in promoting learning; knowledge of ways of interacting with visitors, to actively involve visitors in the visits; and knowledge of resources to support the understanding of content. This type of knowledge allows mediators to have a repertoire of facilitation strategies for didactically appropriate mediation practices, that is, in which the discourse and resources of the exhibition are strategically used to favour the active involvement of visitors and their learning. As in the case of content knowledge, didactic or facilitation knowledge requires structured training instances for its development (Cândido Vendrasco & Marzábal, 2023; Tran & Halversen, 2021).

Context knowledge is composed of four areas: knowledge of the space, knowledge of the objectives of the space, knowledge of the objects in the collection, and knowledge of the dynamics of the collection. This type of knowledge allows contextualizing mediation practices in a specific non-formal space, taking into consideration the space, the objects and their dynamics, and the purposes that visitors are intended to achieve by interacting with them. Given the particular nature of context knowledge, mediators can strengthen this type of knowledge through professional experience as they develop mediation practices, interact with other professionals in the space, and reflect on it (Tran & Halversen, 2021).

Knowledge about visitors and their learning is composed of four domains: knowledge about visitors' interests, about their preconceptions and their role in learning, and about the factors that may influence visitors' different ways of learning. Recognizing the diversity of audiences and understanding the diverse ways in which visitors can engage and learn during visits allows mediation practices to be responsive, that is, to recognize and use the sociocultural characteristics, interests, knowledge, experiences, and perspectives of diverse audiences to foster visitors' interaction with the exhibit more effectively and equitably (Cândido Vendrasco et al., 2024; Cândido Vendrasco & Marzábal, 2023).

Although this is still an emerging framework, it is recognized that mediators influence visitor learning in different ways depending on the application of these knowledge components in their mediation practices (Busch et al., 2022; Berdnikoff & Silva, 2020; Tran & King, 2011). However, there is still little empirical evidence on how this professional knowledge is acquired and applied by mediators and how mediators acquire, articulate and develop it. In this regard, it can be assumed that mediators' knowledge is dynamic and constantly evolving and reorganizing. It is then suggested that there is a need to understand the development of mediators' professional knowledge over time as they move from novice practitioners to experienced mediators (Busch et al., 2022).

16.2.3 Mediator's Professional Development and Profile

Over the past few decades of professionalization in the non-formal science education field, the vast majority of mediators enter the field from a discipline other than museum studies or pedagogy, and while academic success in science provides a strong foundation for some aspect of non-formal education work, it rarely prepares one to understand the realities of the visitor experience (Falk & Dierking, 2016). Some studies have examined the backgrounds of mediators and found them to be a diverse group of individuals in terms of training and background. This diversity of backgrounds can be advantageous, as they must interact with an equally diverse range of visitors and program participants, but it can also be problematic, as the lack of shared professional experience has led to a lack of common language, differing perspectives on their roles and on pedagogy and mediation models.

In this sense, mediators' professional knowledge and its development are strongly anchored to their initial training, to the professional training opportunities they participate in, their experience and reflection on practice (Busch et al., 2022; Berdnikoff & Silva, 2020; Tran & King, 2011).

First of all, the initial training of mediators is identified as an important factor in their professional development (Gomes & Cazelli, 2015). Research in different countries has shown that most mediators in non-formal educational settings have a disciplinary background related to the sciences, and to a lesser extent, they have initial training related to pedagogy (Carletti & Massarani, 2015; Ennes et al., 2020; Massarani et al., 2021).

Secondly, continuous training within the same space is identified, carried out through modeling or accompaniment by more experienced educators (Ennes et al., 2020; Gomes & Cazelli, 2015; Pedersoli & Docters, 2020) or through training opportunities designed for the mediators of each institution in each institution (Macías-Nestor et al., 2020; Pedersoli & Docters, 2020; Batista et al., 2020). The modeling or accompaniment format is the most frequent in Latin American non-formal education spaces (Gomes & Cazelli, 2015; Pedersoli & Docters, 2020), it consists in the voluntary accompaniment of an experienced mediator by the novice mediator, often without a directed and reflective process (Ennes et al., 2020; Pedersoli & Docters, 2020). On the other hand, the characteristics of training programs designed for mediators, even if they are mostly focused on disciplinary content, are very diverse and depend on the context in which they are developed (Tran & Halversen, 2021).

Third, it highlights their own experiences as mediators and the reflective processes around their own practice that each professional engages in (Patrick, 2017). Although this reflective practice can foster the progressive development of educators' mediation practices, this process does not always occur naturally and should be intentional in the training processes mentioned above (Patrick, 2017; Tran, 2007; Tran & Halversen, 2021).

In addition to training and experience, some studies have sought to identify the profiles of these professionals, in relation to age, gender and the role they play in the space. A study conducted in the United States showed that most of the respondents were described as predominantly female, between the ages of 25 and 44, with degrees concentrated in the natural sciences, science education, or other educational disciplines such as elementary education or museum education (Rende et al., 2021). In Brazil, most of the respondents were also women with degrees in the natural and exact sciences, but between the ages of 18 and 25 (Carletti & Massarani, 2015). Finally, the situation is very similar in Europe, where most of the respondents were women with scientific training, between the ages of 21 and 30 (Richard, 2010).

These studies have also aimed to identify the various roles expected of mediators in different spaces, illustrating a broad spectrum of activities and tasks related to the profession. Some of the functions mentioned in the case of Brazil include guiding the public within the space, serving as a bridge between the public and the exhibition, and welcoming the public upon arrival at the museum (Carletti & Massarani, 2015). In Europe, in addition to the mentioned roles, there is also mention of the role of entertainment or animator, leading science workshops and shows, training other mediators, and designing activities and project (Richard, 2010).

Despite the large differences reported in the literature in relation to the professional education of mediators, due to the wide variety of mediators' professional profiles and the scarce opportunities for training and intentional reflection on their practices available to them (Bevan & Xanthoudaki, 2008), we assume that there is a sufficient level of convergence among mediators' professional development processes to recognize certain key stages in their trajectory (Bevan & Xanthoudaki, 2008; Norberto Rocha & Marandino, 2020; Keller-Schneider et al., 2020).

The identification and characterization of these stages could contribute to overcoming the heterogeneous view that has predominated so far, facilitating the development of appropriate models of professional development and training (Tran & King, 2011).

16.3 Research Questions

What are the stages of professional development of mediators?

What are the characteristics of these stages in relation to the development of the mediator's knowledge?

16.4 Methods

16.4.1 Participants

To achieve the research objective, we conducted a three-stage sequential sampling. The first consisted of a purposive criterion sampling (Patton, 2007), in which we selected non-formal education institutions that met the following criteria: (a) they were located in any region of Brazil (b) were defined as science museums, science centers, zoos, aquariums, botanical gardens or similar (planetariums, butterfly gardens), (c) had human mediation in their educational programs. These institutions were selected from an extensive database created from previous surveys in the literature (Massarani et al., 2015), searches of international associations and contacts with key informants. All of these institutions were invited, stating the objectives and expectations of the project in which this study is framed. In this first stage, 294 Brazilian non-formal science education institutions were invited, and 31 of them agreed to participate in the study. It is worth noting that the number of participating institutions is mainly due to the fact that the data collection was conducted in 2020 and 2021, i.e., in the context of the pandemic, and most of the educational spaces were not fully operational.

Subsequently, in the second stage all the mediators who agreed to participate in the research were included, without applying any exclusion criteria at this stage in order to obtain the broadest and most representative sample possible. A total of 78 responses of the questionnaire were obtained from different educators. Of the 78 responses obtained, 60% corresponded to female mediators and 40% to male mediators. Likewise, 51% of the participants were between 18 and 25 years old, according to the mediator profile previously collected by Massarani et al. (2015), and 18% were between 26 and 30 years old. Finally, with regard to years of experience in non-formal education, 27% had less than 1 year, 60% had between 1 and 10 years of experience and 13% had more than 10 years.

Finally, for the third stage, based on a maximum variation sampling (Patton, 2007), using as a criterion the years of experience answered in the questionnaire, we invited 8 participating mediators to a semi-structured interview. Table 16.1 characterizes the participants of the final stage.

16.4.2 Data Collection

In order to access mediators' perceptions of their professional knowledge, a questionnaire was designed and validated to be applied in online format. The construction of the instrument had the following stages: (a) review of referents (Berdnikoff & Silva, 2020; Tran & King, 2011), (b) definition of the dimensions, (c) construction of the items, (d) validation of the instrument.

Identifying and assessing the professional knowledge of mediators is a complex task for three reasons. First, due to the diversity of the types of non-formal educational spaces and their specific characteristics (Tran & King, 2011; Rennie, 2014). Secondly, due to the heterogeneity of initial and continuous training profiles of these professionals (Tran & King, 2011; Rennie, 2014) and, thirdly, given the variety of professional experiences associated with mediation (Tran & King, 2011). The above constitute important barriers to the construction of a single instrument that identifies the knowledge of mediators from different spaces. That is why, according to previous studies (Ennes et al., 2020), the items were constructed focused on the

Table 16.1 Characterization of interview participants

Mediator	Years of experience	Age range	Gender
M42	Less than 1	21 to 25	Female
M49	Less than 1	21 to 25	Male
M71	2 years	26 to 30	Female
M07	3 years	21 to 25	Male
M04	4 years	26 to 30	Female
M64	9 years	31 to 35	Female
M31	10 years	31 to 35	Male
M51	16 years	36 to 40	Male

mediators' perception of their own knowledge, in a generic way that could be applicable to mediators from all types of settings. At the same time, it was decided to identify their perception of their knowledge, given that this would play a critical role in their decision-making, objectives and behaviors in their professional development and could enhance their mediation skills and practices (Bandura, 2004; Ennes et al., 2020; Jegede et al., 2000; Kuntze et al., 2013).

Thus, the questions were structured in a closed format and with response alternatives on a Likert-type rating scale with four categories: Strongly Agree; Agree; Disagree; and Strongly Disagree. Regarding the controversy over the use of a scale with even or odd numbers of responses, the choice of a scale with four categories is selected to avoid the tendency to respond indifferently or in the middle (Asún et al., 2016) and facilitate the positioning of the participants by eliminating the idea of neutrality (Bisquerra Alzina & Pérez Escoda, 2015).

The validation of the questionnaire was carried out by a panel of experts with 4 expert judges (2 from Brazil, Brazil, 1 from Chile and 1 from Mexico) in science education and non-formal education, who evaluated the clarity, relevance and coherence of the questions, pertinence and coherence of the questions with their respective dimensions. After that, we incorporated the suggestions and piloted the questionnaire with 2 educators with similar characteristics to the participating sample.

Finally, a semi-structured in-depth interview was conducted with the mediators (Creswell, 2007), with the objective of identifying their perception of their professional knowledge in greater depth and to identify their views on their role in the guided visits, on their mediation structure and the actions they carried out during the visit. The interview questions were posed from the literature review and the previously applied questionnaire, but there was also room for questions emerging from a previously established script of themes, dimensions and categories (Zhai & Dillon, 2014). The interviews were conducted immediately after observing the mediators conducting a guided visit, so that the questions could be focused and adapted to the characteristics of the mediation during the visit and based on evidence from the observation (Zhai & Dillon, 2014). The interviews lasted approximately 45 min and were audio-recorded for later analysis.

16.4.3 Data Analysis

After a complete reading of all the responses, the questionnaire data were ordered and analysed. The questions are processed on a Likert-type scale, and scores are attributed to each answer (1: strongly disagree; 2: disagree, 3: agree and 4: strongly agree). From this, we identify the average score of mediators with different years of experience for each of the knowledge categories: content knowledge, didactic and facilitation knowledge, context knowledge, and knowledge about the visitor and their learning (Berdnikoff & Silva, 2020; Busch et al., 2022; Tran & King, 2011). Detailed results for each knowledge category can be found in the previous article by

Cândido Vendrasco & Marzábal, 2023. From this, we identify the trends in the stages of professional development.

Finally, we analyzed the interview conducted with each mediator to identify the aspects mentioned by them about their way of conducting mediation. These analyses made it possible to contrast and discuss the aspects highlighted by the mediators with the answers to the questionnaire and the aspects observed during the visits.

Regarding the validity of the coding, it should be noted that the process was carried out independently by two researcher-coders using a constant comparative method, in which the interpretations and coding of each researcher were discussed in an iterative process.

16.5 Results and Discussion

When analysing the questionnaire data on the development of perceptions of content knowledge, didactic and facilitation knowledge, context knowledge, and knowledge about visitors and their learning, it can be observed that they are perceived differently over time, as can be seen in Fig. 16.2.

Figure 16.2 shows the trends in the percentages of mediators who consider that they have a deep knowledge of the content, didactic and facilitation, context, and visitors and their learning according to their years of experience. For the construction of the graph, percentages were used for each type of knowledge, and each interval of professional experience. This representation is elaborated with the purpose of visualizing more clearly the intervals of growth, decrease and continuity in the development of the mediators' perceptions of their knowledge.

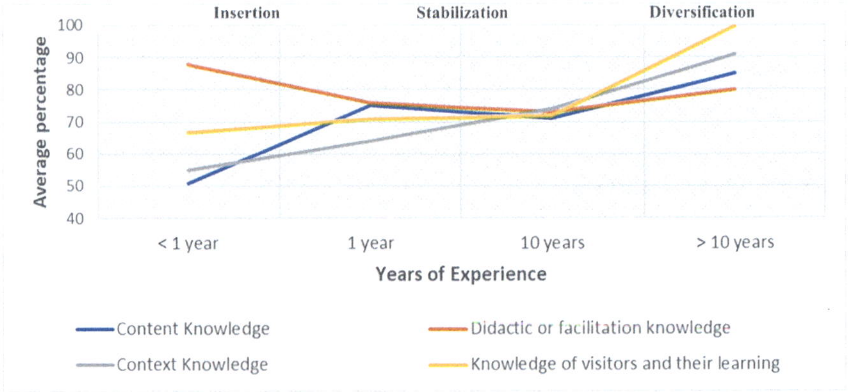

Fig. 16.2 Trends in the perception the mediators on their mastery of the content knowledge, didactic and facilitation knowledge, context knowledge, and knowledge about the visitor and their learning based on the years of experience (n = 78)

Although these knowledges are interconnected and articulated in mediation practices (Busch et al., 2022) they evolve differently over time. These differences seem to respond to each knowledge's nature and how the different professional training opportunities can potentially contribute to their development throughout their careers (Tran & Halversen, 2021).

When looking at each of these dimensions of professional knowledge, we can distinguish three stages in their professional development: a first stage from the beginning of their professional the first year of professional practice, a second stage between 1 and 10 years, and a third stage from 10 years onwards (Fig. 16.2). These stages present distinctive patterns in relation to the various dimensions of mediators' professional knowledge, which allows us to characterize each one of them, and to discuss the potentialities and limitations of mediators' practices in these stages according to their available knowledge (Cândido Vendrasco & Marzábal, 2023).

16.5.1 Insertion

The first stage (<1 year) of professional development of the mediators who participated in this study takes place in their first year of professional practice and represents an initial moment of insertion into mediation in a non-formal science context. In this first stage, the mediators perceive that didactic and facilitation knowledge as the most developed, followed by knowledge about visitors and their learning. On the other hand, knowledge about content and context are the ones they perceive to be the least mastered.

This initial perception could be related, on the one hand, to the fact that the novice mediators recognize less components of their own knowledge (Busch et al., 2022) and, on the other hand, to the fact that a large part of novice mediators in Latin America at this stage are volunteering to accompany an experienced mediator, often without a directed and reflective process about what they are observing (Ennes et al., 2020; Gomes & Cazelli, 2015; Pedersoli & Docters, 2020), so their self-perceived knowledge of didactics and facilitation would be permeated by the experience of accompaniment and by the practices of the experienced mediator, without a more critical perception of their own knowledge. This is reflected in the interview of M42, who mentions in his interview that *"at the beginning of my practice, I received more technical training related to the plants that the garden has in the collection that is displayed to the public. I also had the opportunity to accompany more experienced mediators to see how they explained the guided tour of the garden"*.

In the same way, the perception they have about their knowledge of the visitors and their learning is also related to what they observe in the visits of other mediators or in preconceptions about the learning processes. This is reflected in the interview of M49 who mentioned that *"For me the most important thing is that visitors learn what I explain and become aware of the importance of plants."* Similarly, the mediator M42 when asked about his role in the visits describes their role as guided public attention, seeking to teach about environmental conservation, introducing the importance of the botanical garden, explaining about the plants present and demystifying the ideas brought by the visitors. He also mentions that *"We have a document with the main stopping points of the guided tour and we do the same route with all groups, I just make adjustments in the language depending on the age of the visitors"*.

Considering the perception of the available knowledge, it was expected that mediators at the insertion stage could develop didactically appropriate mediation practices, and taking into consideration the different groups of visitors and their learning (Cândido Vendrasco et al., 2024). However, the interviews shows that even though the mediators of the insertion stage perceive that they have in-depth knowledge about didactics and about visitors and their learning, their practices are still very structured, mostly expository and do not respond to the characteristics of the visiting groups, reinforcing that their perceptions of their knowledge are permeated by what they accompany from more experienced mediators and perhaps by prior ideas of what it means to learn and not so much of a reflective and informed process.

The above is supported by the evidence that in the transition to the next stage the number of mediators who perceive they have in-depth knowledge of teaching strategies and facilitation decreases. At the same time, in the transition it seems that the focus is on the development of the knowledge of content and context. Prioritizing content and context knowledge in this transition would be given, mainly, by the institutions, given that the initial training processes are focused on the conceptual content related to the exhibition and familiarization with the educational setting (Macías-Nestor et al., 2020; Pedersoli & Docters, 2020; Batista et al., 2020). Enriching these training opportunities, incorporating other types of knowledge and enhancing supportive reflective work of more experienced educators would be especially fruitful in this transition of stages, given that the mediation model, and its associated mediation practices, are not yet consolidated (Ennes et al., 2020; Gomes & Cazelli, 2015; Pedersoli & Docters, 2020; Macías-Nestor et al., 2020; Batista et al., 2020).

16.5.2 Stabilisation

In the stabilization stage (1 to 10 years), it seems that there is a consolidation and probable implementation of the mediators' knowledge, since a stable trend is observed in all dimensions. Even so, it is possible to suppose that an integration of knowledge and even reorganizations of this knowledge on a smaller scale may be occurring.

In consideration of the knowledge available (or at least perceived as such), mediators in the stabilization stage could develop disciplinary and epistemologically appropriate mediation practices, contextualized to the space in which they work. However, the now perceived limitations in didactic and facilitation knowledge hinder responsive mediation practices, and would then be structured mediations, with previously designed discourses focused on knowledge transmission, and with little space for interaction, which are replicated independently of the audience, limiting visitors' learning (Cândido Vendrasco et al., 2024).

This is reflected in the interview of M71, when she mentions that "*I have the knowledge that people want to hear, so I think my role is essential, because, although the circuit is one that you can do without a mediator but it's much more enriching when you have someone there, trained and prepared to pass on this information to the visitor*". The ideas about the structuring of the pre-established visit and discourse can be observed in the interview of mediator M07 when he describes that "*we go with a pre-established itinerary, which aims to visit all the points in the park... regardless of the age of the visitors. The only time we select is with very young children, because sometimes it gets too tiring for them to walk the whole way*". These two examples reflect that, even though the mediators' have more experience, mediation processes at this stage are still rather expository and do not necessarily respond to the different groups of visitors. Although it should be noted that mediators M71 and M07 have 2 and 3 years of experience, respectively, so they are still in a transition process from their insertion stage.

Despite the above, at this stage, ideas associated with responsive mediation begin to appear. For example, mediator M07 mentions the importance of the use of questions during mediation "*Generally, the question arises much more with the idea of making the audience feel included than necessarily teaching something, I would say, because educational discourse in general already has a function of teaching something, so the addition of questions is much more about making the person feel included in it. To maintain interest... It's information that I could have just told the person, but they found out on their own—all right, I helped them come up with it—but they understood it on their own, and they feel good about it, they feel included because I asked, because they found out*". Just like mediator M71, when she described "*I'm more of a question person, so I always like to ask, "Have you done this before?" I try my best to bring what they already know from home so that we can reconcile and create a bond there, right? So my mediation is always with questions. I always ask questions, hoping they'll participate, but they don't always, so I think what I expect is participation, even if it's from a few people.*"

When we look at the interviews of mediators M04 and M64, who have 4 and 9 years of experience respectively, some differences can be observed in their reflections that point to a more reflective development of their knowledge. For example, mediator M04 mentioned that "*I try to recapture the previous knowledge that the group already has about the animals, about the zoo's function, and from that I put together my visit and always according to the interest of the class...I like the students to participate, to question, to bring what they already know so that we can often reaffirm it or try to add a new vision to what they already know. I like them to participate at all times, you've noticed that I keep asking them a lot of questions so that they can interact, so that I can find out what they already know*".

As well as the mediator M64 who indicates that "A*ll participants can contribute to the dialog during the visit, regardless of their level of education or social status. The person may live in the countryside, but when they come here, they have a wealth of cultural knowledge and experience. So, names that we use here, they use in a* *different way there...This whole audience is very helpful to us, regardless of age, regardless of social or cultural situation, because each one has their own context, their own perception of all this, so we always try to encourage this participation".*

These two examples show that mediators with more experience in this stabilization stage can mobilize their knowledge in more responsive and visitor-centered mediation practices. Even so, the apparent stabilization of knowledge at this stage and the absence of professional knowledge development reinforces the idea that the mediator's experience alone does not necessarily promote knowledge development. This stabilization would also be explained by the fact that most educational spaces do not have instances of continued training after the mediator insertion stage (Ennes et al., 2020; Richard, 2010). Therefore, this stage presents opportunities for development centred on the experience and intentional reflection associated with their own practice (Patrick, 2017; Tran & Halversen, 2021), which could more effectively promote the development of mediators' knowledge and the mobilization of this knowledge towards didactically appropriate practices that are more responsive to the needs of visitors.

16.5.3 Diversification

In a third stage (more than 10 years) there is a new trend of increase in the mediators' perception of their knowledge in all dimensions. Mainly, an important change is observed in the knowledge about visitors and their learning. At this stage, most of the mediators can recognize the existence of several factors that influence visitors' learning and acknowledge the importance of interest and previous ideas about the exhibition.

Consequently, at this stage, further adjustments of mediation practices around the needs of the audience could be identified, for example, taking into account their previous ideas, interests and knowledge to define the discourse and structure of the visit (Cândido Vendrasco et al., 2024). At the same time, different facilitation practices could be identified, incorporating the use of questions, interaction with the collection, use of resources and encouraging active participation of visitors (Cândido Vendrasco et al., 2024). In sum, at this stage the mediators could develop a mediation that is contextualized, disciplinary, epistemologically and didactically appropriate and also responsive to the audience.

This is reflected in the interviews of the two mediators who were at this stage of development, M31 and M51 with 10 and 16 years of experience respectively.

M31 for example, when asked about his role as mediator mentions that *"we plan this visit according to the visitor's interests...we can always prioritize discussion, the exchange of information and, above all, consider the background of these visitors, whether they are properly planned, identified, which we have the profile all reported, right, or perhaps a visitor who arrives and will have their first contact with the museum, it will be the first contact with the Botanical Garden object, many even as a cultural space, we also receive groups in situations of social vulnerability, so this means that the educator needs to be very dynamic in their activities, right, and since Mediation proposes dialog, listening above all, then we also find this refuge, even if we don't know who is going to be with us on the next visit, but of course we establish an itinerary, and we adapt the language, we adapt the theme itself, the objective of this visit, according to that profile. So I'd say that considering our target audience is the fundamental role for us to be able to act and generate a visit that brings a positive experience, of exchanging information, sharing knowledge with this visitor, whatever they may be, child, adult, person who already has some experience in a museum space, right, or not".*

Regarding the structure of the visit, mediator M51, when asked about how he conducts the guided tours, mentions that *"I work a lot with questions. So, before I start talking, "But what do you know about ferns?". Because I'll start my speech from where they are. Otherwise it becomes an unbearable lecture. I talk, I talk, I talk and the students just listen… I go into the subjects, I do it, now I think about it, I do it all the time. I ask them more questions than they ask me, so if they've seen it, what they've heard about it, it's like an investigation, looking into what they know. Every time we're in a space I get information from them.!".* And also mentions the use of biological material for hands-on activities *"whenever I manage to catch the element, I bring it to the children. So fruit from the forest, I try, I catch it, I show it to them. Leaves, branches, trunks, any insects I can get my hands on, if it's not a dangerous insect. So I try to use these kinds of strategies to encourage the interest and participation of everyone.*

The mediators' perception of their knowledge at this stage coincides with Busch et al. (2022), that more experienced mediators are able to better link the various dimensions of their professional knowledge. Even when knowledge is more developed at this stage, there are still opportunities for professional development in which mediators can in a directed, reflective and participatory way, continue to articulate this different knowledge to enhance innovative mediation practices that are more responsive and visitor-centred, enhancing the purposes of equity and inclusion that these educational spaces have (Cândido Vendrasco & Marzábal, 2023). At the same time, mediators at this stage of development can participate in the training processes of novice mediators, supporting them to be more effective (Ennes et al., 2020; Pedersoli & Docters, 2020).

16.6 Conclusions and Projections

The application of a knowledge perception questionnaire to 78 mediators allowed us to characterize the mediators' perception of their knowledge of content, didactic and facilitation, context and visitors and their learning in relation to their years of professional experience. In this regard, we have seen that these perceptions are characterized in different ways. From the observation of the trends in each category of knowledge, we have been able to recognize three stages in the professional development of mediators: *insertion, stabilization and diversification.* The first stage of insertion, in which mediators perceive that the didactic and facilitation knowledge and knowledge of visitors are the most developed, while they perceive less mastery of content and context knowledge. The perception of the latter two dimensions increases in the transition between the first and second stage, while the perception of facilitation knowledge decreases. In the second stage of stabilization, mediators perceive all their knowledge in a similar way and no significant change is observed in any type of knowledge, but it is assumed that a reorganization and integration of previously developed knowledge could occur. And in the third stage, of diversification, an increase is observed in the mediators' perception of their knowledge in all dimensions, mainly in the perception of the visitors' knowledge and their learning.

From the interviews with mediators in each of the stages, it was possible to better demonstrate how this knowledge is articulated with the mediators' practices. All of the above has important implications for learning in spaces of non-formal science education.

Given that the professional career of mediators in Latin America is generally very short (Gomes & Cazelli, 2015; Massarani et al., 2021), a significant number of mediators would not be able to deeply develop the knowledge dimensions mainly related to the visitors and their learning and also, with the didactic and facilitation strategies, limiting the visitor's experience and the achievement of the educational objectives of each institution. This is because the late development of mediators' didactic and facilitation knowledge would make it difficult to incorporate innovative, didactically appropriate, responsive and visitor-centred mediation practices, maintaining visit formats in which the mediator has the role of delivering disciplinary knowledge, and in that the public has a low level of involvement and participation.

This implies that, on the one hand, we need to promote a more extensive professional career for mediators in Brazil and Latin America, in which they feel valued and have work projections within their educational space. On the other hand, we need to promote intentional training that promotes and anticipates this process of developing diverse knowledge.

Regarding the training of mediators, we consider that there are three fundamental aspects in their professional development: (a) initial training, (b) continuous training within the same field, and (c) experience and reflection on practice. We recognize that educational settings can primarily influence continuing education and reflection on practice. Therefore, we suggest that these processes are integrated in an intentional manner and from early stages, allowing mediators to be trained within the educational environment and that this training is accompanied by continuous reflection on their experience and practice starting with their incorporation into the educational space. However, it is crucial that these formative processes are grounded in theory (Robinson, 2019) and, although they must take into account the contextual aspects of each setting, it is essential to promote a common framework of knowledge to be addressed in this formative process to foster good mediation practices (Tran & Halversen, 2021).

In terms of future research, we could benefit from longitudinal studies that follow mediators over extended periods, looking specifically at the development of their different skills and the various factors that may influence this process, paying particular attention to the formative processes in which they participate. In this way, we could delve deeper into the different stages of their professional development and identify the best way to enhance this process. On the other hand, it would be interesting to explore how the different skills relate to each other and how this relationship is linked to mediation practices, which would allow us to better understand the mediators' decision-making processes during guided tours and their possible effects on the visitors' experience (Cândido Vendrasco & Marzábal, 2023).

16.7 Limitations

It is recognized that some methodological aspects limit the scope of this research. First, the focus of the study is on the knowledge of mediators at each stage of their professional development, so the data and analysis focused on this. However, we can recognise that there are many other factors that could influence or affect the professional development of these professionals, such as salaries, working conditions (paid versus voluntary), educational policies of the space in which they work and others. In future research on the topic, other relevant factors need to be incorporated in order to understand more broadly the developmental stages of mediators.

Furthermore, given that it is uncommon for mediators to have career trajectories of 10 years or more, i.e. to reach the stage of diversification, studying the transformation of their knowledge would necessitate longitudinal studies, which is not the case in our research. Nevertheless, given the dearth of research on the careers of mediators in Latin America, this study makes a valuable contribution to the existing literature by providing insights into the professional development of mediators in the region.

On the other hand, we understand that 78 mediators is a small sample considering the population of mediators that exists in Brazil, so we suggest caution in generalising the results of this research, even though it may shed important light to open the discussion on the subject. Therefore, it is planned to extend this research to more mediators in different types of non-formal educational spaces and even to other countries in the region.

Despite these limitations, it is thought that the justification of these methodological decisions and the theoretical basis of the proposed discussions enabled rigorous and significant research to be conducted on a topic that has been little explored and is directly related to the training of individuals who act as mediators in non-formal spaces of science education. It is therefore believed that the data collected provide relevant information on the research questions and allow further areas of research to be identified.

Disclosure It should be noted that part of this chapter was published in an article in Portuguese that can be found at https://orcid.org/0000-0003-3932-2579 and gave rise to the publication of the ESERA Conference 2023, which was selected for this book. In this chapter, we elaborate on the previous findings by incorporating interviews with 8 selected mediators of different experiences.

Acknowledgments We thank all the mediators who participated in the research. We also thank the peer reviewers who provided valuable feedback on the initial version of this chapter.

This work was supported by the following projects: CONICYT-PFCHA/Doctorado Nacional/2019-21190181 (Chile) and National Council for Scientific and Technological Development (CNPq)/409899/2021-6 (Brazil).

References

Asún, R. A., Rdz-Navarro, K., & Alvarado, J. M. (2016). Developing multidimensional Likert scales using item factor analysis: The case of four-point items. *Sociological Methods & Research, 45*(1), 109–133. https://doi.org/10.1177/0049124114566716

Batista, M. F. A., Gonzalez, A. C. D. S., Oliveira, D., & da Silva Barros, H. (2020). A formação de mediadores no Museu da Vida: múltiplas vivências. *Journal of Science Communication América Latina, 3*(2), A05. https://doi.org/10.22323/3.03020205

Bisquerra Alzina, R., & Pérez Escoda, N. (2015). Pueden las escalas Likert aumentar en sensibilidad?. REIRE. *Revista d'Innovació i Recerca en Educació, 2015, 8*(2), 129–147. https://doi.org/10.1344/reire2015.8.2828

Bandura, A. (2004) "The growing primacy of perceived efficacy in human self-development, adaptation and change" En: M. Salanova, R. Grau, E. Cifre, S. Llorens y M. García-Renedo (eds). Nuevos horizontes en la investigación sobre la autoeficacia. : Universitat Jaume I Servicio de Comunicación y Publicaciones.

Berdnikoff, A. N., & Silva, J. R. S. (2020). Os saberes museais dos mediadores. *Revista de Ensino de Ciências e Matemática, 11*(6), 856–876.

Bevan, B., & Xanthoudaki, M. (2008). Professional development for museum educators: Unpinning the underpinnings. *Journal of Museum Education, 33*(2), 107–111.

Busch, K. C., Kudumu, M., & Park, S. (2022). Pedagogical content knowledge for informal science educators: Development of the ISE-PCK framework. *Research in Science Education, 1-22*, 253–274. https://doi.org/10.1007/s11165-022-10055-9

Carletti, C., & Massarani, L. (2015). Mediadores de centros e museus de ciência: um estudo sobre quem são estes atores-chave na mediação entre a ciência e o público no Brasil. *Journal of Science Communication, 14*(2), 1–17. https://doi.org/10.22323/2.14020201

Cândido Vendrasco, N., & Marzábal, A. (2023). Percepção dos mediadores de espaços de educação científica não formal sobre seus conhecimentos nas diferentes etapas de desenvolvimento profissional. *Investigações em Ensino de Ciências, 28*(2), 308–331.

Cândido Vendrasco, N., Marzabal, A., & Pugliese, A. (2024). Towards responsive mediations in guided visits to non-formal science education settings. *Studies in Science Education, 60*(1), 45–74.

Creswell, J. (2007). Philosophical, paradigm, and interpretive frameworks. Cap. 2. In *Qualitative Inquiry & Research Design. Choosing among five approaches* (2nd ed.).

Daza-Caicedo, S., Ariza, A. P., Falla, S., Múnera, A., & Quiroga, J. (2020). Encuentros y desencuentros de quienes median y sus públicos. Reflexión sobre el rol de la mediación desde un museo de ciencias en Colombia: el Museo Interactivo Maloka. *Journal of Science Communication, América Latina, 3*(2), A06.

Dillon, J., Rickinson, M., & Teamey, K. (2016). The value of outdoor learning: Evidence from research in the UK and elsewhere. In Dillon, J. Ed., Towards a convergence between science and environmental education (pp. 193–200). Routledge.

Ennes, M., Jones, M. G., & Chesnutt, K. (2020). Evaluation of educator self-efficacy in informal science centers. *Journal of Museum Education, 45*(3), 327–339. https://doi.org/10.1080/10598650.2020.1771993

Falk, J. H., & Dierking, L. D. (2012). Lifelong science learning for adults: The role of free-choice experiences. In *Second international handbook of science education (pp. 1063–1079)*. https://doi.org/10.1007/978-1-4020-9041-7_70

Falk, J. H., & Dierking, L. D. (2016). The museum experience revisited. *Routledge*. https://doi.org/10.4324/9781315417851

Geveke, C. H., Steenbeek, H. W., Doornenbal, J. M., & Van Geert, P. L. (2017). Attractor states in teaching and learning processes: A study of out-of-school science education. *Frontiers in Psychology, 8*, 299.

Gomes, I., & Cazelli, S. (2015). Formação de mediadores em museus de ciência: saberes e práticas. *Ensaio Pesquisa em Educação em Ciências (Belo Horizonte), 18*(01), 23–46. https://doi.org/10.1590/1983-21172016180102

Hernandez, C. M., Morales, A. R., & Shroyer, M. G. (2013). The development of a model of culturally responsive science and mathematics teaching. *Cultural Studies of Science Education, 8*(4), 803–820. https://doi.org/10.1007/s11422-013-9544-1

Jegede, O., Taplin, M., & Chan, S. L. (2000). Trainee teachers' perception of their knowledge about expert teaching. *Educational Research, 42*(3), 287–308. https://doi.org/10.1080/001318800440614

Keller-Schneider, M., Zhong, H. F., & Yeung, A. S. (2020). Competence and challenge in professional development: Teacher perceptions at different stages of career. *Journal of Education for Teaching, 46*(1), 36–54. https://doi.org/10.1080/02607476.2019.1708626

Kuntze, S., Siller, H. S., & Vogl, C. (2013). Teachers' self-perceptions of their pedagogical content knowledge related to modelling–an empirical study with Austrian teachers. In *Teaching mathematical modelling: Connecting to research and practice* (pp. 317–326).

Macías-Nestor, A. P., Reynoso Haynes, E., & Torreblanca-Navarro, O. (2020). Formación de mediadores en los museos y centros de ciencias de la Universidad Nacional Autónoma de México. *Journal of Science Communication América Latina, 3*(2), A03. https://doi.org/10.22323/3.03020203

Marandino, M. (2008). *Educação em museus: a mediação em foco*. Geenf · Grupo de Estudos e Pesquisa em Educação Não-formal e Divulgação em Ciência/FEUSP.

Massarani, L. (2020). Museus de ciência e a mediação entre ciência e sociedade. [Science museums and the mediation between science and society]. *Journal of Science Communication América Latina, 3*(2), editorial. https://doi.org/10.22323/3.03020501

Massarani, L., Alvaro, M., Norberto Rocha, J., Vieira de Abreu, W., Silveira, F., Falla Morales, S., et al. (2021). Mediadores de centros e museus de ciência: um estudo sobre os profissionais que atuam na América Latina. *Museologia e Patrimônio-Revista Eletrônica do Programa de Pós-Graduação em Museologia e Patrimônio–Unirio, 14*(1).

Massarani, L., Leon-Castella, A., Aguirre, C., Reynoso, E., Lindergaard, L., & Fernandez, E. (2015). *Guia de centros e museus de ciência da América Latina e do Caribe*. RedPOP-UNESCO e Museu da Vida.

Norberto Rocha, J., & Marandino, M. (2020). 'O papel e os desafios dos mediadores em quatro experiências de museus e centros de ciências itinerantes brasileiros' [The role and challenges of mediators in four experiences of itinerant Brazilian museums and science centers]. *Journal of Science Communication América Latina, 03*(02), A08. https://doi.org/10.22323/3.03020208

Patton, M. Q. (2007). *Sampling, qualitative (purposeful)*. The Blackwell encyclopedia of sociology. Wiley.

Patrick, P. G. (2017). Informal science educators and the nine dimensions of reflective practice. In *Preparing informal science educators* (pp. 41–65). Springer.

Pedersoli, C., & Docters, M. L. (2020). Formar para la mediación pedagógica en el museo de ciencias: una experiencia en la Universidad Nacional de La Plata, Argentina. *Journal of Science Communication América Latina, 3*(2), 1–2. https://doi.org/10.22323/3.03020204

Peters, T. (2019). ¿Qué es la mediación artística? Un estado del arte de un debate en curso. *Corima, Revista de Investigación en Gestión Cultural, 6*(4).
Rende, K., Fromson, K., Jones, M. G., & Ennes, M. (2021). The privilege of low pay: Informal educators' perspectives on workforce equity and diversity. *Journal of Museum Education, 46*(4), 430–440.
Rennie, L. J. (2014). Learning science outside of school. In Handbook of research on science education, Volume II (pp. 134–158). Routledge.
Richard, O. (2010). Report on the profile of European explainers, PILOTS Project, D3.3.URL: http://www.ecsite.eu/sites/default/files/project_docs/D3_3_Report_on_the_profile_of_European_explainers_0.pdf (acedido em 20 de diciembre de 2023).
Robinson, C. (2019). Thoughts on the history of professional development programs in museums. *Journal of Museum Education, 44*(2), 123–130.
Specht, I., & Loreit, F. (2021). Empirical knowledge about person-led guided tours in museums: A scoping review. *Journal of Interpretation Research, 26*(2), 96–130.
Tran, L. U. (2007). Teaching science in museums: The pedagogy and goals of museum educators. *Science Education, 91*(2), 278–297. https://doi.org/10.1002/sce.20193
Tran, L. U., & Halversen, C. (2021). *Reflecting on practice for STEM educators: A guide for museums, out-of-school, and other informal settings*. Routledge.
Tran, L. U., & King, H. (2011). Teaching science in informal environments: Pedagogical knowledge for informal educators. In *The professional knowledge base of science teaching* (pp. 279–293). Springer. https://doi.org/10.1007/978-90-481-3927-9_16
Weiler, B., & Smith, L. (2009). Does more interpretation lead to greater outcomes? An assessment of the impacts of multiple layers of interpretation in a zoo context. *Journal of Sustainable Tourism, 17*(1), 91–105. https://doi.org/10.1080/09669580802359319
Zhai, J., & Dillon, J. (2014). Communicating science to students: Investigating professional botanic garden educators' talk during guided school visits. *Journal of Research in Science Teaching, 51*(4), 407–429.

Natália Cândido Vendrasco has a PhD in Education. During the last 10 years, she worked as an educator in various non-formal science education settings in Brazil and Chile and as a teacher trainer at different universities. Currently, she is an assistant professor at the Pontificia Universidad Católica de Chile.

Ainoa Marzábal has a PhD in Science Education. Her area of interest focuses on discursive practices and modelling-based science education. Currently, she is an associate professor at the Pontificia Universidad Católica de Chile.

Chapter 17
Examining Teachers' Perceptions of Authentic Science Practices and Practices-Based Curriculum

Kathleen M. Hill, Tiffany Lewis, and Matthew M. Johnson

17.1 Introduction

The Next Generation Science Standards (NGSS) in the United States call for engaging students in authentic science through the practices of scientists and engineers (NRC, 2013). To a practicing scientist, authentic science is a body of complex content and discipline-dependent methodologies, analyses, and social interactions with the goal of constructing or refining scientific knowledge (Wong & Hodson, 2010). Thus, authentic science in the classroom involves engaging students in investigations that seek to answer scientific questions in ways that mirror or closely approximate those of scientists in today's world (Crawford, 2012, p. 113). However, in the typical science classroom, students engage in "simple inquiry tasks" (Chinn & Malhotra, 2002, p. 176) that do not reflect the day-to-day activities of practicing scientists. Also referred to as school science, these tasks typically include contrived activities where students are following steps to verify already known information (Crawford, 2012) that include limited or no student agency and require only basic reasoning (Chinn & Malhotra, 2002).

One strategy to increase teachers' knowledge of science and engineering practices is to immerse them in a research experience (e.g., Research Experience for Teachers—RET). While some positive effects have been reported from prior studies, interventions associated with RET programs have been understudied (Krim et al., 2019). This paper reports on a RET program intended to develop teachers' understanding of science and engineering practices through a targeted series of professional development (PD) activities embedded in an immersive research experience. The theoretical framework is grounded in situated cognition, because teachers

K. M. Hill (✉) · T. Lewis · M. M. Johnson
Center for Science and the Schools, Pennsylvania State University, University Park, PA, USA
e-mail: kmm173@psu.edu; tzs80@psu.edu; mmj125@psu.edu

G. Cakmakci, M. F. Tasar (eds.), *Connecting Science Education with Cultural Heritage*, Contributions from Science Education Research 15,
https://doi.org/10.1007/978-3-031-90944-3_17

learn about the science and engineering practices through their cognitive apprenticeships (Brown et al., 1989) in social situations like a master-apprentice relationship. A goal of the RET program was to have teachers experience the ways in which scientists create knowledge and reflect on how these practices can be translated into K-12 classrooms. This study focused on teachers' perceptions of science and engineering practices and how they prepare curriculum for students to investigate and ultimately explain a particular phenomenon or solve a problem. Our research questions included: (1) In what ways did teachers characterize a range of classroom activities in terms of authentic science and engineering practices? (2) To what extent did teachers include the science and engineering practices in their curriculum plans based upon their summer research experience?

17.1.1 What Is Authentic Science in the Classroom?

Fortunato et al. (2018) characterized features of contemporary science to include: intentional novel problem selection, interdisciplinary work through team science, and political considerations of career trajectories and funding structures. These activities require professional scientists to have sophisticated content knowledge and methodologies with established cultural practices to answer questions or solve problems in the natural and designed worlds. Critics of bringing authentic science to the K-12 classroom argue that it is not realistic to think that science as a research discipline can be directly translated to the K-12 classroom because the goals and epistemologies of the two endeavors are different (Chinn & Malhotra, 2002). What is considered authentic for adults is not the same as what is authentic for students (Burgin, 2020). How students engage in authentic science is going to be different from the work of professional scientists; however, students should still be given the opportunity to practice science beyond memorization and learning the "rules" of science. Students should be enculturated into scientific communities of practice over time, from a young age to university level, as they gain more sophisticated understandings of science as a discipline (Burgin, 2020).

Burgin (2020) argues that students can engage in activities that are authentic, but the activities that the students carry out can lack authenticity in how they are situated in the classroom. For example, titrations are a common chemistry laboratory activity for students. Students determine the unknown concentration of a strong base by titrating the base into a strong acid containing an indicator solution. This is an authentic activity that scientists may engage in, but typically the context and significance to science is not apparent to students as they carry out the lab. For an activity to be authentic for students, the teacher must attend to epistemological underpinnings of the activities by engaging students in the practices of scientists and engineers (Burgin, 2020).

Building on the work of Burgin (2020), we identify three components of learning experiences for authentic science in the classroom: (1) scaffolded engagement in science and engineering practices, (2) student agency, and (3) purpose-driven activities (Johnson & Hill, 2023). This includes providing the opportunity for students to act as agents of their own learning as they collaboratively engage in purpose-driven activities through the science and engineering practices to make sense of phenomena. This view of authenticity concedes that authentic science activities should be rooted in the work of scientists and engineers, while recognizing that the goal of science education is not for students to engage in novel scientific investigations. The goal is for students to adopt epistemological habits of mind important to the process of science. Students' engagement in authentic activities should be scaffolded over time, shifting from teacher-driven activities to student-driven activities (Johnson & Hill, 2023).

17.1.2 Next Generation Science Standards (NGSS): Promoting Authenticity in the Science Classroom

Science educators can create developmentally and culturally appropriate experiences for students in the classroom. Current science education reform in the United States, articulated by the Next Generation Science Standards (NGSS; NRC, 2013), attempt to bridge the gap between science content and enacting authentic science. NGSS advocates for teaching science concepts through engaging students in the science and engineering practices (NRC, 2012). Forty-nine of fifty states have adopted NGSS or a state-based version of NGSS (Fig. 17.1), thus serving as a framework for our PD that reaches teachers across the United States.

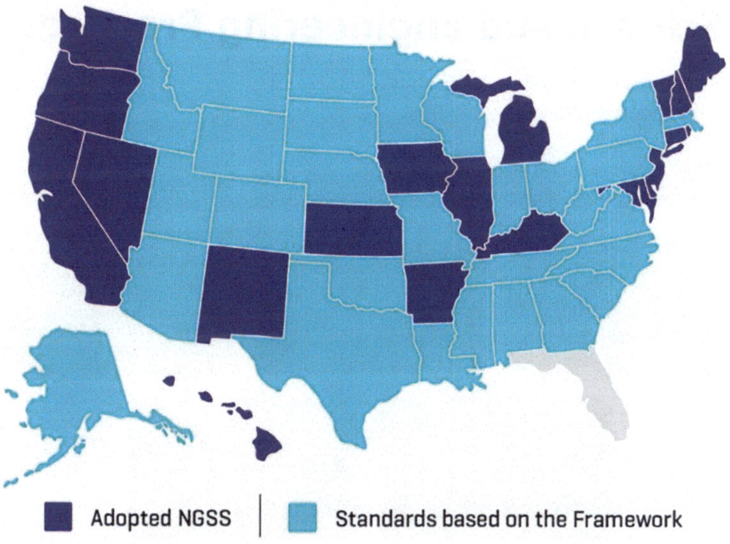

Fig. 17.1 States that have adopted NGSS or have written standards based on NGSS (NSTA, 2024)

Previous science education reform, National Science Education Standards, focused on the use of inquiry-based teaching to bring the scientific method to life as a skill that was separate from learning the content (NRC, 1996). Moving beyond inquiry-based science education with stand-alone investigations, NGSS promotes students learning science content through engaging in eight science and engineering practices (Fig. 17.2; NRC, 2012). Developed in conjunction with professionals from the fields, these practices include activities that describe the daily work of scientists and engineers such as developing and using models, engaging in argument from evidence, and constructing explanations and defining solutions (NRC, 2012). This shift to the practices is an entry point for a more epistemologically aligned view of science and provides an epistemic context through with students can learn science.

17.1.3 Teacher Research Experiences Promote Authentic Science in the Classroom

The difficulty of shifting towards reform-based teaching practices as described by the *Framework* has been widely described (Patchen & Smithenry, 2014) and is not a new realization for science education. Traditional teacher education programs do not require preservice teachers to engage in scientific or engineering research (Burgin, 2020; Windschitl, 2004). This puts secondary science teachers at a disadvantage as most teachers lack expertise in the day-to-day practices of scientists that is required to effectively integrate these practices into their science curriculum

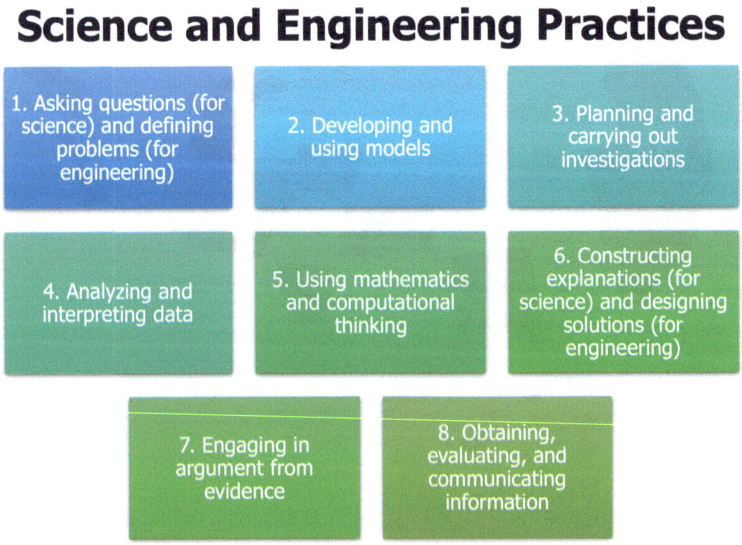

Fig. 17.2 Eight NGSS science and engineering practices (NRC, 2012)

(Antink-Meyer & Meyer, 2016; Davidson et al., 2022). One way for teachers to reduce this gap in knowledge is through participating in an immersive science PD experience, such as a Research Experience for Teachers (RET) program (Davidson et al., 2022).

RET programs embed teachers in laboratories alongside scientists and engineers to engage in a research project, which allows teachers to learn the practices and cultures of science in the context of authentic research. Studies of RET program outcomes suggest that extended participation in research experiences have helped teachers develop deeper understanding of science principles and epistemologies (Blanchard et al., 2008; Davidson et al., 2022; Miranda & Damico, 2015). However, RET programs differ in terms of the ways in which they engage teachers in the practices of scientists and engineers. "Practices are learned through participation and often entail extended interactions with members already familiar with the ways that practices are recognized as socially significant" (Kelly & Licona, 2018); therefore, teachers must be engaged in the work that contributes to the research of their lab and not a contrived project developed by the lab simply for the teacher (Blanchard & Sampson, 2017). Working with scientists and engineers for several weeks positions teachers to develop better disciplinary understandings of science (Davidson et al., 2022) with the aim that teachers translate these practices back to the classroom using their pedagogical expertise.

In a study on the long-term outcomes of an RET, Pop et al. (2010) found that teachers reported that their participation in the RET program created changes in their teaching, including student-centered teaching practices and adding activities where students applied science learning to real life situations. Participating in a research experience also helped teachers gain more insight into how recent reforms of science education frame disciplinary understandings of science based on epistemic practices (Davidson et al., 2022). However, teachers participating in RET programs often struggled to effectively translate their research experiences into classroom practices (Krim et al., 2019).

To address this challenge, some RET programs have provided critical supports for teachers to develop curriculum that mirrors or closely approximate the work of STEM experts in ways that are appropriate for precollege learners (Hill et al., 2017a; Hill et al., 2017b; Hill et al., 2018). An important feature of an impactful RET program is the science education-focused PD that helps teachers unpack how they have been engaging in the practices and how they can translate the practices to their science classrooms. Studies of RETs that include pedagogical professional development suggest that the intervention of a research experience for teachers can have a positive impact on teacher knowledge of inquiry (Herrington et al., 2012; Southerland et al., 2016) but modest to few influences on the changes in teacher practice with RET programs that lack science education PD (Enderle et al., 2014; Sadler et al., 2010). These findings make it important to consider how an RET program influences teacher practice and student impact (Blanchard & Sampson, 2017).

17.2 Methods

The theoretical framework of situated cognition guided our study, focusing on teachers' knowledge development being a product of the activity, context, and culture in which the teachers, teacher educators, and researchers operated (Brown et al., 1989). Learning takes place within a social context using authentic practices, much as an apprentice would learn from their master (Brown et al., 1989). The goal of the RET program is not to develop *knowledge producers* (Feldman et al., 2013), but to build teachers' content knowledge for performing science and engineering research and provide critical supports for translating their research experience into curriculum appropriate for secondary-level students in their school context.

This qualitative study was designed to characterize teachers' perceptions of SEPs as they engaged in science and engineering research and how their learning manifested in the curriculum they developed based on their research experience.

17.2.1 Description of the Summer Research Experience Program

During the summer of 2021, the Center for Science and the Schools (CSATS) at the Pennsylvania State University led a RET program in which participating teachers conducted authentic research projects remotely under the direction of a science or engineering faculty member and graduate students for seven weeks. The program engaged teachers in research 80% of the time and PD 20% of the time. Each week, the teachers participated in six hours of PD designed to support them in translating their research experience into the classroom setting. Four hours of PD per week focused on practices-based pedagogy and two hours on understanding the practices of researchers. The pedagogy PD sessions were held virtually due to pandemic restrictions and teachers' locations across the United States.

Authentic Science Activity Continuum: during the fourth week of the PD session, teachers were asked to place example science and engineering classroom activities (Table 17.1) along an author-developed continuum of authentic science activities (Fig. 17.3). These example activities were identified by the researchers as activities that are typically found in science and engineering classrooms in the United States. Four possible placements were included on the continuum ranging from 1 (most contrived) to 4 (most authentic). The science activity was considered to be more authentic when: (1) students engage in science and engineering practices and (2) the activity is situated in a research project.

The science activities were presented with a short description and pictures to represent the activity to initiate a discussion about what counts as an authentic activity with students. The continuum was provided to focus and provoke discussion about authenticity in the classroom; therefore, there were no "correct" placements of the activities on the continuum. The teachers were purposely provided with

Table 17.1 Examples of activities placed along the activity authenticity continuum

Common Science Classroom Activities	Group Placements on the Continuum			
	Group 1	Group 2	Group 3	Group 4
1. Drawing pictures of objects viewed through a microscope	1	1	1	1
2. Students draw a landscape with multiple geologic features and describe possible evolution of the landscape	2	2	2	2
3. Wearable device design project that solves a problem identified by students (or the teacher)	4	3 or 4	4	4
4. Hydroponic system design and comparison to soil-based growing systems to maximize yield	4	2	4	3
5. Creating circuits on a breadboard	2	1	3	2
6. Creating an aluminum boat to hold the most pennies	2	1	3	1
7. Mixing diet coke and Mentos	1	1	1	1

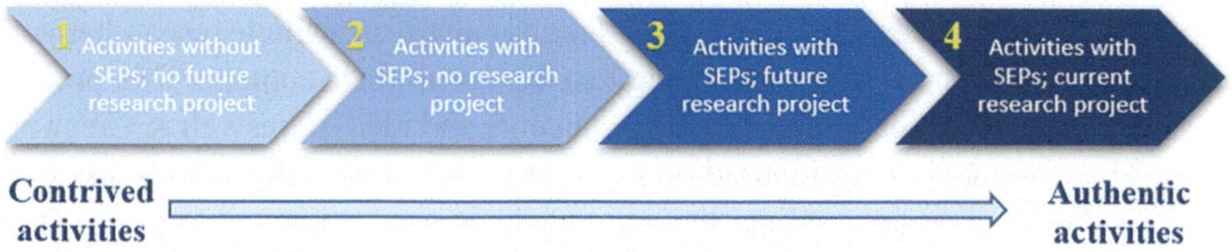

Fig. 17.3 Authentic Science Activity Continuum used during the professional development

limited context about the enactment of the activity to drive the discussion. For example, an activity was described for teaching about interactions between materials: Menthos candy was added to a bottle of Diet Coke, resulting in an eruption of the drink from the container. No specifics were provided in terms of the role of the teacher or student in the activity. By limiting the contextualization, the teachers were able to generate possible classroom scenarios for enactment of the activities and negotiate the level of authenticity.

17.2.2 Study Sample and Data Collection

The study sample included 12 secondary STEM teachers who applied and were selected to participate in the CSATS summer RET program. Selection criteria included teaching assignment, years of teaching experience, and stated interested and anticipated benefits described in the application. The teachers were from 8 different states in the United States. Four of the teachers have an undergraduate degree in a science discipline and ten with graduate degrees (master's level) in education. The teachers spanned the secondary grades, with nine at the high school level and three at the middle school level. Additionally, seven of the teachers taught in schools with an underserved population of students with high diversity (>50% students of color) and/or low socioeconomic status (>50% of students receive free or reduced lunch), as identified by each school districts' demographic reports.

Multiple sources of data were collected for the study. Teachers' responses to the authenticity continuum and resulting small and large group discussions were recorded on Zoom and transcribed. In addition, the teachers created two artifacts as part of the RET program: (1) a technical research poster and (2) a "CLassroom Research Project" (CLRP) plans to engage their students in authentic investigations related to their summer research. These artifacts were collected from all participating teachers.

17.2.3 Data Analysis

To determine teachers' perceptions of the authenticity of science and engineering classroom activities, open coding and thematic analysis were employed to identify key patterns and themes. The open coding process involved a single researcher generating initial codes for segments of the text based on their content. This process allowed for the identification of recurring concepts and ideas (Creswell & Creswell, 2018). The generated codes focused on any text related to describing the classroom activities. Following open coding, thematic analysis was conducted to group these codes into broader themes (Saldaña, 2021).

In analyzing the data from the teachers' artifacts created during the RET program, a priori coding and comparative analysis were utilized. A priori coding involved applying predefined codes of the SEPs to identify the presence of specific science and engineering practices in their technical research and CLRPs (Creswell & Creswell, 2018). This method ensured consistency and alignment with the study's objectives. Comparative analysis was then employed to examine the similarities and differences across the artifacts (Creswell & Creswell, 2018). By comparing the technical research posters and CLRP plans, patterns and variations were identified in how teachers integrated their summer research into classroom activities.

17.2.4 Limitations

Several limitations exist in the study. First, teachers self-selected to take part in the RET program, so the teacher sample is skewed to teachers that are apt to participate in professional development they chose, which limits generalizability to larger groups of teachers that do not select to participate in professional development. Another limitation is that we did not collect lessons that teachers' intended to replace with their new curriculum plans. This would have been useful to measure teachers' ideas of authentic science prior to participation in the RET program. Finally, we were only able to collect data on teachers' curriculum plans, not on the enactment of those plans. Due to the teachers' locations in eight states across the

United States, we were unable to visit classrooms to observe teachers in their classrooms to determine if students engaged in the science and engineering practices as planned.

17.3 Results

17.3.1 Research Question 1: Characterizing Classroom Activities in Terms of Authenticity

During the week 4 discussion focused on rating the authenticity of science activities, all four groups came to similar consensus for four out of seven of the activities along the continuum (Table 17.1). Based upon coded discussion responses, teachers' main determining factor for authenticity was whether the students were driving the process of the activity. The teachers mentioned that an activity was more authentic when the students designed a system, collected their own data, or identified an engineering problem. Three out of seven of the activities produced more varied responses by the teachers: hydroponic system design, creating circuits on a breadboard, and creating an aluminum boat. These three activities produced a common theme of discussion regarding how the activity was scaffolded or situated in the classroom. Two teachers mentioned that an activity is more authentic when it is tied to the science content. Even though science content was not part of the continuum, the teachers discussed tying science content to practices, shifting the activity from simply being "hands-on" to students considering more complex ideas.

The activities that produced the most discussion among the teachers and variation in continuum rating was based on the importance of how the activity was scaffolded. For example, in a discussion of the activity where students created a circuit with a breadboard, the teachers provided different examples for each level of the continuum based on how the activity was enacted by the teacher.

> If it's just like following directions to connect things exactly how they prescribed…then it would be a level one. But if you are giving the ability for the students to construct the circuit in a way that they're designing, that's when I think you move to a level two.—RET Teacher

The hydroponics activity also generated quite a bit of discussion about where the development of a hydroponics vs. soil growth system should fall on the continuum. One teacher argued that the activity was a 4, "If it was dealing with the current real-world situation of feeding the world, it's definitely a current research project." This is opposed to if the students were just creating the systems, but not testing them, the activity would be a three because the students could be working towards a future research project (Fig. 17.4).

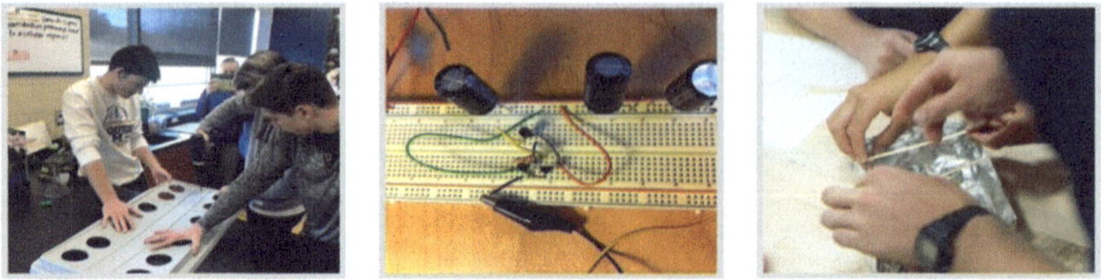

Fig. 17.4 Examples of common science classroom activities: (**a**) Students create a hydroponics system (own photo); (**b**) Students create circuits using a breadboard (Roberts, 2010); (**c**) Students create aluminum foil boats to determine how much weight in pennies can be held by the boat (U.S. Coast Guard Academy, 2013)

17.3.2 Research Question 2: Translating Science and Engineering Practices to Curriculum Plans

In terms of their own research practices, the teachers' technical posters provided evidence of multiple investigations that they conducted to answer a research question or design an engineering solution. All the teachers engaged in seven to eight of the science and engineering practices outlined in the NGSS, and their CLRPs included multiple investigations that incorporated many of the practices of their research experiences (Tables 17.2 and 17.3). Teachers' understanding of some practices were aligned with the actual work of scientists and engineers; however, some teachers expressed a view of practices, such as "asking questions" and "developing and using models," that differs from those of technical experts or described in the NGSS document.

17.4 Discussion and Conclusions

Research shows that most teachers do not engage their students in authentic science activities. Providing teachers with immersive opportunities to work alongside practicing scientists is an important way for teachers to better understand the practices of the scientific community. This study addresses the gap in the literature of examining the impact of an RET program that included science education-focused PD. The CSATS RET program, which combined technical research with targeted professional development (PD), significantly influenced teachers' understanding and implementation of science and engineering practices (SEPs) as outlined by the NGSS (NRC, 2013). These findings are consistent with previous research indicating that immersive research experience can deepen teachers' knowledge of content and practices involved in technical research (Blanchard et al., 2008; Davidson et al., 2022). By focusing on the SEPs and their application in classroom settings, the PD sessions supported teachers to refine their understanding of these practices and

Table 17.2 Translation of teacher research into curriculum plans

Teacher Research Technical Poster Topic	Teacher Classroom Research Project Topic
1. Perceptive robots for field-oriented industry	Robotics automation with biological vital sign sensors
2. Light optimization for visual quality and conservation of artwork	Light optimization in school's learning spaces to improve student's performance
3. Indoor contaminants on PM2.5 and long-term health issues	Using sensors to improve energy efficiency and indoor air quality in our school
4. Using thermal imaging to detect energy leakage in existing windows	How can we use infrared imaging to develop a way to stay cool while keeping a bearded dragon in the classroom?
5. Using PEWI to model and evaluate the tradeoffs between crop yield, nitrous oxide emissions, and nitrate leaching from land uses and best management practices	Mitigating the leaching of nitrate and nitrous oxide emissions from agricultural landscape
6. Quantifying annual daylighting for LEED certification	Design solutions to add optimal daylight to a new school building addition
7. Can tributary stations be used as proxies for the Mainstem: A case study in the Chesapeake Bay	Biogeochemical cycling in our pond/stream system on campus
8. A micro wearable device that increases accessibility of healthcare by extending the point-of-care	How can we use STEM to create a solution to health care?
9. Development of human breast phantom for portable ultrasound and photoacoustic imaging of breast cancer	Using biotechnology to detect probability of inherited cancer
10. Earthworm abundance as a soil health indicator in agro-ecosystems	How and why do organisms interact with their environment and what are the effects of these interactions?
11. Using modeling to manage organic residues and maximize soil carbon benefits	Sustainability and impact on agrosystems
12. Influence of STEM identity on computational design thinking	Understanding the design implications of minimizing building daylighting, energy use, and structural weight through dynamic building design in the context of computational design thinking

incorporate them into curriculum plans, which is consistent with other studies that highlight the importance of PD in enhancing the impact of research experiences on teaching practices (Herrington et al., 2012; Southerland et al., 2016).

One of the key insights from this study is that teachers' development of practices-based curriculum was closely tied to their own research experiences. This was evident in the way teachers incorporated SEPs into their Classroom Research Project (CLRP) plans, reflecting the specific practices they engaged in during their summer research projects. This finding supports the notion that authentic research experiences can help teachers translate complex scientific practices into classroom activities (Pop et al., 2010; Davidson et al., 2022); however, the practices are limited to those included in the teacher's technical research project.

Table 17.3 Translation of science and engineering practices into curriculum plans

CLRP Topic	Science and Engineering Practices[a] Present in CLRP Plans							
	1	2	3	4	5	6	7	8
1. Robotics automation with biological vital sign sensors			X					X
2. Light optimization in school's learning spaces to improve student's performance	X	X		X	X	X	X	X
3. Using sensors to improve energy efficiency and indoor air quality in our school	X		X	X	X	X	X	X
4. How can we use infrared imaging to develop a way to stay cool while keeping a bearded dragon in the classroom?	X	X	X	X		X	X	
5. Mitigating the leaching of nitrate and nitrous oxide emissions from agricultural landscape	X	X	X	X	X	X	X	X
6. Design solutions to add optimal daylight to a new school building addition	X	X	X	X	X	X	X	X
7. Biogeochemical cycling in our pond/stream system on campus	X	X		X	X	X	X	
8. How can we use STEM to create a solution to health care?	X		X	X		X		X
9. Using biotechnology to detect probability of inherited cancer	X	X	X	X			X	X
10. How and why do organisms interact with their environment and what are the effects of these interactions?		X	X	X		X	X	X
11. Sustainability and impact on agrosystems	X		X		X			X
12. Understanding the design implications of minimizing building daylighting, energy use, and structural weight through dynamic building design in the context of computational design thinking	X	X	X	X	X			X

[a]List of science and engineering practices is included in Fig. 17.2

The study also revealed a discrepancy between teachers' understanding of authenticity in classroom activities and their inclusion of SEPs in their curriculum plans. While teachers effectively integrated SEPs into their CLRPs, their characterization of authentic classroom activities during week 4 of the PD was primarily based on student agency rather than engagement in SEPs. This could be because the authenticity continuum did not specifically include the practices in the rating criteria, but the CLRP template specifically asked for teachers to identify the practices included in their curriculum plans. Also, one component of our definition of authenticity is student agency, which may have focused teachers' attention on this concept early in the continuum rating activity. While teachers recognize the importance of student-driven learning, they may need further support to fully integrate SEPs into their concept of authenticity (Burgin, 2020).

In conclusion, this study highlights the potential of RET programs to enhance teachers' understanding and incorporation of SEPs in classroom activities, thereby promoting more authentic science learning experiences for students. Teachers need more authentic experiences accompanied with associated pedagogical professional development to help them think about moving beyond contrived activities to provide more authentic experiences for students. The findings suggest that combining immersive research experiences with targeted PD can effectively bridge the gap between scientific research and classroom practice. Future efforts should focus on providing ongoing support for teachers to refine their understanding of authenticity and fully integrate SEPs into their teaching practices. More collaborations between STEM researchers and pedagogical experts are needed to bring authentic STEM learning experiences to precollege classrooms (Hill et al., 2017a; Hill et al., 2017b; Hill et al., 2018).

References

Antink-Meyer, A., & Meyer, D. Z. (2016). Science teachers' misconceptions in science and engineering distinctions: Reflections on modern research examples. *Journal of Science Teacher Education, 27*(6), 625–647. https://doi.org/10.1007/s10972-016-9478-z

Blanchard, M. R., Southerland, S. A., & Granger, E. M. (2008). No silver bullet for inquiry: Making sense of teacher change following an inquiry-based research experience for teachers. *Science Education, 93*(2), 322–360.

Blanchard, M. R., & Sampson, V. D. (2017). Fostering impactful research experiences for teachers (RETs). *EURASIA Journal of Mathematics, Science and Technology Education, 14*(1). https://doi.org/10.12973/ejmste/80352

Brown, J. S., Collins, A. M., & Duguid, P. (1989). Situated cognition and the culture of learning. *Educational Researcher, 18*(1), 32–42.

Burgin, S. R. (2020). A three-dimensional conceptualization of authentic inquiry-based practices: A reflective tool for science educators. *International Journal of Science Education, 42*(9), 1465–1484. https://doi.org/10.1080/09500693.2020.1766152

Chinn, C., & Malhotra, B. (2002). Epistemologically authentic inquiry in schools: A theoretical framework for evaluating inquiry tasks. *Science Education, 86*(2), 175–218. https://doi.org/10.1002/sce.10001

Crawford, B. A. (2012). Moving the essence of inquiry into the classroom: Engaging teachers and students in authentic science. In K. Tan & M. Kim (Eds.), *Issues and challenges in science education research* (pp. 25–42). Springer. https://doi.org/10.1007/978-94-007-3980-2_3

Creswell, J. W., & Creswell, J. D. (2018). *Research design: Qualitative, quantitative, and mixed methods approaches* (5th ed.). SAGE Publications.

Davidson, S. G., Jaber, L. Z., & Southerland, S. A. (2022). Cultivating science teachers' understandings of science as a discipline. *Science & Education, 31*(3), 657–683. https://doi.org/10.1007/s11191-021-00276-1

Enderle, P., Dentzau, M., Roseler, K., Southerland, S., Granger, E., Hughes, R., Golden, B., & Saka, Y. (2014). Examining the influence of RETs on science teacher beliefs and practice. *Science Education, 98*(6), 1077–1108. https://doi.org/10.1002/sce.21127

Feldman, A., Divoll, K. A., & Rogan-Klyve, A. (2013). Becoming researchers: The participation of undergraduate and graduate students in scientific research groups. *Science Education, 97*(2), 218–243.

Fortunato, S., Bergstrom, C. T., Börner, K., Evans, J. A., Helbing, D., Milojević, S., Petersen, A. M., Radicchi, F., Sinatra, R., Uzzi, B., Vespignani, A., Waltman, L., Wang, D., & Barabási, A.-L. (2018). Science of science. *Science, 359*(6379), eaao0185. https://doi.org/10.1126/science.aao0185

Herrington, D. G., Luxford, K., & Yezierski, E. J. (2012). Target inquiry: Helping teachers use a research experience to transform their teaching practices. *Journal of Chemical Education, 89*(4), 442–448. https://doi.org/10.1021/ed1006458

Hill, K. M., Johnson, M., Smith, A., & Ward, A. (2017a). *Challenges of teachers developing and implementing practices-based student investigations*. National Association for Research in Science Teaching.

Hill, K. M., Johnson, M., Smith, A., & Ward, A. (2017b). *Supporting teachers to translate research experiences into practices-based instruction*. European Science Education Research Association.

Hill, K. M., Smith, A., Johnson, M., & Ward, A. (2018). *Using the MASTER model framework to support teachers in incorporating research practices in the secondary classroom*. Association of Science Teacher Educators.

Johnson, M. M., & Hill, K. M. (2023). Revolutionizing STEM education with broader impacts. In M. M. Johnson & S. B. Vassmer (Eds.), *General proceedings of the ARIS summit 2023*. Baltimore.

Kelly, G. J., & Licona, P. (2018). Epistemic practices and science education. In *History, philosophy and science teaching: New perspectives* (pp. 139–165). Springer.

Krim, J. S., Coté, L. E., Schwartz, R. S., Stone, E. M., Cleeves, J. J., Barry, K. J., et al. (2019). Models and impacts of science research experiences: A review of the literature of CUREs, UREs, and TREs. CBE—Life sciences. *Education, 18, ar65*, 1–14.

Miranda, R. J., & Damico, J. B. (2015). Changes in teachers' beliefs and classroom practices concerning inquiry-based instruction following a year-long RET-PLC program. *Science Educator, 24*(1), 23–35.

National Research Council. (1996). National science education standards. *National Academies Press*. https://doi.org/10.17226/4962

National Research Council (NRC). (2012). *A framework for K-12 science education: Practices, crosscutting concepts, and core ideas*. National Academies Press.

National Research Council (NRC). (2013). Next generation science standards: For states, by states. *National Academies Press*. https://doi.org/10.17226/18290

NSTA. (2024). *K-12 science standards adoption across the U.S [infographic]*. National Science Teachers Association. https://www.nsta.org/science-standards

Patchen, T., & Smithenry, D. W. (2014). Diversifying instruction and shifting authority: A cultural historical activity theory (CHAT) analysis of classroom participant structures. *Journal of Research in Science Teaching, 51*(5), 606–634. https://doi.org/10.1002/tea.21140

Pop, M. M., Dixon, P., & Grove, C. M. (2010). Research experiences for teachers (RET): Motivation, expectations, and changes to teaching practices due to professional program involvement. *Journal of Science Teacher Education, 21*(2), 127–147. https://doi.org/10.1007/s10972-009-9167-2

Roberts, D. (2010). *miniSADbot025* [Photograph]. Openverse. https://www.flickr.com/photos/50516268@N03/4813681809

Sadler, T. D., Burgin, S., McKinney, L., & Ponjuan, L. (2010). Learning science through research apprenticeships: A critical review of the literature. *Journal of Research in Science Teaching, 47*(3), 235–256. https://doi.org/10.1002/tea.20326

Saldaña, J. (2021). *The coding manual for qualitative researchers* (4th ed.). SAGE Publications.

Southerland, S. A., Granger, E. M., Hughes, R., Enderle, P., Ke, F., Roseler, K., Saka, Y., & Tekkumru-Kisa, M. (2016). Essential aspects of science teacher professional development: Making research participation instructionally effective. *AERA Open, 2*(4), 233285841667420. https://doi.org/10.1177/2332858416674200

U.S. Coast Guard Academy. (2013). *Swab summer cadets participate in the transport effectiveness exercise in McAllister hall [photograph]*. Openverse. https://www.flickr.com/photos/43228104@N07/9368346500

Windschitl, M. (2004). Folk theories of "inquiry:" how preservice teachers reproduce the discourse and practices of an atheoretical scientific method. *Journal of Research in Science Teaching, 41*(5), 481–512.

Wong, S. L., & Hodson, D. (2010). More from the horse's mouth: What scientists say about science as a social practice. *International Journal of Science Education, 31*(11), 1431–1463. https://doi.org/10.1080/09500690903104465

Kathleen M. Hill is the Director of the Center for Science and the Schools (CSATS) and a professor at the Pennsylvania State University. Her primary research interests include science teacher education with an emphasis on translating science and engineering practices into precollege learning experiences. She holds multiple grants to design and implement teacher professional development programs that bridge cutting-edge science and engineering research and K-12 education.

Tiffany Lewis is STEM Education Specialist and lead of CSATS Summer RET program at the Pennsylvania State University. Her primary research interests include science teacher education, focusing on the impact of immersive STEM research experiences on teachers' conceptions of authentic science pedagogy.

Matthew M. Johnson is Professor of Practice for Engagement and Outreach with a partial appointment in the Center for Science and the Schools at the Pennsylvania State University. His research and teaching interests focus on supporting elementary teachers in rural schools to leverage local funds of knowledge to engage their students in engineering practices.

Chapter 18
Representation of the Nature of Science in Exemplary Inquiry Teaching

Shani Zur and Tali Tal

18.1 Introduction

In recent decades, there has been a notable increase in public engagement with science-related issues, which often carry personal, economic, or public implications. Consequently, science education is pivotal in equipping students with the skills to navigate scientific discourse to make informed decisions (Feinstein & Waddington, 2020; Osborne, 2023). This emphasis on fostering scientific literacy is underscored by Roberts and Bybee's Vision II of scientific literacy, which emphasizes that scientific literacy competencies should encompass an understanding of scientific procedures and an awareness of the broader social and human contexts in which science operates to make informed decisions (Roberts & Bybee, 2014). Therefore, educators should cultivate students' awareness and engagement and discuss the epistemic aspects of scientific work to make informed decisions and take action on science-related issues of social, political, or environmental significance (Chinn et al., 2021; Lederman & Lederman, 2020). Thus, teachers should not only teach scientific content and promote procedural understanding but rather support students' epistemic vigilance concerning the scientific process and the scientific community as a community of practice (Chinn & Duncan, 2021; Osborne, 2023).

The questions to be asked are: How could educators support students in gaining content and procedural knowledge as well as expanding their epistemic vigilance?

S. Zur (✉)
Education in Science and Technology, Technion Israel Institute of Technology, Haifa, Israel
e-mail: Shani.zur@technion.ac.il

T. Tal
Education in Science and Technology, Technion Israel Institute of Technology, Haifa, Israel

Samuel Neaman Institute, Haifa, Israel
e-mail: rtal@technion.ac.il

G. Cakmakci, M. F. Tasar (eds.), *Connecting Science Education with Cultural Heritage*, Contributions from Science Education Research 15,
https://doi.org/10.1007/978-3-031-90944-3_18

One approach for cultivating students' understanding of how scientists carry out scientific research and achieve consensus is through inquiry-based teaching, a well-established pedagogical approach through which teachers can offer students opportunities to participate and reflect on the scientific endeavor. In inquiry learning, students are asked to ask questions, investigate, explain, and reason (Duschl et al., 2007), all of which are grounded in scientists' thought processes and activities that yield scientific ideas and knowledge (Erduran & Dagher, 2014). Productive inquiry teaching should promote both procedural and epistemic understanding of the scientific enterprise (Duncan et al., 2021); it should foster students' understanding of how scientists carry out investigations and their work as a community that holds shared values and norms. Encourage students to think critically, judge evidence and sources of information, and discuss and reflect on the measures scientists take to pursue accuracy and reliability (OECD, 2023). Encouraging students to reflect on the intricacy of the Nature of Science (NOS), which encompasses cognitive-epistemic and social institutional dimensions, could support their competence in communicating science-related issues and increase their trust in science (Erduran & Dagher, 2014; Dagher & Erduran, 2023).

Inquiry learning is embedded within the curriculum across various science disciplines. However, it often emphasizes how knowledge is constructed while focusing less on the ontological characteristics of scientific knowledge or the Nature of Scientific Knowledge (Lederman & Lederman, 2020). Evidence indicates that inquiry teaching in schools is frequently used to provide students with a procedural understanding and lacks a substantiated view of the epistemic aspects of science and an understanding of scientists' work as part of the scientific community (Osborne, 2014; Schwartz et al., 2023). There seem to be two main challenges in inquiry enactment: First, it often presents a teacher-oriented approach that primarily focuses on examining the relationships between variables in a laboratory setting. Employing highly controlled inquiry projects that primarily emphasize procedural-technical aspects hinders students from engaging in epistemic thinking and fails to facilitate a comprehensive understanding of NOS (Duschl & Grandy, 2013). Secondly, teachers hold limited views about inquiry and NOS instruction (Capps & Crawford, 2013). They also mentioned challenges in designing and enacting inquiry and encountered difficulties related to the teachers' personal and working contexts (Zur & Tal, 2021).

Despite the broad literature on the implementation of inquiry-based teaching and NOS instruction, we do not know enough about how teachers could best utilize inquiry teaching to enhance their students' reflection about the NOS and about scientific inquiry as a collective endeavor of the scientific community (Dagher & Erduran, 2023; Schwartz et al., 2023). Drawing back to Roberts and Bybee's (2014) Vision II, we need to further understand how inquiry could support students in adopting a more sophisticated view that would help them discuss and make informed decisions and actions.

To narrow this gap, we intended to learn from teachers who perform excellent inquiry teaching. Learning from exemplary cases is a well-known approach that could yield valuable insights from individuals who faced and overcame challenges and obstacles to reach their desirable goals (Tal et al., 2006). For example, a case study on exemplary teaching of argumentation examined teachers' views and knowledge about argumentation, as well as their perceptions, constraints, and successes in implementing it. By articulating the teachers' vision, the authors highlight key aspects that can support sustainable educational reform (Erduran & Dagher, 2007).

18.1.1 Exemplary Inquiry Teaching

Excellent inquiry teaching could vary across disciplines and teaching approaches. Nevertheless, as McComas et al. (2020) indicated, "...in an arena as varied as the school classroom in a domain as complex as science teaching, there cannot be one best practice. However, if we allow that there are small groups of procedures and orientations that are more effective than others in science practice and pedagogy, we can feel quite justified in calling such practices exemplary."

We wish to outline some aspects of design and implementation that we considered EIT. The idea of EIT is based on the literature about productive inquiry, and especially adopted and adjusted Duncan, Av Shalom, and Chinn et al.'s (2021) principles for effective inquiry learning from teachers' perspective:

1. Enhance student engagement in multiple scientific practices and utilize the experiences to negotiate the complex interactions of successful inquiry.
2. Support and scaffold epistemic practices and guide students to consider the aims and values of these practices. Such as constructing complete arguments and critically evaluating them.
3. Give students the autonomy to inquire (epistemic agency); invite students to raise questions that hold personal, social, or environmental value, participate in planning the investigation; and engage them in scientific practices.
4. Discuss and reflect on scientific endeavors to support students' meta-cognitive understanding of the practices that scientists pursue to collect evidence and reasoning.
5. Foster social communities of inquiry; appreciate the scientific process scientists undertake to be persistent and critically evaluate their work as they aim for trustworthiness.

The EIT outline clarifies that teachers should support their students to be familiar with the NOS. They should discuss and reflect on how scientific knowledge is constructed and verified, considering the intricacy of the scientific endeavor which\ encompasses cognitive, social, financial, cultural, and political aspects.

Furthermore, based on our assumption that NOS is important to foster students' epistemic understanding of science and inform them in communicating science-related issues, we sought to learn how NOS is presented and discussed by teachers who carry out EIT. Specifically, we ask in what ways and contexts teachers who implemented EIT present and discuss the NOS?

18.2 Method

We followed a phenomenological approach, carrying out a thematic analysis to identify which NOS aspects were addressed in the interviews and whether the teachers intentionally reflected the NOS. We also adopted and followed Erduran and Dagher's (2014) expanded version of the Family Resemblance approach (FRA), which offers a coherent and holistic framework for the NOS in the context of science education. To examine the aspects of NOS that are reflected by the teachers whom we identified as "exemplary", we focused on:

1. The aspects of NOS that emerge in the EIT.
2. The contexts in Which the EIT teachers discussed the NOS, and Whether the discussion of the NOS was intentional

18.2.1 Participants

To select participant-teachers who carry out EIT, we sought recommendations from science education leaders such as chief superintendents from the Ministry of Education, researchers familiar with these teachers and their practices, and teacher educators. We approached only teachers for whom we received two different recommendations indicating that their enactment could be considered EIT. Altogether, we received recommendations for 15 exceptional teachers known for incorporating excellent inquiry-based learning. Subsequently, we conducted brief conversations with each of these teachers to learn about their implementation of inquiry-based teaching. After careful consideration, we selected 12 teachers whose practice appeared to align with the criteria for EIT. Most participants were teaching as a second career. Recognizing that teachers' working context, including the school's curriculum and student backgrounds, can influence instructional practices, we deliberately selected teachers working in public schools that serve students from average to low-average socioeconomic backgrounds. Participants' demographics are detailed in Table 18.1.

18.2.2 Data Collection and Analysis

We conducted in-depth interviews with each of the 12 teachers. In the interviews, we asked teachers to describe, in detail, how their inquiry projects were carried out and to point out their initial goals. For example, we asked why you designed and taught using scientific inquiry and what inquiry-based learning is in your view. All interviews were audio-recorded and transcribed, and then transcripts were divided into utterances. We also collected relevant documents such as teachers' inquiry teaching plans, instructions for students, and their artifacts used for triangulation

Table 18.1 Participant demographics

Name	Degree	Background prior to teaching	Teaching subject
Ori	Ph.D.	Academia and industry	Biology
Dina	M.Sc.	Academia	Biology
Tzipi	M.Sc.	Academia and industry	Middle school science
Nimrod	M.Sc.	Academia and industry	Chemistry
Tom	M.Sc.	Academia	Physics
Amit	M.Sc.	Academia and industry	Physics
Nili	M.Sc.	Academia	Biology
Roth	B. A	–	Middle school science
Riva	M.Sc.	Industry	Chemistry
Jordan	M.Sc.	–	Biology
Genny	M.Sc.	Academia	Biology
Ollie	M.Sc.	–	Biology

teachers' declared statements. We examined which and how the following FRA categories were reflected in teachers' descriptions:

- Cognitive-Epistemic aspects: aims and values (AV), scientific practices (SP), methods (M), and scientific knowledge (SK).
- Social-Institutional aspects: professional activities (PA), scientific ethos (SE), social certification and dissemination (SCD), social values of science (SV), social organizations and Interactions (SOI), political power and structures (PPS), and financial systems (FS).

Every utterance highlighted in one of the FRA categories was marked 1. Utterances that did not explicitly refer to any FRA category were marked 0. In each case, we examined the distribution of the categories within either the epistemic-cognitive or the social-institutional domains and summed the number of NOS-related statements.

Then, we reviewed the data corpus to identify how teachers presented a considerable reference to NOS in the EIT cases. By considerable, we mean that both cognitive-epistemic and social-institutional aspects were reflected and that at least six categories of the FRA were represented overall.

Finally, we reviewed all teachers' statements again to identify the context in which they addressed the NOS and whether these references were intentional. We considered an intentional, clear statement made by the teacher that mentioned the NOS as pivotal to the inquiry learning process.

18.3 Findings

Nine out of the 12 teachers we selected considerably addressed NOS aspects referring to cognitive-epistemic and social-institutional dimensions (see Fig. 18.1). Interestingly, they did not address political powers or economic factors in the context of science.

The remaining three teachers indicated less than six categories of the FRA and, thus, did not reflect the broad spectrum of the NOS. Although these teachers expanded the concept of inquiry and focused on various scientific practices, they neither drew clear connections nor discussed how the student learning experiences reflect the scientific process or refer to how scientists work.

Overall, in nine cases, 223 statements were identified as congruent with the FRA categories (As detailed in the method section). Notably, the cognitive-epistemic aspects were given more consideration as they were mentioned twice as much as the Social-institutional aspects (66% and 34%, respectively). The most common category mentioned by the teachers was scientific practices (SP), which appeared in 26% of all NOS-related statements. The least discussed categories were in the social-institutional domain—professional activities (PA), social values (SV), and social certification and dissemination (SCD), which received only 4% out of all NOS statements each.

Additionally, we examined the context in which teachers emphasized the NOS and how the NOS was reflected. We found that teachers discussed the NOS as they described:

- Their teaching goals and objectives and how they planned inquiry teaching.
- Their encounters with students during inquiry teaching and student-teacher dialogues.

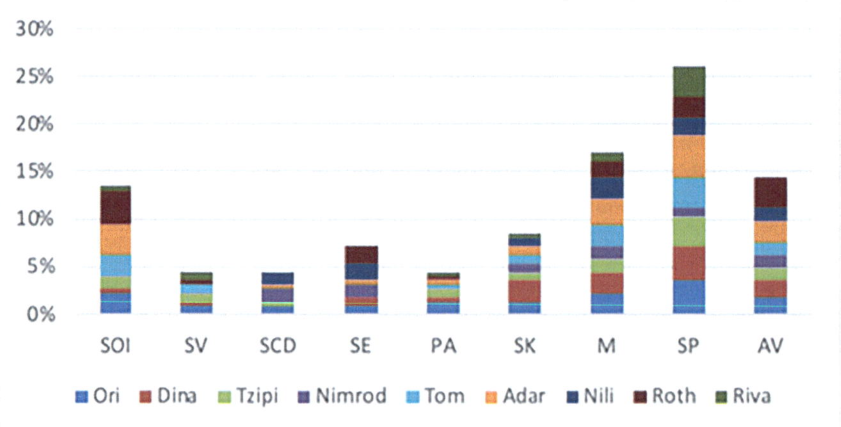

Fig. 18.1 Percentage of the represented FRA categories as represented in EIT

Although our findings represent all nine teachers, we provide a detailed account of Ori's case to illustrate the aspects we identified. Ori is a biology teacher who chose to teach as a second career after gaining a Ph.D. in science, completing post-doctorate research, and working for several years in various research laboratories. He created an inquiry-based biology curriculum that the Ministry of Education acknowledged as an alternative to the high school biology requirements. Ori's "Investigate Biology" program required the students to experience various scientific practices and then conduct their own research.

18.3.1 NOS Representation in Teachers' Objectives and Planned Instruction

As teachers discussed why they designed and implemented inquiry-based teaching in various ways, they mentioned two primary objectives:

1. Engage students in inquiry and support their understanding of scientific processes. Teachers wished to educate their students about the scientific enterprise and familiarize them with the scientific process and the work of scientists. They emphasized the cognitive-epistemic domain, such as utilizing methods and repeating experiments, so students will better understand how scientists pursue a rigorous investigation.
2. Utilize inquiry teaching to address personal and/or social relevance issues to enhance students' competency to inquire into real-world issues or make informed decisions. They planned to provide students with opportunities to practice scientific thinking and engage them in practices that would support them in acquiring the skills to inquire about everyday problems and attributes to become thoughtful individuals. The teachers discussed their personal experience in inquiry as a meaningful aspect that helped shape who they are as individuals. They also mentioned aiming to provide their students with similar experiences. Their goal was to use inquiry as a catalyst to reflect on scientists' systematic thinking.

For instance, Ori discussed his inquiry learning experience as a high school student, which inspired him to teach "thought inquiry" and commit to writing a whole biology curriculum focusing on students' inquiry. When he was asked what teaching thought inquiry meant to him, he responded:

> Inquiry is a thinking process, and every aspect of life is a process of exploration… inquiry is a form of systematic thinking, not entirely open-ended thinking. What I constantly try to convince my students is that we investigate in life all the time. When the light in my room does not turn on, I start to ask what happened; I ask questions and hypothesize whether it is only a light bulb or a broader problem; I then investigate. It is a thinking process. I also wish to educate my students about scientific critique. That is always to ask questions, try to obtain answers, and make reasoned conclusions based on evidence. Scientific critique never crosses the threshold as a given, nor is it an acceptance of the obvious.

> I conducted inquiry activity [to enhance their] scientific critique about the efficacy of the Pfizer vaccine and whether it should be encouraged. We continuously examined sources [of information] and, above all, consistently posed questions. We asked all sorts of questions, and the students eventually had to inquire and then formulate and present a scientific poster.

Ori views scientific inquiry as a thinking process. His primary goal is to adopt scientific thought to enhance student thinking about everyday problems. By utilizing scientific practices, he aims to encourage students to "think like scientists," but he does not talk about educating students about scientists' work. For instance, he indicates that students should present a poster but does not mention the value of students reflecting on how scientists communicate information. Nevertheless, Ori purposefully aims to discuss the NOS ideas that stem from personal or social contexts relevant to students' lives, such as vaccination. He uses scientific inquiry to enhance his students' epistemic awareness of the scientific process and their decision-making process.

In addition to looking at NOS in teachers' goals, we found that teachers incorporated the NOS in their planned inquiry instruction. By planned instruction, we refer to teachers indicating how they intend to guide their students, such as by discussing relevant information, supporting students in acquiring necessary skills or practices, and creating opportunities to discuss and reflect.

According to the above statement, for example, Ori planned for his students to engage in scientific practices and reflect on the scientific endeavor. By indicating that "We have continuously examined sources [of information]," he emphasized his intention to collaborate and support his students while investigating.

As teachers unfolded their inquiry plans and detailed how the inquiry teaching was implemented, they mentioned several NOS aspects, primarily scientific practices (SP), utilizing various methods (M), and communicating the inquiry process to reflect scientific community professional activities (PA). They planned to ask their students to communicate the inquiry process and their discoveries in conference-like events where students were asked to prepare and present scientific posters and presentations and critically discuss scientific information with peers. The NOS aspects were reflected in teachers' statements when they emphasized the verbal guidelines they provide to their students, as well as in the written assignments.

Ori has planned a long-term inquiry process. He planned to allow students to choose what to investigate and provide them with opportunities to experience and master scientific practices and reflect on the process. For example, he formulated worksheets that asked his students to read and critically evaluate scientific articles that expanded their knowledge about the phenomena they chose to study. The instructions Ori formulated were divided into assignments; in each assignment, he emphasized what students were required to do and the aims and value of participating in the learning process. Here is an example of the second task students were given:

> Work in pairs, discuss, and write four questions that came to mind considering the article you read. The questions could include ideas or concepts you are unfamiliar with, or subjects and ideas based on the question you are interested in investigating more about. In this

> assignment, you will practice critical scientific thinking, and you will also develop the ability to ask questions; both are key aspects of scientists' work (Ori's written instructions)

In the tasks Ori prepared, he expected to engage his students in scientific practices and in communicating with peers. He explicitly drew students' attention to the fact that these practices reflect scientists' work. Thus, his planned instruction guides students to exercise procedural processes and draws their attention to the fact that the experience reflects how scientists think and work.

18.3.2 NOS Representation in Students and Teacher-Student Dialogues

The NOS was also indicated as teachers discussed situations that occurred while students were investigating. The encounters were usually given as examples of teacher-student dialogue or student-student interactions. Overall, teachers addressed the NOS while supporting their students' investigations or during discussions that resonated with the scientific endeavor. Evidently, teachers aim to model scholarly behavior or support their students' learning while addressing the expected values and norms. Nevertheless, although the teachers referred to the NOS, it was implicit. They did not directly point to the broader context beyond the specific example.

Across cases, it was evident that teachers have changed the traditional roles or power relations between the teacher and their students, shifting away from positioning themselves as the main source of knowledge and offering students scaffolding to study as active and independent learners. Teachers act as in-formed learners alongside their students while giving them the autonomy to lead the inquiry and discuss and interact, similarly to how scientists communicate. For example, Ori described a dialogue with two of his students who approached him at the end of the class. He mentioned the students being frustrated that they had not finished their assignments.

> A pair of students approached me at the end of class, both with tears in their eyes. [I asked] what happened? [They said] for the whole class, we only argued about one question, so we did not finish the rest of the task. I told them 'wonderful'! They were shocked. Just shocked. They thought that accomplishing the list of tasks was the most important thing. I told them [you did] fantastic work. Do you know what I saw the whole class? I sat and looked at you, and I was just thrilled; I was so excited. This is the greatest joy for a teacher [to see his students work]. The whole time you were negotiating, each brought her own arguments, and you tried to contradict the other's [claim] by turning to different sources of information. This exercise is ten times more important than completing all the tasks. We are not here for quantity. We do not have an exam at the end; we are in pursuit of satisfying our curiosity and the urge to learn, which is a great privilege. (Ori, interview)

In Ori's description of his dialogue with his students, he emphasized the importance of argumentation and reasoning: "They were constructing arguments and critically evaluating them". This discursive practice is part of the scientific practice that scientists exercise. Ori's applause to his students could indicate the importance he attributes to norms represented in the scientific ethos. Nevertheless, Ori mentions that the reason for facilitating these practices is to "satisfy our curiosity and the urge to learn." He does not deliberately draw the students' attention to how these practices are used in the scientific community and to their value in scientific investigations.

18.4 Discussion and Conclusions

We examined exemplary cases of inquiry teaching to understand how the NOS is exemplified in best-practice inquiry teaching. We identified nine teaching cases that aligned with the criteria of EIT and presented considerable indications for NOS.

In the nine cases we identified as EIT, teachers seemed to pay considerable attention to the idea of the NOS as they designed and guided the inquiry. This indication aligns with scholars who argue that inquiry teaching could be a valuable approach to encouraging students to consider the epistemological aspects of science (Duncan et al., 2021) and promoting their understanding of the NOS (Duschl & Grandy, 2013).

Nevertheless, we found that teachers primarily centered on cognitive-epistemic aspects while the social-institutional aspects received less emphasis in teacher descriptions. The different considerations of the cognitive vs. social aspects may stem from the place these aspects have in curriculum requirements and inquiry learning—reform (e.g., NRC, 2000). Inquiry teaching addresses the procedural and epistemic aspects of how scientific knowledge is constructed and a lesser emphasis on the science community as a milieu that holds social and cultural norms and values (Osborne, 2023; Lederman & Lederman, 2020).

Interestingly, even though teachers highlighted that one of their primary objectives was to educate their students to make informed decisions, which aligns with Roberts and Bybee's (2014) vision II of scientific literacy, focused more on the process that aimed to construct scientific knowledge rather than the social aspects of science that navigate how science is discussed and critically evaluated within the scientific community.

There is broad agreement about the importance of reflecting and discussing science in a social context to enhance students' epistemic vigilance and support them in engaging in informed science-related discussions (Duncan et al., 2021; Feinstein & Waddington, 2020). Thus, our findings highlight that a stronger emphasis should be placed on reflecting science as a community, especially on the norms and values that scientists pursue to achieve accuracy and reliability.

To enhance deep reflection on the NOS in inquiry teaching, it is imperative to acknowledge that teachers' pedagogical approaches are rooted in their individual perspectives and knowledge of NOS (Dagher & Erduran, 2023). Hence, it is essential to provide teachers with opportunities to discuss and reflect on scientific issues in social contexts and the epistemic process as they are reflected in these issues as an inherent part of their professional development (Zion et al., 2020). We suggest involving teachers in meaningful experiences and discussions and encouraging reflection on the pertinent aspects of this knowledge in science education, which could encourage them to reflect on these dimensions in their inquiry-based teaching.

Additionally, we found that teachers were intentional about reflecting on scientific work in their initial aims and planned instruction. Nevertheless, while teachers described their dialogues with their students, they did not seem to address some NOS ideas in a broader context. For example, they discussed how utilizing scientific practices and methods drew students' attention to the fact that different methods could enhance the reliability of the results. The contribution of incorporating explicit NOS instruction to promote a profound understanding of NOS has been previously discussed (Duschl & Grandy, 2013; McComas et al., 2020). Our findings further emphasize that teachers should be better aware of reflecting the idea of the NOS to guide their students' inquiries.

This study holds several limitations. Methodologically, our analysis of teachers who implemented EIT was based on their lived experiences, relying on teachers' descriptions and written documents. Consequently, we did not study how teachers' instruction, student interactions, and dialogues were carried out. Additional research is needed to address whether Co-learning strategies could foster inquiry teaching, including intervention and assessment. We recommend further research to include in-class observations to further indicate how the NOS manifests as teachers guide their students in EIT.

References

Capps, D. K., & Crawford, B. A. (2013). Inquiry-based instruction and teaching about nature of science: Are they happening? *Journal of Science Teacher Education, 24*, 497–526. https://doi.org/10.1007/s10972-012-9314-z

Chinn, C. A., Barzilai, S., & Duncan, R. G. (2021). Education for a "post-truth" world: New directions for research and practice. *Educational Researcher, 50*(1), 51–60. https://doi.org/10.3102/0013189X209406

Chinn, C., & Duncan, R. G. (2021). Inquiry and learning. In R. G. Duncan & C. A. Chinn (Eds.), *International handbook of inquiry and learning* (pp. 1–14). Routledge.

Dagher, Z. R., & Erduran, S. (2023). To FRA or not to FRA: What is the question for science education? *Science & Education, 32*(5), 1247–1264. https://doi.org/10.1007/s11191-023-00425-8

Duncan, R. G., Av-Shalom, N., & Chinn, C. (2021). Inquiry and learning in science. In R. G. Duncan & C. A. Chinn (Eds.), *International handbook of inquiry and learning* (pp. 327–344). Routledge.

Duschl, R., & Grandy, R. (2013). Two views about explicitly teaching nature of science. *Science and Education, 22*(9), 2109–2139. https://doi.org/10.1007/s11191-012-9539-4

Duschl, R., Schweingruber, H., & Shouse, A. (2007). Taking science to school: Learning and teaching science in grades K-8. In R. A. Duschl, H. A. Schweingruber, & A. W. Shouse (Eds.), *Taking science to school*. National Academies Press.

Erduran, S., & Dagher, Z. R. (2007). Exemplary teaching of argumentation: A case study of two science teachers. In *Contributions from science education research* (pp. 403–415). Springer. https://doi.org/10.1007/978-1-4020-5032-9_31

Erduran, S., & Dagher, Z. R. (2014). *Reconceptualizing the nature of science for science education* (Vol. 43). Springer Netherlands.

Feinstein, N. W., & Waddington, D. I. (2020). Individual truth judgments or purposeful, collective sensemaking? Rethinking science education's response to the post-truth era. *Educational Psychologist, 55*(3), 155–166. https://doi.org/10.1080/00461520.2020.1780130

Lederman, N. G., & Lederman, J. (2020). Nature of scientific knowledge and scientific inquiry. In V. L. Akerson & G. A. Buck (Eds.), *Critical questions in STEM education* (pp. 3–20). Springer International Publishing.

McComas, W. F., Clough, M. P., & Nouri, N. (2020). Nature of science and classroom practice: A review of the literature with implications for effective NOS instruction. In W. F. McComas (Ed.), *Nature of science in science instruction, science: Philosophy, history and education* (pp. 67–111). Springer. https://doi.org/10.1007/978-3-030-57239-6_4

Organization for Economic Cooperation and Development (OECD). (2023). *PISA 2025 Science Framework*. Oxford University Press.

Osborne, J. (2014). Teaching scientific practices: Meeting the challenge of change. *Journal of Science Teacher Education, 25*(2), 177–196. https://doi.org/10.1007/s10972-014-9384-1

Osborne, J. (2023). Science, scientific literacy, and science education. In *Handbook of research on science education: Volume III* (Vol. 3, pp. 785–816). https://doi.org/10.4324/9780367855758-30

National Research Council (NRC). (2000). *Inquiry and the national science education standards: A guide for teaching and learning*. Center for Science, Mathematics, and Engineering Education/National Academy Press.

Roberts, D., & Bybee, R. (2014). Scientific literacy, science literacy, and science education. In *Research in science education* (pp. 545–558). Routledge.

Schwartz, R. S., Lederman, J. S., & Enderle, P. J. (2023). Scientific inquiry literacy. In *Handbook of research on science education* (Vol. 3, pp. 749–782). Routledge. https://doi.org/10.4324/9780367855758-28

Tal, T., Krajcik, J. S., & Blumenfeld, P. C. (2006). Urban schools teachers enacting project-based science. *Journal of Research in Science Teaching, 43*(7), 722–745. https://doi.org/10.1002/tea.20102

Zion, M., Schwartz, R. S., Rimerman-Shmueli, E., & Adler, I. (2020). Supporting teachers' understanding of nature of science and inquiry through personal experience and perception of inquiry as a dynamic process. *Research in Science Education, 50*(4), 1281–1304. https://doi.org/10.1007/s11165-018-9732-9

Zur, S., & Tal, T. (2021). "We know our students": Teacher views of inquiry-based learning implementation in light of their working context. *Proceeding of Singapore National Academy of Science, 15*(2), 91–104. https://doi.org/10.1142/S2591722621400093

Shani Zur holds a PhD in Education from the Technion – Israel Institute of Technology. Her main research interests include science education and inquiry-based teaching, focusing on the teacher's identity and teacher understanding of the Nature of science. Currently, she is a leading curriculum developer and a teacher mentor at the National Teacher Center of Environmental Education and Sustainability and at the Taking Citizen Science to School Center.

Tali Tal, the Ruben and Tallu Academic Chair of Science is a professor of science and environmental education at the Faculty of Education in Science and Technology, Technion – Israel Institute of Technology, and a Senior Research Fellow at the Samuel Neaman Institute for National Policy Research. She served as president of NARST (2019–2022) and as the Faculty Dean (2020–2024). Currently, she chairs the Technion's Pre-University Education Center. Her research focuses on out-of-school science education, inquiry-based learning, and environmental education.

Chapter 19
Collaborative Network Literacy in Science Education Leadership: A Case Study of the Danish Academy of Science Education (NAFA)

Tina Maria Brinks and Gitte Miller Balslev

19.1 Introduction

In this chapter, we expand upon our initial presentation at ESERA 2023, which focused on how to facilitate networked science education leadership to promote an open and sustainable educational culture. The original discussion was based on a Danish case study of a science program in lower secondary education. Here we proposed *distributed leadership* as a particularly effective approach to foster interdisciplinary collaboration in science education. While the original case of *Tåsingeskolen* (the school of Tåsinge) addressed school leadership within a specific science education program at a single school, this chapter focus on leadership in science education faculty. Although mostly theoretical, the discussion analyses these two cases based on case descriptions created for the purpose of this chapter. The two cases, one about (formal) leadership and the second about professional learning communities of science teacher educators in the Danish Academy of Science Education (NAFA) (2020). We discuss the two cases from a theoretical perspective of how interdisciplinary perspectives can help science education leadership facilitate complex and dynamic challenges in science education.

Despite lifting our discussion into the specific context of a far-reaching intervention in science education teaching faculty in Denmark, our underlying perspective remains consistent and unchanged: modern science education leadership is highly dynamic and must adapt to address complex challenges. Facilitating relevant and necessary changes over time requires a deep understanding of these dynamics. We understand *science education leadership* as more than formalized leadership

T. M. Brinks (✉) · G. M. Balslev
STEM Education Research Center (FNUG), Department of Mathematics and Computer Science, University of Southern Denmark (SDU), Odense, Denmark
e-mail: tinabrinks@imada.sdu.dk; gm@sdu.dk

G. Cakmakci, M. F. Tasar (eds.), *Connecting Science Education with Cultural Heritage*, Contributions from Science Education Research 15,
https://doi.org/10.1007/978-3-031-90944-3_19

positions, but as leadership, which can both be the types of roles that are formalized, e.g. formal leadership of an educational institution, but which can also be enacted in collaborative structures within education (self-leadership, autonomy, etc.), e.g. groups of teachers with distributed leadership. Rather than attempting to map NAFA as an organization or as a specific intervention, we use the NAFA case as a point of departure for discussing collaboration in science education practices.

We offer a critical perspective on educational leadership in science education, advocating for greater emphasis on the collaborative processes in which science teaching is embedded and developed. This type of leadership is multifaceted, contextually situated, and grounded in an awareness of professional diversity, collaboration, and trust. Such characteristics contribute to creating a sustainable school culture where collaboration is essential for designing engaging education with strong connections to real-world applications (Díaz-Gibson et al., 2014). From distributed leadership, we proposition *networked leadership* as a complementary framework, characterized by leaders who successfully facilitate interdisciplinary collaboration in education. A type of leadership, which is multifaceted, contextually situated, and grounded in an awareness of professional diversity, collaboration, and trust. Consequently, leadership transcends traditional top-down or bottom-up paradigms. Instead, it reflects a "bottom-linked" approach (context-responsive and aware) (Balslev, 2012) which is neither strictly centralized nor decentralized but is instead highly contextualized. From a pragmatic and interdisciplinary perspective, we explore how the concepts of distributed leadership and networked leadership can contribute to fostering sustainable institutional collaborative cultures.

To support this argument, we begin by mapping the theoretical landscape of the two NAFA cases, introducing key theories from organizational learning about leadership. These include the concepts of distributed leadership, networks as organizational constructs, and professional learning communities as frameworks for collaboration.

19.2 Science Education and its Complex Nature: and What Is Science Education Ledership Anyway?

Although important and inevitable as an underlying premise for the development of teaching practices (i.e. *content knowledge*), nature of science (NoS as in Lederman, 2023) is not in itself sufficient to comprehend aspects such as collaboration and leadership. As Treagust and Won (2023) points to, science education is much more complex, as it branches off into other traditional research and practice fields, thus becoming a multifaceted field in its own right. Science teacher *education* and the teacher educators who teach future science teachers navigate the overlapping spheres of their own teaching practice as well as teaching future (pre-service) science teachers to develop their own teaching practices, as in *PCK* (Dolin et al., 2022; van Driel et al., 2023). Science teacher educators play an important role in the communication, while navigating the difficult task of teaching and communicating complex and dynamics realities (i.e. climate change). At the same time studies have shown that academic achievement, such as good grades, does not necessarily correlate with student motivation to learn science (Osborne & Dillon, 2008).

Science *leadership* as a concept, which conceptually primarily draws upon established research fields in organizational learning and education research about, we find, lacks a clear definition of how it differentiates itself from educational leadership in general. This chapter initiates this discussion through the two cases.

19.3 Educational Leadership and Collaborative Networks

Science education is struggling with a cultural educational heritage grounded in a silo-oriented system that inhibits school-community collaboration (Brinks & Balslev, 2023; Díaz-Gibson et al., 2014). Potentially, this heritage narrows learning opportunities for students as well as for professionals. According to Reinholz and Apkarian (2018) institutional leadership is essential to promote systemic change in STEM-departments. In this sense, an important task for institutional leadership is equally to address the institution's culture and its ability to widen the scope of educational collaboration, by engaging with the local community.

19.3.1 Distributed and Humanistic Leadership in Education

Research on formal educational leadership shows that models based on hierarchical authority, isolation, and control, are shifting to more distributed, collaborative, and networked forms (Díaz-Gibson et al., 2014; Liou & Daly, 2021). This approach understands schools as *learning ecosystems* that embrace collaborative action within and across the community to increase social capital and collective learning, and the innovative and disruptive component as a central focus. In educational settings bottom-linked leadership (Balslev, 2012, 2021; Kristensen, 2011) is described as the deliberate linking of political understandings of a given program and the situated understandings in the actual educational settings (Balslev, 2012, 2021). In general education as well as in science education this active linking of various perspectives is crucial for enhancing local teaching and building culture in schools and educational institutions. Hence, the educational leadership team undertakes an important role in bridging and translating between top-down, formal educational visions (sense-giving) and bottom-up, normative understandings in teacher practices (sensemaking). We argue in the following that this type of educational leadership is distributed and collaborative in nature.

When it comes to *distributed leadership*, the concept has gained traction as an alternative perspective on school leadership, challenging the traditional bureaucratic paradigm with an emphasis on individual leaders, often depicted as "heroic" principals (Spillane, 2005). Instead of focusing solely on personal traits and actions, distributed leadership recognizes that leadership emerges from the collaborative interactions among members and various resources within the school or specific units such as a science program or department. It emphasizes that leadership is a result of collective and networked effort where multiple organization members work interdependently. By shifting the focus away from individual leaders, distributed leadership changes roles and relations, and highlights the importance of tasks, practices and learning culture that are shared among personnel within the given educational institution. In practice, distributed leadership often manifests through teamwork and sharing leadership across the organization, for example through activities around collaborative translation of national science curriculum into local teaching practices. Distributed leadership works through teams of teachers sharing ideas and resources for innovation and group development, thereby exemplifying distributed leadership in action.

However, the concept of distributed leadership, as evidenced by Woods (2016), Spillane (2005), Youngs (2017) and Raae (2021), carries inherent risks associated with power struggles and conflicting authority dynamics. This is particularly notable in organizations characterized by entrenched paradigms, such as bureaucracy alongside innovation, as discussed by Brunsson (2002). The tension between traditional, profession-founded organizational models and innovative, politically driven project structures highlights the limitations of distributed leadership. This is especially true when both the design and expectations of leadership are rooted in its distributed nature.

In sectors like education, where public sector innovation programs often operate, there are implicit expectations of distributed leadership from project founders, steering groups, or philanthropic funders (Balslev, 2012). Managing these dual paradigms demands careful attention from project leadership. Despite the evident advantages, mitigating the potential downsides of distributed leadership necessitates meticulous planning, effective communication, and continuous monitoring. This ensures that the distributed leadership structure aligns with the organization's objectives while upholding accountability, coordination, and inclusivity. In other words, distribution of leadership calls for a collaborative focus that can be found in leadership approaches such as humanistic leadership.

Humanistic leadership, as delineated by Andersen et al. (2023), is a collaborative leadership style founded on ethical imperatives that prioritize the cultivation and nurturing of trust within the organization. This type of leadership is exercised in collaboration, "where we must take care of the trust that is given as the basis for human interaction" (Andersen et al., 2023). Humanistic leadership is socially responsible leadership, which acknowledges that problems should be addressed at the level they arise and not pushed down into the system—based on an understanding that all problems unfold on the relational level and therefore risk infiltrating and poisoning the relationships necessary for task completion regardless of the organizational level. Leadership scholars within humanistic leadership call for consensus-seeking and inclusive leadership, where trust is the prerequisite for collaboration (ibid.). This morally responsible leadership is responsive and continuously involves taking leadership feedback seriously—including the less tangible ones, based on intuition, perceptions, and feelings. It is a form of leadership that actively engages in sense-making as interpretations rather than objective knowledge and evidence. And generally speaking, as a responsible and contextually responsive leader, one must both manage and negotiate intentional direction in connection with a specified and delineated task as in bottom linked leadership. Leading entails a distinct yet more flexible process. The identification or clarification of the task holds paramount importance in the undertaking. This interpretation of the leadership role encompasses an intention to convince others of a specific comprehension of the task while simultaneously considering their perspectives in shaping the evolving definition of the task.

Thus, this type of leadership is a dynamic collective process of sense-giving and sensemaking e.g. around the translation of the national science curriculum. Put more dramatically, leadership becomes a battle of defining (Raae, 2021). Leadership is hence a set of behaviors used to help people align their collective direction, to execute strategic plans, and to continually renew an organization which is therefore situated in a collaborative setting.

19.3.2 Collaborative Practices in Networks, and its Influence on Culture

Whereas leadership is one of the structural conditions for educational practices we also point to the fact that a humanistic leadership style is sometimes the fuel to constructive cross-professional and cross-institutional collaborations. Moreover, we argue that collaborative practices must be interpreted with an understanding of the contexts that they take place in. We therefore introduce some important theoretical concepts to help us understand the relationship between collaborative practices, network as a concept, and the meaning of culture in this specific context.

Changing collaborative practices is a difficult endeavor within educational institutions with centuries long traditions and cultural beliefs. Nonetheless, it is important to maintain and nurse cultures as any attempts towards improvement under these circumstances may be "doomed to tinkering" (Stoll et al., 2006, p. 235) instead of creating lasting effects. According to Lave and Wenger (2001) the participants in a community of practice such as a PLC *are absorbed* and *absorb into* a so called 'culture of practice', which means "giving them exemplars, leading to shared meanings, a sense of belonging and increased understanding" (Stoll et al., 2006, p. 234).

The concept of professional learning communities (PLC) is a well-described and well-researched concept of *collaboration*. A literature review done in the context of NAFA, however, found that from 2018–2021 there is little new international research about PLCs specifically about science education (Albrechtsen et al., 2022). According to the review NAFA describes a PLC as being "binding and systematically investigative by nature" and that they consist of communities with a group of educators who share experiences and knowledge from their practices (ibid., p. 8). This is done through investigative and reflective dialogues centered around concrete teaching and the students' learning. Professional learning networks (PLN) as a concept builds upon PLCs and stems from the "growing focus on the collaborative power of inter-school social capital" (Poortman et al., 2022, p. 1). Whereas a PLC is usually delimited to whole-school PLCs or networked, whole-school PLCs, a PLN is a construct of several educators from different institutions who collaborate for student outcome purposes (ibid., p. 3) and conduct informal leadership in regard to their shared purpose. According to Brown and Poortman, a PLN can be described as "any group who engage in collaborative learning with others outside of their everyday practice in order to improve teaching and learning" (2018, p. 1).

PLCs and PLNs are based around collaboration and knowledge-sharing in nature and requires shared visions and goals to work as efficient and autonomous teams. Professional leadership is thus implicitly, or explicitly, distributed. We discuss the implications of this within the frame of NAFA in the latter part of the chapter.

Whereas *collaboration* is a theoretical concept, network research covers an extensive field of both theory, method and methodology. *Social* network research centers around patterns of relationships by looking at the nodes of actors in a network, and the ties that connect them (Liou & Daly, 2021). Any given phenomenon can be examined and observed as a network if it consists of some kind of relationship between its entities (Borgatti & Foster, 2003). While network analysis provides a formal framework for studying relationships within a network, such as NAFA, *relational sociology* (e.g. Mische, 2011) emphasizes the significance of social relationships and interactions in shaping individuals' behaviors and outcomes. Bridging these fields involves conceptualizing social structures and dynamics in terms of both relational concepts (e.g., trust, reciprocity) and network measures (e.g., centrality, density). Mische explores the connection between culture and relationships in networks. She argues that it is necessary to "craft an approach to theory and research that views relations, interpretations, and actions as mutually generative" (ibid., p. 90). While network methodology provides insight into what structural opportunities, constraints, and dilemmas actors are confronted with, these are factors that are always interwoven with the cultural context which the complex relationships are embedded in (Fuhse, 2021).

In shifting focus from individual leaders, distributed leadership alters roles and relationships, emphasizing the shared tasks, practices, and learning culture among personnel within the educational institution. This shift is often observed in practice through teamwork and the sharing of leadership responsibilities across the organization, with informal teacher teams collaborating to share ideas and resources for innovation and group development, thus demonstrating distributed leadership in action. Crafting an approach to theory and research that views relations, interpretations, and actions as mutually influential becomes necessary. Consequently, leadership is seen as a collection of behaviors utilized to align collective direction, execute strategic plans, and continuously renew an organization, all within a collaborative setting. Cross-institutional networks must therefore be understood and analyzed in the context in which they are situated.

The theoretical perspectives on distributed and humanistic leadership, along with collaborative practices, highlight decentralized power structures, prioritize relationships, and seek to create resilient, inclusive, and adaptive educational systems, forming a foundation for analyzing science education leadership. In the following, we delve deeper into these human-centered, networked approaches in relation to the two selected cases.

19.4 Methodology

While this chapter is primarily a theoretic discussion and an attempt to build theory and bridge concepts and research fields, it is grounded in the case of NAFA (practically two cases but discussed as a single case (Yin & Campbell, 2018)). The cases were described specifically for the purpose of this chapter but is informed by data from ongoing research. Each of the two cases are rooted in NAFA as educational and organizational practice.

The case of teacher education in NAFA is based on empirical data collected in the PhD project (2023–2026). The PhD project is a part of the larger research project LAVIN—Longitudinal Study of Knowledge Ecologies in NAFA—which is a NAFA research project that examines how new knowledge of science education is created, how it emerges, and how it flows between actors and collaboration structures (Svabo et al., 2022). The PhD project applies a mixed methods-design to collect data through surveys (once a year) as well as qualitative data; NAFA project documents and formal descriptions and observations from NAFA-organized meetings. The PhD project also apply social network analysis to map the collaboration, and to investigate the distinction between professional learning communities and professional learning networks.

The other case is an illustration of leadership of the NAFA program, and is based on qualitative empirical data, journal entries, project descriptions, meeting minutes and dialogues. This study explores and analyzes cultural dynamics, meanings and the design of distributed leadership in the NAFA program examining experiences within the context of broader cultural and social frameworks, aiming to uncover deeper insights into individual and/or collective identities, beliefs, and practices.

To overcome ethical concerns and overgeneralization when applying the qualitative case study method, we have chosen to draw on feedback from peers in the validation our interpretations (Levitan et al., 2017). Furthermore, we refrain from generalizations in our conclusions and acknowledge that our findings are situated in relation to the NAFA-case. However, findings from this chapter can be applied to reflect on *what* science leaderships is in a given contextual setting and *how* science education leadership can help inform and develop science teaching practices.

The two cases described below are used as important and current points of discussion in regard to how distributed leadership emerges in an organization such as NAFA.

19.5 The Danish Academy of Science Education (NAFA)

According to Danish educational policy children and young people should experience and feel the fascination of exploring and understanding scientific phenomena to a greater extent, and they should experience strengthened scientific mastery to be applied in their roles as individuals, citizens, and professionals (NAFA, 2020). NAFA is a 7-year (2022–2028) collaboration which includes four Danish universities as well as all six university colleges, where teacher education takes place in Denmark. The vision is to ultimately create science teaching that motivates children and young people. NAFA is implemented through three key strategic branches: coordinated initiatives in schools, teacher education, and research. This requires a certain level of cultural and organizational transformation in teaching methods throughout the education system.

The teacher education initiative is largely centered around the science teacher educator faculty (population size of 70) at the university colleges and is centrally coordinated vertically across the three initiatives and horizontally in the participating educational institutions. Collaboration is thus coordinated centrally in NAFA to develop teaching practices as well as systematically implement new scientific knowledge. Development of teaching practices in teacher education is operationalized through intra- and inter-institutional collaborative teams. Together they create a network of professional learning communities (PLCs), which operate within their own university colleges as well as across them. This source of new knowledge creation and systematic collaboration is analyzed in the next part of the chapter.

19.5.1 Case 1: Science Teacher Educators and Their Organization

All participants in the teacher education initiative are all teacher educators within the natural science subjects (biology, nature and technology, physics and chemistry, and geography) taught in Danish teacher education institutions (university colleges). The initiative is then operationalized through different types of collaboration structures as formalized PLCs. The theoretical framework which NAFA operates on a basis of, is largely structured around the work in the PLCs and identified through six dimensions as showed in Fig. 19.1.

The formalized collaboration structures are described as respectively *local PLCs*, which are the PLCs located at each of the six university colleges, and *national PLCs*, where the PLCs consists of teacher educators in each of the four natural science subjects. NAFA illustrates this interconnectivity in Fig. 19.2. The PLCs are largely self-organized as they are and leadership is therefore distributed, in which participants collaborate to create new knowledge of how to teach science education to eventually enhance the quality of science teaching in the school system by improving and developing upon the teaching practices across science teacher

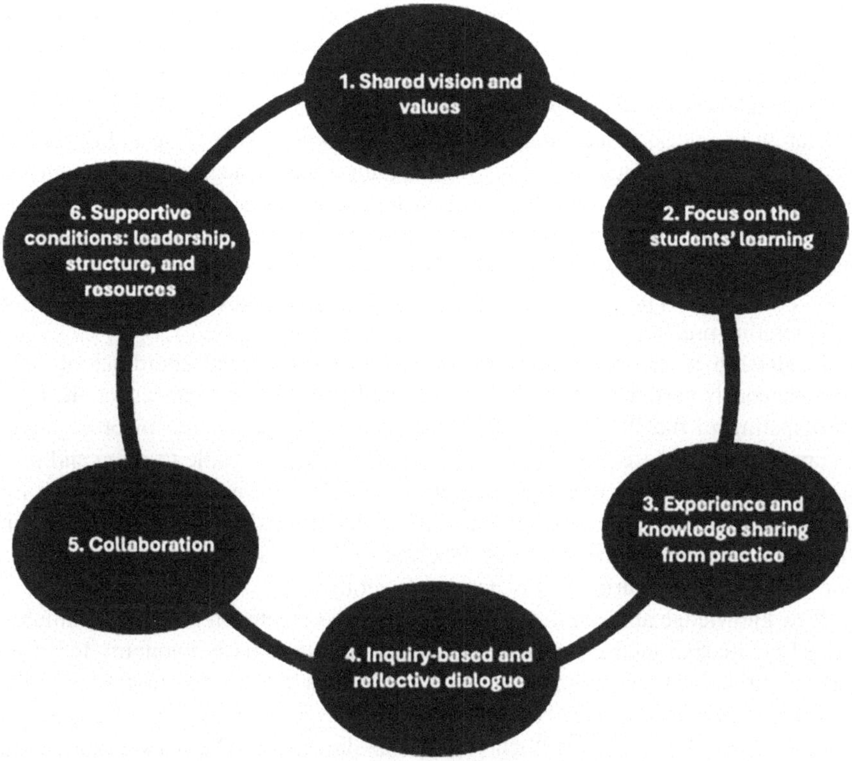

Fig. 19.1 The six dimensions in a professional learning community (OLC), translated and adapted from the NAFA program description (NAFA, 2020, p. 22)

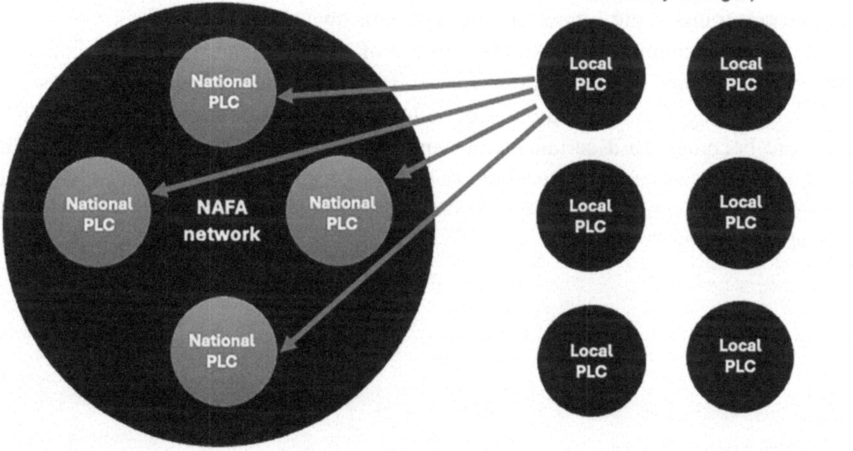

Fig. 19.2 Professional learning communities in NAFA, translated and adapted from the NAFA program description (NAFA, 2020, p. 22)

education institutions. This initiative is centrally organized in NAFA on a leadership level, but at the same time central coordinators are appointed from each PLC at the university colleges.

The teacher education initiative in NAFA is structured around the PLCs as collaboration practices, as a conduit for creating new knowledge for teaching student teachers in science education. This organization finds its legitimacy through the purpose of NAFA and is therefore not an initiative that comes directly from the teacher educators who participate in the PLCs, although collaborative practices such as PLCs are not new concept in Danish teacher education (e.g. Rahbek Thyssen et al., 2024). NAFA provides time and funding for the teacher educators to enact this collaboration practice.

Leadership is therefore partly distributed to these central coordinators, who simultaneously participate in the PLCs on equal terms as their colleagues. Besides participating in the PLCs, the teacher educators also participate in other larger recurring events where different actors in NAFA meet to be able to share and possibly create new knowledge. For example, NAFA-leadership organize meeting for the PLC coordinators to increase knowledge sharing and to provide a forum for cross-institutional organization, and PLCs at different UCs are provided with an opportunity to meet during yearly network meetings.

New knowledge and connected practices are thus created in NAFA and simultaneously recreated, taking new forms. Teacher educators have autonomy to create meaningful collaboration processes and science content for education. One of the knowledge practices (social capital) which is considered as a type of consensus principle in NAFA is "den dobbelte didaktiske dimension" (Ellebæk et al., 2022), directly translated to 'the double dimension of *didaktik*'[1] (e.g. Krogh et al., 2021), which semantically translates to an equivalent of second order knowledge, which conceptually describes the dual role of the teacher educator; the teacher educator is creating new knowledge (concepts, models, instructional modus operandi, etc.) for the teaching of teacher students who then in turn translate (and/or reproduce, recreate, reform, transfer, etc.) these practices to their own teaching practices.

Some of the characteristics of the purpose of NAFA as an organization is that it inherently becomes an organization that exists and operates simultaneously as the university colleges in which the teacher educators have their daily practice. NAFA therefore becomes, to a certain extent and in various degrees, an intervention in itself which integrates in the teacher educators' already existing practices. We will discuss the implications for culture and cross-institutional collaboration in the last part of the chapter.

[1] 'Didaktik' in the nordic countries is different than the concept of 'didactics' and is therefore often spelled differently for clarity. For reference we can refer to Krogh et al. (2021).

19.5.2 Case 2: Leadership in NAFA

The NAFA program has a complex governance structure with multiple layers of interest groups serving various functions, including a steering group, a national management group, and a reference group composed of representatives from various unions. This governance structure is designed to support the daily management and ongoing development of the program. The management group is recruited from geographically diverse university colleges to ensure broad representation and to incorporate diverse institutional interests. However, this structure also embeds institutional political tensions. In addition to the governance structure, the NAFA organization encompasses three focus areas, a Nordic board of researchers, and 10 partner institutions, each with its own institutional priorities. The primary goal of this multifaceted governance model is to establish a clear common direction, foster local ownership, and involve the strongest experts.

Overall, the NAFA-program can be viewed as a case of educational public sector innovation (Balslev, 2012; Jensen et al., 2008) on a national scale. Its complex governance structure presents challenges to leadership, decision-making processes, and implementation efforts, as it involves inherent institutional political tensions. Leadership within the NAFA program is inherently distributed across its various initiatives, with each unit having its own leadership. Distributed leadership recognizes that leadership emerges from collaborative interactions among members and the diverse resources within a group or organization. This is particularly evident in the NAFA-program, where leadership facilitates the initiatives of multiple members working independently across different institutional settings. However, our analysis focuses on the barriers to distributed leadership in a complex project organization like NAFA.

The NAFA program management group comprises a national management group representing three key NAFA initiatives: teacher education, school development, and research. Consequently, leadership is distributed both within and beyond the NAFA leadership team. Our analysis identifies six areas of risk associated with distributed leadership, as outlined in the figure below. In the following section, we explore these six aspects individually.

In the case of the NAFA research initiative—CESE[2]—the distributed leadership task is to initiate and anchor interdisciplinary and cross-institutional collaborative research projects. *Coordination challenges* are evident in this case, particularly due to the conflicting priorities and approaches of indivuals from different institutions participating the research projects. Distributed leadership can be impeded in larger cross-institutional settings where team members face these conflicts. As Liou and Daly (2021) highlight, a lack of interconnectivity among collaborating organizations can foster internal competition, resulting in inefficiencies and difficulty in achieving consensus. For example, *conflicts* can be identified in the form of differences in opinions, values, and leadership styles among research team members as

[2] CESE is an abbreviation for Center for Excellence in Science Education (see also NAFA, 2020).

well as in the NAFA management group. These differences, if unattended, may hinder collaboration and innovation within distributed leadership structure (Fig. 19.3).

As Albæk (2010) among other points out, conflicts—regardless of the level they occur at—will play out in relational spaces. In the case of NAFA, historical conflicts arisen at the organizational level have played out at the relational level in program management as well as in CESE. In this regard, NAFA leadership have put effort into task-focusing and have attempted to constructively and with a conflict-resolving approach, relate to events in the institutional political spaces (e.g. Sløk, 2018). Unfortunately, when institutional policy overshadows and affect constructive collaboration at the relational level there seems to be a danger of reduced *innovation* and risk-taking, as has been the case in the program management in regard to addressing the historical institutional conflicts that affect the distributed leadership in the team. In environments like NAFA where distributed leadership is heavily relied upon, a tendency to prioritize consensus and risk avoidance over innovation and experimentation can be detected. This stifles creativity and limits the program's ability to recognize the need for change or to adapt to change accordingly.

CESE management is obliged to "lead upwards in complex politically governed organizations" (cf. job advertisement) and based on humanistic leadership (Andersen et al., 2023). CESE has tried to create and be mindful of trust both upwards towards the program manager and personnel leader, sideways in relation to the program management group, downwards towards CESE's development group as well as research projects, and outwards towards partners. With this bottom-linked leadership (Kristensen & Voxted, 2009) which connects the complex program, CESE has tried to translate between the many different levels in the NAFA program organization.

However, this form of translational leadership (Balslev, 2012) has not fully resonated in the program management group. Despite repeated attempts at

Fig. 19.3 The Risks of Distributed Leadership in the NAFA program

capacity-building initiatives, division of labor, and cross-level leadership meetings, it seems that the system has not succeeded in transitioning from a predominantly principal-agent-oriented and low-trust-based implementation to a relation-oriented and trust-based and situated realization (Balslev & Raae, 2019). All in all, the distributed leadership model is at risk of collapsing with the lack of clear *accountability* with leadership responsibilities distributed among multiple individuals. This ambiguity can lead to confusion and difficulty in identifying who is ultimately responsible for specific decisions or outcomes.

This examination of distributed leadership within NAFA suggests the presence of alternative leadership approaches and awareness that could effectively unite collaboration and propel a complex program like NAFA forward facilitating innovation characterized by curiosity, responsiveness and sensemaking. The concept of responsible and humanistic leadership may serve as a source of inspiration for future program management during NAFA's concluding and foundational phases. In the last part of the chapter, we discuss the types of leadership which we have conceptualized through the two practice-based cases.

19.6 Collaborative Network Literacy

We initially asked how interdisciplinary perspectives can support science education leadership in addressing complex and dynamic challenges within science education. Several forms of collaborative practice exist in NAFA, and this chapter has examined these through two lenses: the NAFA leadership perspective and the NAFA teacher educator perspective. Leaders in programs like NAFA must develop the ability to monitor the culture of the network or organization. This involves recognizing subtle cues such as "the spirit, the twinkle in the eye, the changed behavior, the bodily posture, the slight concern, the unintended exposure, [and, ed.] the quiet uncertainty" (Hansen & Thorsted, 2022).

This study identifies *network literacy* as an essential skill for inclusive leadership, which requires leaders to address cultural expressions and emotions with empathy. A humanistic perspective on leadership fosters humility, embraces doubt, and acknowledges life's meaning in the context of large programs like NAFA. This approach to leadership is genuine, transparent, and self-aware. Humanistic leaders lead with integrity and honesty, which facilitates trust-building and credibility. Such leaders inspire team members through sincerity and vulnerability, fostering a sense of connection and loyalty. They prioritize collaboration and consider the needs of program partners, interpreting and responding to subtle, often imperceptible signals (Andersen et al., 2023).

This leadership approach represents a fundamental shift from the principal-agent model, which emphasizes hierarchical authority. Instead, humanistic leadership focuses on *collaboration* and *relational dynamics*. The research themes explored in this chapter contribute to the fields of leadership and interdisciplinary approaches in science education innovation. They highlight the importance of raising awareness about the processes of creating, embedding, and reforming practices while considering cultural factors.

In the context of NAFA, distributed leadership is a foundational expectation. However, it faces limitations due to tensions between traditional bureaucratic structures and innovative political organizations (Brunsson, 2002). This tension exposes distributed leadership to challenges, such as inconsistency between what the organization says, what it does, and what it actually achieves or intends. The paradoxical nature of roles and relationships between controlled development and distributed development warrants further research to fully understand the processes of educational transformation in science education.

As highlighted by leadership literature and our analysis of NAFA, distributed leadership requires a program management approach that fosters trust, responsibility, and motivation among all participants. Effective leadership must mobilize the potential of established teams (Wiseman & McKeown, 2010), whether they are leadership teams, steering groups, or other collaborative bodies.

19.6.1 Collaborative Practices in Teacher Education Initiatives

Within the teacher education initiatives, NAFA acts as an intervention in the existing daily practices of teacher educators across university colleges. These colleges have established traditions of collaboration, but NAFA introduces changes that require adaptation. For teacher educators, professional learning communities (PLCs) serve as collaborative structures designed to facilitate practice-based, reflective dialogues. These dialogues aim to produce new knowledge about science education while sharing experiences and insights, as outlined in NAFA's theoretical framework (cf. Figure 19.1).

PLCs are enacted through various activities (cf. Figure 19.2), including meetings and tasks that align with the educators' daily practices. These activities are either self-organized or coordinated locally by PLC coordinators, thereby contributing to horizontal institutional structures. Simultaneously, national PLCs introduce vertical collaboration across university colleges, transcending existing educational practices and institutional cultures.

Leadership in these contexts is distributed among PLC coordinators and the PLCs themselves, where teacher educators function as autonomous members of these structures. However, empirical data (e.g., from the LAVIN project) suggest that defining NAFA's collaborative structures solely as PLCs may oversimplify their complexity. Since these structures operate cross-institutionally, a more accurate conceptualization would be to view NAFA as a professional learning network (PLN). This framing acknowledges NAFA as a diverse network where participants collaborate autonomously but align toward common goals and shared purposes.

For teacher educators, this requires a willingness to change, commitment to the program's objectives, and cultural and structural conditions that enable development and innovation. Complex collaborations like NAFA demand responsiveness, awareness, and sensitivity to two overarching aspects: (1) the knowledge generated within these collaborations, and (2) the *collaborative network literacy* that enables participants to navigate cross-institutional boundaries. This literacy involves an understanding of diverse cultures, structural conditions, institutional policies, and prerequisites. It encompasses awareness, knowledge, and competence, all of which are crucial for facilitating effective collaboration.

19.7 Conclusions, Research Implications and Further Explorations

The cases analyzed in this chapter highlight the importance of understanding not only how practices are enacted and re-enacted to create new knowledge in science education but also the second-order knowledge that surrounds daily teaching practices. Without this understanding, there is a risk of perpetuating collaborative patterns that fail to address the complexity and context of contemporary and future science education.

This study concludes that science education leadership is becoming increasingly complex and requires collaboration characterized by cultural traits such as curiosity, responsiveness to change, sensemaking, and a focus on knowledge practices that facilitate institutional culture. Constructive conditions, including the capacity for collaboration and relational leadership, are essential to this process.

Further research is needed to explore how institutional culture can be developed and how to inspire the "spirit" and "twinkle in the eye" necessary for transforming science education practices and leadership. The cases analyzed here underscore the significance of fostering *collaborative network literacy* in both formal and informal educational leadership. This approach aims to avoid repeating the missteps of past educational transformations while paving the way for meaningful and sustainable change.

While this is a case study grounded in a Nordic science education wide-spread intervention, we recognize that there are limitations in terms of generalizability. However, the primary objective was to bridge theoretical concepts within a context of science education at a faculty level. With an initial conceptualization of *collaborative network literacy* this discussion contributes to the field of nature of science education research and to continue to investigate further what the nature of science education leadership is. More research is needed to understand this from different perspectives and from different parts of the educational chain. Furthermore, more perspectives from outside of a Nordic context is needed in the future. In this discussion we found it most important to discuss in-depth instead of a broader perspective.

The PhD project continues to investigate how network research can develop understandings of what collaboration in science teaching looks like and what their implications are. In the longitudinal research project LAVIN (Albrechtsen, T. R. S.,

Svabo, C., Kiær, K. & Brinks, T. M., 2022) we investigate how knowledge flows within NAFA and hope to contribute further to this research field, and that we can expand further on collaborative practices in science education.

References

Albæk, M. (2010). Ét liv Én tid Ét menneske: Hvordan vi glemte at leve et meningsfuldt liv. *Gyldendal.* https://www.saxo.com/dk/et-liv-en-tid-et-menneske_morten-albaek_haeftet_9788702260069

Albrechtsen, T. R. S., Brinks, T. M., Bennedsen, K., & Svabo, C. (2022). *Professionelle læringsfællesskaber: et forskningsoverblik (2018–2021).* https://nafa.nu

Andersen, E., Fannikke, L., Klausen, K. K., & Pedersen, K. M. (2023). *Ledelse i mellemrum: Et humanistisk perspektiv på offentligt lederskab.* Syddansk Universitetsforlag.

Balslev, G. M. (2012). *Omdannelse & uddannelse: innovation, læring og samarbejde.* Roskilde Universitet.

Balslev, G. M. (2021). Learning throughout the innovation journey: A new dimension to learning as a search process. *International Journal of Innovation in Education, 7*(2), 86. https://doi.org/10.1504/IJIIE.2021.10041595

Balslev, G. M., & Raae, P. H. (2019). *REFORMulering–manipulation eller metastyring?:En caseanalyse af reformarbejde på skoleområdet i en dansk kommune* (Vol. 34, pp. 207–228).

Borgatti, S. P., & Foster, P. C. (2003). The network paradigm in organizational research: A review and typology. *Journal of Management, 29*(6), 991–1013. https://doi.org/10.1016/S0149-2063(03)00087-4

Brinks, T. M., & Balslev, G. M. (2023). Facilitating networked science education leadership to promote an open and sustainable school culture. *ESERA, 7*(2).

Brown, C., & Poortman, C. L. (2018). Networks for learning: Effective collaboration for teacher, school and system improvement. In *In 2018* (1st ed.). Routledge.

Brunsson, N. (2002). The organization of hypocrisy: talk, decisions and actions in organizations (2. ed.). Copenhagen Business School Press.

Díaz-Gibson, J., Civís-Zaragoza, M., & Guàrdia-Olmos, J. (2014). Strengthening education through collaborative networks: Leading the cultural change. In *School leadership & management.* (Vol. 34, Issue 2, pp. 179–200). Routledge. https://doi.org/10.1080/2F13632434.2013.856296

Dolin, J., Ellebæk, J. J., & Daugbjerg, P. (2022). *Dannelse og kompetence i læreruddannelses dobbeltdidaktiske perspektiv: med særligt henblik på uddannelse af naturfagslærere til grundskolen.*

Ellebæk, J. J., Dolin, J., & Daugbjerg, P. (2022). Den dobbelte didaktiske dimension: i læreruddannelsens naturfag. *Naturfagsakademiet.*

Fuhse, J. (2021). Social networks of meaning and communication. In *Oxford scholarship online, 2021, 1.* Oxford University Press. https://doi.org/10.1093/2Foso/2F9780190275433.001.0001

Hansen, F. T., & Thorsted, A. C. (2022). At tænke med hjertet: En grundbog i eksistentiel praksisfænomenologi. *Klim.* https://vbn.aau.dk/en/publications/at-t%C3%A6nke-med-hjertet-en-grundbog-i-eksistentiel-praksisf%C3%A6nomenolo

Jensen, K. E., Jensen, J. P., Digmann, A., & Bendix, H. W. (2008). Principper for offentlig innovation. Børsens Forlag.

Kristensen, C. J. (2011). *Medarbejderdrevet innovation i offentlige organisationer–et teoretisk blik* [Roskilde Universitet]. https://forskning.ruc.dk/da/publications/medarbejderdrevet-innovation-i-offentlige-organisationer-et-teore

Kristensen, C. J., & Voxted, S. (2009). Innovation: medarbejder og bruger. In *Hans Reitzel.* Hans Reitzel.

Krogh, E., Qvortrup, A., & Graf, S. T. (2021). Didaktik and curriculum in ongoing dialogue. In *2021, 1. udg.* Routledge. https://doi.org/10.4324/2F9781003099390

Lave, J., & Wenger, E. (2001). Situated learning. In *Learning in doing, 2001, Reprinted*. Cambridge University Press.

Lederman, N. G. (2023). Nature of science: Past, present, and future. In N. G. Lederman, D. L. Zeidler, & J. S. Lederman (Eds.), *Handbook of research on science education: Volume III (Vol. 3, pp. 831–879)*. Routledge.

Levitan, J., Carr-Chellman, D., & Carr-Chellman, A. (2017). Accidental ethnography: A method for practitioner-based education research. *Action Research, 18*(3), 336–352. https://doi.org/10.1177/1476750317709078

Liou, Y. H., & Daly, A. J. (2021). Obstacles and opportunities for networked practice: A social network analysis of an inter-organizational STEM ecosystem. *Journal of Educational Administration, 59*(1), 94–115. https://doi.org/10.1108/JEA-02-2020-0041

Mische, A. (2011). Relational sociology, culture and agency. In P. J. Carrington & J. Scott (Eds.), *Social network analysis* (pp. 80–98). SAGE Publications Ltd..

NAFA. (2020). *Naturfagsakademi NAFA: Et nationalt program for kvalificering af naturfagsundervisning i Danmark.*

Osborne, J., & Dillon, J. (2008). Science Education in Europe: Critical Reflections: A Report to the Nuffield Foundation.

Poortman, C. L., Brown, C., & Schildkamp, K. (2022). Professional learning networks: A conceptual model and research opportunities. *Educational Research, 64*(1), 95–112. https://doi.org/10.1080/00131881.2021.1985398

Raae, P. H. (2021). Udviklingsledelse i gymnasiet: Mellemledelse og udvikling. In *2021, 1. udgave*. Bogforlaget Frydenlund.

Rahbek Thyssen, M., Dahl Madsen, C., & Lund Nielsen, B. (2024). Skoleudvikling, synergi og meningsskabelse:—hvordan bygges videre på tidligere naturfagsprojekter? *MONA - Matematik- Og Naturfagsdidaktik.*

Reinholz, D. L., & Apkarian, N. (2018). Four frames for systemic change in STEM departments. In *International journal of STEM education* (Vol. 5). Springer. https://doi.org/10.1186/2Fs40594-018-0103-x

Sløk, C. (2018). Har en leder altid ansvaret, og aldrig skylden? Om lederes læring ved undervisning i ansvar og ledelse. *Public Governance Research, 4*, 23–41.

Spillane, J. P. (2005). Distributed leadership. In *The educational forum (West Lafayette, Ind.)* (Vol. 69, pp. 143–150). https://doi.org/10.1080/2F00131720508984678

Stoll, L., Bolam, R., McMahon, A., Wallace, M., & Thomas, S. (2006). Professional learning communities: A review of the literature. *Journal of Educational Change, 7*(4), 221–258. https://doi.org/10.1007/s10833-006-0001-8

Svabo, C., Albrechtsen, T. R. S., Kiær, K., & Brinks, T. M. (2022). *Længdestudie af videnøkologier i NAFA: projektbeskrivelse.*

Treagust, D. F., & Won, M. (2023). Paradigms in science education research. In N. G. Lederman, D. L. Zeidler, & J. S. Lederman (Eds.), *Handbook of research on science education: Volume III (Vol. 3, pp. 3–27)*. Taylor and Francis. https://doi.org/10.4324/9780367855758-2

van Driel, J. H., Hume, A., & Berry, A. (2023). Research on science teacher knowledge and its development. In *Handbook of Research on Science Education* (Vol. 3, pp. 1123–1161). Routledge. https://doi.org/10.4324/9780367855758-41

Wiseman, L., & McKeown, G. (2010). *Multipliers* (1st ed.). HarperBusiness.

Woods, P. A. (2016). Authority, power and distributed leadership. In *Management in education.* (Vol. 30, Issue 4, pp. 155–160). SAGE Publications Ltd.. https://doi.org/10.1177/2F0892020616665779

Yin, R. K., & Campbell, D. T. (2018). *Case study research and applications* (6th ed. (Issue Mi)). SAGE Publications, Inc.

Youngs, H. (2017). A critical exploration of collaborative and distributed leadership in higher education: Developing an alternative ontology through leadership-as-practice. In *Journal of higher education policy and management.* (Vol. 39, Issue 2, pp. 140–154). Routledge. https://doi.org/10.1080/2F1360080X.2017.1276662

Tina Maria Brinks is a PhD research fellow at the STEM Education Research Center (FNUG) at the University of Southern Denmark and in the Danish Academy of Science Education (NAFA). Her research interests focus on organizational learning in science education, collaborative practices within teacher education, conceptualization of teacher knowledge, and social network analysis as data visualization method for teacher professional development.

Gitte Miller Balslev holds a PhD in collaborative innovation in public sector education and is an assistant professor and co-director of the STEM Education Research Center (FNUG) at the University of Southern Denmark as well as SDU-liaison and researcher in the national program University Science Education (USE). Furthermore, she is a former leader of the national Center for Excellence in Science Education (CESE) at Copenhagen University College, Denmark. Her research interests include STEM-related educational innovation on educational change through national reforms, educational leadership in reform realization, collaborative innovation, and educational ecosystems from a network perspective.

Chapter 20
"Gaijatra": Connecting Science Education with Cultural Heritage in a Professional Development Workshop in Nepal

Bhaskar Upadhyay, Cuc Kim Vu, and Stephanie Erickson

20.1 Introduction

Tara Yosso (2005) argues that cultures that are beyond the mainstream cannot be stereotypically imagined as myths and misaligned conceptions that do not contribute to learning. Yosso, drawing from Bourdieu and Passeron (1977) and Valenzuela (1999), also asserts that most standard textbooks and curriculum promote culture and socialization skills and practices that are from middle class mainstream dominant groups. Hence, making schools optimal places to reinforce only dominant practices and knowledge. In this chapter we recognize the concerns of Yosso and fully support that Nepal's science curriculum and national standards are dominated by Western science and practices with some sprinkling of local examples rather than valuing local knowledge at par with current science knowledge and practices (Upadhyay & Sadykova, 2024). In a country like Nepal where, according to the National Statistics Office [NSO] (2023), national population and housing census of 2021 shows that there are 142 casts/ethnicities and 123 different mother tongues spoken. Therefore, there are immense opportunities to bring cultural heritage as science teaching and learning contexts in almost any science classrooms, curriculum, and textbooks.

The science education community has been advocating for science teaching and learning that bridges science contents, skills, values, and ideas to the local cultural

B. Upadhyay (✉)
Organizational Leadership, Policy, and Development, University of Minnesota, Minneapolis, MN, USA
e-mail: bhaskar@umn.edu

C. K. Vu · S. Erickson
Curriculum and Instruction, University of Minnesota, Minneapolis, MN, USA
e-mail: vuxx0023@umn.edu

G. Cakmakci, M. F. Tasar (eds.), *Connecting Science Education with Cultural Heritage*, Contributions from Science Education Research 15,
https://doi.org/10.1007/978-3-031-90944-3_20

contexts of students and their lived experiences. Bouillion and Gomez (2001) show that fifth grade teachers were successful in connecting the issues of local river pollution in a community to what students learned in school science. Upadhyay et al. (2020) assert that Indigenous students in Nepal were more attuned to inquire, explore, and engage with science when socio-cultural issues provided contexts for learning science. Similarly, science teachers in semi-urban schools in Nepal, who were ambivalent about marginalized students' knowledge from home, struggled to support active student learning (Koirala, 2023) and in a study of an elementary teacher in Pakistan, Upadhyay et al. (2005) show that the local issues and culture when linked to science, made the teacher feel more effective in teaching science to students from marginalized groups. These studies continuously show that science taught through culturally sustaining/relevant pedagogy tends to be personally and collectively more meaningful and satisfying to students and in bringing success to learning. Despite the importance of culture in science teaching and learning, there is a persistent challenge in Nepal in providing appropriate resources and skills to science teachers to be well prepared in making their teaching culturally relevant (Koirala, 2021). This paper will utilize a qualitative methodology approach to explore how a workshop helped teachers consider local cultural heritage useful in science teaching and learning and what would be some of the challenges of culturally relevant science teaching classrooms. Before moving on we provide some key contexts for this study.

20.2 A Brief History of Gaijatra

Gaijatra started in Kathmandu during the reign of Pratap Malla (1641–1671 A.D.) to console his queen after their son's untimely death (Regmi, 2007). To console the queen, he ordered every home in the kingdom who had lost a family member to take out a procession in front of the palace, so the queen saw many citizens had experienced loss as she had done. The queen remained unconsolable. The king then announced that people could dress up in any manner and use any means to bring joy to the queen. He also announced rewards to people if the queen felt relieved from sadness of losing their son. All the homes where a family member had deceased that year, took out processions dressed up in outlandish, absurd, and silly costumes to imitate, impersonate, and mock various aspects of social, and cultural customs. These processions made the queen laugh and helped her come out of the sadness of her loss. The king then decreed that every year on that day there would be a celebration of Gaijatra. Gaijatra is a ritualistic performance celebrated by the Newar ethnic group across Nepal. 'Gai' means 'cow' and 'Jatra' means 'festival'. Thus, when people dress up, they somehow depict a cow in some form in their attire. Most of the time, it is the mask or a headgear that carries some feature of a cow.

Since Kathmandu Valley is the heart of the Newars, Gaijatra is celebrated more intensely in the valley. Every year, families who have lost someone will bring out a procession where family members are dressed up in funny clothes and masks. The procession goes around various communities as desired. A key feature of Gaijatra and how people are dressed is that each person must have some kind of connection to a feature of a cow. Most of the time a mask or a headgear tends to show that. Over time, Gaijatra has also symbolized and permitted people to lampoon and critique social, cultural, political, economic and other injustices and systems of injustices. Therefore, this cultural heritage has a much bigger significance in the Nepali society rather than just a ritualistic performance by the Newars.

In this chapter we draw from Gaijatra as the cultural heritage context to engage teachers at a school in Nepal in learning about culturally relevant science teaching and learning in their classes and the potential challenges of it in a high-stakes external assessment system.

20.3 Education in Nepal

In Nepal, there are three groups of schools, public (government funded), private (for-profit and non-profit school without any government funding), and religious schools. In this chapter, we discuss public schools only. The education system in Nepal at the school level is categorized into three levels based on grade levels.

All students attend pre-primary school for 1 year. After completing pre-primary school, students attend basic education level school from the 1st grade to 8th grade. After that, they then attend high school from the 9th grade to 12th grade. According to the Center for Education and Human Resource Development [CEHRD] (2024), the student teacher ratio was the highest in the secondary grade (46:1) and the lowest at the grades 1–5 (23:1). However, a much greater challenge is that many teachers are hired as long term 'relief' teachers to teach subjects such as mathematics, science, and English (CEHRD, 2024). All relief teachers are untenured and most of them are out-of-subject area teachers who have both pedagogical and content area mastery challenges (CEHRD, 2024). Therefore, many schools struggle to hire science teachers who have degrees in science (Upadhyay et al., 2010). Furthermore, CEHDR's report shows that resources that handicap teachers in quality and culturally relevant teaching are curriculum, reference materials and teachers' guide to effective teaching. Only 63% schools have these materials readily available to teachers. Thus, professional development workshops become highly valuable in enhancing teachers' pedagogical skills, science contents, and design skills.

20.4 Himal School

This study was conducted at Himal School (pseudonym), a public school, where about 10% of the teachers are tenured (permanent) and the rest are working under temporary or relief teacher status. In the case of relief teachers, they can be terminated at any time. Thus, there is a greater number of science teacher turnout because many of them are relief teachers. Furthermore, teacher professional development opportunities are fewer and far between for many of the teachers because regular teacher professional education programs are in policy documents but hardly implemented on a regular basis to benefit most teachers (Borg & Consult, 2023). Himal school has about 650 students from pre-school to high school. However, about a quarter of the students come from internal migrant families who work in the local brick factories. Almost 90% of these students return to their village during monsoon with their parents and other relatives when brick factories are closed. Many of these families return after the end of the monsoon season to the brick factories increasing the school student population. As these students come from many cultural, linguistic, and geographic differences, Gaijatra gives an appropriate context for teachers to gain ideas, skills, and ways to learn and understand how cultural heritage could be a part of science teaching and learning. The goal of this chapter is to document and understand the value of the workshop for teachers in potentially utilizing cultural heritage context in their teaching. Additionally, to also understand what challenges teachers perceive when considering culturally relevant pedagogy in science teaching and learning. To answer these questions, we draw from the theories of culturally relevant pedagogy to understand teachers' perceptions about teaching science using cultural heritage context.

20.5 Literature Review

We briefly present scholarly works on science and cultural connections. Further, we also present the idea of culture and cultural heritage and their value in science teaching and learning.

20.5.1 Value of Culture in Science Education

There are two kinds of cultural heritage—tangible and intangible. In our case, both types of cultural heritage are factors in contextualizing cultural heritage for science teaching and learning. We use tangible cultural heritage such as the monuments, architectures, arts, clothes, and other aspects of tangible cultural heritage in the teacher professional development workshops to build science activities around. We specifically focused on mask making, the most popular depiction of Gaijatra, in the workshop. However, we also recognized the value of intangible cultural heritage to better connect the workshop between science and culture. For example, we encouraged teachers to consider "practices, representations, knowledge, and expressions" (UNESCO, 2022, Article 2, p. 5) as they engaged in workshop activities. An important component of cultural heritage is that many of these skills and values are "transmitted from generation to generation [and] constantly recreated by communities"

(UNESCO, 2022, Article 2, p. 5) over time these skills, values, knowledge, and artifacts become essential culture and identity of a community.

On the other hand, culture is "a social domain that emphasizes the practices, discourses, and material expressions, which, over time, express the continuities and discontinuities of the social meaning of a life held in common" (James et al., 2015, p. 53). Both cultural heritage and culture clearly articulate that community practices, artifacts, knowledge, languages, and skills are central part of individual and community identities. Bang et al. (2013) and Tolbert (2015) specifically focus on the challenges of cultural mismatch between Indigenous students' culture and science culture. They argue that when there is a disconnect between Indigenous epistemology and knowledge and scientific epistemology and knowledge, teachers and students both struggle to productively engage in science teaching and learning. Fortney et al. (2019) express that teachers who teach students from marginalized communities must recognize and actively and critically draw from the cultural assets of the communities of the students to make science teaching more equitable and transformative for students. Upadhyay (2009) shows that when immigrant and other marginalized students' knowledge, skills, and culture are integrated and valued during science teaching and learning, students tend to find greater connections and value in learning science. Thus, the value of cultural heritage is unquestionably important for better science teaching as well as learning. Therefore, the workshop provided an appropriate and valuable opportunity for teachers to learn how to connect science with the local cultural heritage "Gaijatra".

20.6 Culture and Cultural Experiences in Science Teaching and Learning

The idea of culturally sustaining pedagogy (McCarty & Lee, 2014) in conjunction with culturally relevant pedagogy (Ladson-Billings, 2014) provides a strong rationale for connecting culture with science teaching in attaining academic success, building positive feelings and connection to one's cultural heritage, sociocultural awareness, and helping sustain local cultural values and skills. In science education, researchers have articulated that science teaching and learning not only brings relevance between science content and the culture of students, but it also evokes emotions in teachers and students during classroom interactions (Meylan, 2014) because personal and communal relationships are disturbed or supported like in Gaijatra. When students' culture is used in teaching science, students seem to enjoy and excel in active classroom participation and find science more meaningful. Similarly, studies have shown that when teachers are prepared and given opportunities to infuse cultural heritage in science, they seem to be more willing to make science activities aligned with local sociocultural activities and issues (O'Neill et al., 2023; Upadhyay, 2009).

20.7 Culture and Science Teachers in Nepal

Science teachers are constantly encouraged to engage in science teaching that brings students' cultural experiences to support learning so that science is more meaningful to students. However, in the context of Nepal, science teacher preparation and professional development efforts during in-service years still lacks in building teachers' pedagogical capacities to become more culturally inclusive. Koirala (2021) shows in his study of classroom teachers in Nepal that the teachers lacked sufficient knowledge and pedagogical tools to make science teaching relevant to students' linguistic and home cultural diversity. Similarly, Upadhyay et al. (2018) showed that when Tharu Indigenous teachers explicitly contextualized science teaching with the local cultural heritage of 'Maghee', teachers found students more enthusiastic about learning science. Yet, there is a lack of study in this area that has explored how teacher development workshops provide support and space to explore local cultural heritage in classroom science activities. Furthermore, many Nepali science teachers continue to believe that knowing the facts about science ideas devoid of local cultural contexts is a hallmark of good pedagogy because this produces good results in externally administered high-stakes tests.

Yet, multicultural and culturally relevant science pedagogy generally tend to happen in many public schools in Nepal through multilingual communication and instructional events (Koirala, 2023; Upadhyay et al., 2020). However, students' experiences in other parts of their lives such as household activities, history, and social issues seem to be less prevalent in science instruction except with some teachers who are willing to take the risk and value science teaching and learning as a social justice activity (Upadhyay et al., 2021). Thus, there is a gap in understanding how and why teachers in Nepal could draw from local cultural events to build a more culturally relevant pedagogical skills in science teaching and learning. This chapter shares some of the early outcomes from a professional development program.

In this chapter, we seek to explore the following questions: (1) In what ways do science teachers find the workshop help them consider local cultural heritage useful in science teaching and learning? (2) What challenges do science teachers perceive in culturally relevant science teaching?

20.8 Brief Context for Workshop

The data for the paper comes from a 6-week-long collaborative partnership between a high school in Nepal and the authors. Nepal has centrally designed curriculum standards with some flexibility for the local education departments. The academic success of the students is measured by passing the regional and national exams. Therefore, there is pressure on teachers to teach-to-the test rather than connections to cultural relations and critical thinking. However, recently there has been recognition that cultural connections to science teaching could help academic success and build value in knowledge and skills from local cultural heritage. We provided two professional development workshops for more than 40 teachers focusing on integrating science education and culture. Since the school wanted most teachers to build pedagogical skills and tools to draw from local cultural contexts, the school decided that all teachers would benefit from our workshops. This chapter shares results from one workshop.

20.9 Methodology and Methods

We used qualitative methodology (Miles et al., 2014), which provided us guidance to document and understand the nuanced value of "Gaijatra" as the context for cultural connection with science contents and teachers' sense of the value of local culture in science teaching. A qualitative methodology is a methodology that allows us to document in-depth participants' experiences, perceptions, beliefs, values, and the meanings they make of their experiences.

We kept reflective field notes to document the activities during the four-hour long workshop. We documented observations of the activities, teachers' conversations, questions, and interactions during the workshop to capture what they said, how they connected Gaijatra, to science, their feelings, and ideas to connect to their contents such as social studies, mathematics, science, English, and Nepali language.

We recorded informal conversations with 12 teachers during the workshop. We asked questions such as what connections they saw between science and Gaijatra, designing the masks, the cultural value of what the masks represented for them, and how mathematics and other skills like questioning and design changes they did connect to their designing of masks. We interviewed three science and one mathematics teacher and audio recorded them. During the 60–90 min long one-on-one interviews, we asked questions such as how much value they found in using Gaijatra context in teaching their contents; how they made sense of cultural heritage and science or mathematics connections; in what ways Gaijatra would be valuable for them in teaching their contents; how they saw personal and cultural connections in this workshop valuable.

20.10 Analysis

We analyzed data thematically using the approach described by Saldaña (2012). We first generated open codes looking for similarities and dissimilarities in their perceptions, values, challenges and opportunities in using cultural heritage as context of science teaching and learning. After that we collapsed open codes into larger groups of similar codes, reducing the number of open codes. These larger codes were then developed into themes capturing participants' ideas, perceptions, challenges, and opportunities. We established the following two themes as findings for the purposes of this chapter—mask making is a cultural activity that connects cultural heritage with science contents and there is an ambivalence of science and culture connection.

20.11 Findings

20.11.1 Mask-Making: Connecting Cultural Heritage with Science Contents

The researchers, after the input from the teachers decided that Gaijatra would be a suitable cultural context to frame science (STEM) with the local cultural heritage, Gaijatra. The teachers shared with us that "each local community has its own cultural and historical variation on mask-making." Since most of the teachers were from the Bhaktapur area, they decided to use that local variation in their masks. In each group (five groups), there were teachers from all content areas, so they leveraged contributions from the respective contents to make the masks as culturally relevant as possible. During the making of the masks, teachers drew from mathematics (measurements), science (strength of the materials, mixing colors), history (variations in designs), design components such how a mask fits the facial contours of a person and reshaping it as needed, and language arts (communication of the process and connections to cultural heritage and how to scientifically describe the processes).

During the sharing of each of the group's masks, some of the comments from the teachers were: "we can see how easily I can teach some of the math(s) and science like units and measurements, light and color,(...)"[mathematics teacher and science teacher1]; "I recognize that I can use this in how students can describe science activities [using] science language" [elementary teacher12]; "I actually can teach about pressure [strength of the materials used to make mask] to high school class" [high school science teacher3], "I really like how I or social studies teachers can take other cultural activities where we use masks to depict deities to teach science" [principal]; "parents and students will really like this and might help in understanding and engaging in science and maths [rather than] struggling" [assistant principal].

These interactions during the workshop showed that teachers were seeing connections between the mask making aspects of Gaijatra and how they could utilize this context to teach their contents. Furthermore, during the interview the science and the math teachers found several ways to make Gaijatra a valuable context to connect science, math, and design processes in their respective classes. The following excerpt represents the cultural heritage and science connections as perceived by two science teachers.

> Science Teacher2: We [most teachers] don't think about festivals and culture having connections with science, mathematics, engineering. So, the workshop was a good example of science, maths, and design [engineering] coming together...
>
> I also liked how you all connected geometry, proportions, and measurements in making mask[s]. (...) I can see connecting it to modeling similar and congruent shapes, (...) symmetry, etc. (...). We have other cultures and heritage like Machindranath making wooden chariot which can connect to mathematics, science. The workshop was really helpful because it made us think using other cultural heritages in Nepal. Festival is a part of that, [and] gives lot of opportunities for teachers [and students] to explore in science too...I think students will see various sciences in these kinds of activities as useful.
>
> Science Teacher 1: I think students, Newars and others also, feel that they can understand science concepts like color mixing, tensile strength, measurements, confirmatory tests. They also get to explore other things openly because Gaijatra as cultural heritage has so many differences in the three cities in the Valley...We hardly thought about cultural heritage in this way (...) makes our teaching method more locally connected.

The excerpts indicated to us that science teachers could envision cultural heritage as a springboard to contextualize and introduce various ideas, concepts, skills, and practices in their teaching. Additionally, both the teachers also recognized other cultural heritages that could pedagogically enhance their teaching to bring cultural connections during science teaching.

20.11.2 The Ambivalence of Science and Cultural Connections

Teachers saw value in learning Western science as explained in their standard textbooks. Yet, they were very frustrated and, at times, overwhelmed by a sense of hopelessness that the local knowledge had no place in the science they were teaching. Most teachers would agree what one teacher said, "we need to consider if our students pass the external exam (...) the school looks bad if majority don't pass." Thus, indicating a persistent dilemma for the teachers to either spend time in teaching science contents as facts and ideas or to make science more culturally relevant. The researchers heard from the teachers both the value of culture in science teaching and learning but also doubted if that would support student learning—specifically passing the high-stakes external exam. For example, during the workshop, one teacher said, "This is really helpful to connect with the local culture students know but there is a *danger* of not having enough time to teach what is asked in the exams." Another teacher (math), "I know this [workshop] is helpful, but the textbooks don't give any example of Gaijatra and other cultural events." An elementary teacher said, "I see the value in the integration of subjects, but I worry that [culture becomes] an object in science rather than a deeper understanding of cultural value[s], and we can lose the culture (...) and [embedded] knowledge and skills in it as well as the emotional value to us." Most teachers agreed that there is a lot of "sadness and worry" we are "losing our culture", but on the other hand, "joy that we can see [in this workshop] how our culture Gaijatra could be preserved." A statement from an elementary teacher captures the value of culture but also challenges:

> I know this and other cultural event(s) can easily [be] taught in history and English. (…) We talk about masks like Lakhe dance and makes easy for students to understand. (…) We (teachers) like making things and elementary students also like them; it's fun for them and [also] learn about culture. (…) I now can make connections to science and maths too (…) we [teachers] have to be careful about finishing curriculum, so students pass the external exam (…) parents don't want children to fail either.

The sense of connections to culture, science, and mathematics brings opportunities but at the same time creates a challenge of how not to get carried away in cultural connections without doing science and mathematics or teaching other content areas such as English language.

Similar sentiments were expressed by the science teachers we interviewed.

> Science Teacher 3: I agree that the workshop showed a way to connect cultural heritage and some science contents we teach in different grades (…) as a Newar, I'm proud of our cultural heritage connected, but 10th grade external exams or 11th and 12th grades exams don't ask questions like you all [in the workshop] engaged us—brining different ideas together.
>
> Science Teacher 1: I teach 8th grade, and I grew up with Gaijatra and other jatras, so I saw many connections to science ideas and connected to health and environment (…) but the district external exams don't ask questions to critical[ly] think and apply concepts (…) I'm very indebted to the workshop for showing us how our local culture has so much to contribute to our teaching.

During the interview, the science teachers showed interest and recognition in the value of cultural heritage in their pedagogy. They also agreed that the workshop illuminated how and why cultural heritage could provide opportunities to make teaching and learning culturally connected. However, they were equally concerned about the required external exams that students had to pass to get promoted to the next grade. They attested the value of local cultural heritage and the connections it provided to contents and communities. Thus, there was an ambivalence in their reflections about cultural heritage and science teaching and learning. These ambivalences were real and the teachers were continuously managing them to make learning useful for multiple purposes so that it supported effective pedagogy and better learning in Nepali schools.

20.12 Discussions and Implications

In this chapter we showed that science teachers in Nepal could draw from the local cultural heritage to make their science teaching and learning connected to local culture and heritage (Upadhyay et al., 2018). They see that there are opportunities to link culture to science teaching and learning but there are also some systemic constraints such as the external exams. Teachers also felt cultural connections generated challenges or ambivalence in balancing culturally relevant pedagogy for better learning opportunities and the high-stakes exams that all students must pass. The implications of the findings of this study are that culture and heritage provide a multitude of opportunities for science teachers to leverage local culture for active and engaged science teaching and learning environments. Another implication is that teachers could envision varied cultures and cultural experiences from the lens of knowledge and skills parity with science-based knowledge and knowledge from cultural heritage. For example Indigenous knowledge of the environment such as the “Māori ancestral sayings that describe the pollination of *harakeke* (New Zealand flax [*Phormium tenax*]) by native *kākā* parrots (*Nestor meridionalis*)” (Wehi, 2009 as cited in Jessen et al., 2021, p. 95) and similarly Tharus in Nepal using hollowed out and dried gourd to carry water indicate to us that cultural heritage carries knowledge at par with science rather that a myth (Yosso, 2005).Furthermore, when science teachers are more knowledgeable about the local culture, they can infuse the cultural skills and knowledge in their teaching with many benefits to students and make the curriculum richer and more authentic. There are several key take home messages for the readers from this chapter. First, cultural heritage is an authentic and viable context to connect science ideas and values with that of culture and heritage. Second, cultural heritage can be utilized as an authentic and culturally relevant resource for science teachers, including all other STEAM fields, from all grade levels. Third, science teachers, including elementary teachers, in the Global South countries need continuous professional development opportunities to learn and enhance skills to how to integrate cultural heritage and its knowledge in their teaching and learning. Fourth, there is a real and justifiable challenge to teachers when external pressures such as doing well in the high-stakes exams supersede culturally relevant teaching for understanding and positive identity building among students and teachers. Fifth, some cultural heritage could generate emotions or feelings of stress among teachers and students, such as Gaijatra because it is directly connected to the loss of a beloved family member, friend, or neighbor. Therefore, teachers need to be cautious in how they balance between teaching and learning and managing emotions of their own and that of the students. Finally, we assert that cultural heritage is not only an authentic context for science teaching and learning, but it is also a representation of valid and reliable knowledge holder and a means of transferring invaluable local knowledge to the next generation. Therefore, science and STEAM teachers should seriously and deeply consider utilizing this resource in their teaching and learning for more culturally relevant learning opportunities for students and their identity enhancement.

References

Bang, M., Warren, B., Rosebery, A. S., & Medin, D. (2013). Desettling expectations in science education. *Human Development, 55*, 302–318. https://doi.org/10.1159/000345322

Borg, S. & Consult, V. (2023). *The status of teacher education and development in Nepal*. British Council. Retrieved on December 10, 2024, from https://www.britishcouncil.org.np/sites/default/files/the_status_of_teacher_education_and_development.pdf

Bouillion, L. M., & Gomez, L. M. (2001). Connecting school and community with science learning: real world problems and school–community partnerships as contextual scaffolds. *Journal of Research in Science Teaching, 38*(8), 878–898. https://doi.org/10.1002/tea.1037

Bourdieu, P., & Passeron, J. (1977). *Reproduction in education, society and culture*. Sage.
Center for Education and Human Resource Development [CEHRD]. (2024). Flash report 2080 (2023/2024). Ministry of Education, Science, and Technology. Retrieved September 21, 2024, from https://dang.edcu.gov.np/upload_file/files/post/1724905049_287811443_FLASH%20REPORT%202080.pdf
Fortney, B., Morrison, D., Rodriguez, A., & Upadhyay, B. (2019). Rethinking equity in science teacher education: Introduction to special issue. *Cultural Studies of Science Education, 14*, 259–263. https://doi.org/10.1007/s11422-019-09943-w
James, P., Magee, L., Scerri, A., & Steger, M. (2015). *Urban sustainability in theory and practice: Circles of sustainability*. Routledge.
Jessen, T.D., Ban, N.C., Claxton, N.X., & Darimont, C.T. (2021). Contributions of Indigenous Knowledge to ecological and evolutionary understanding. *Frontiers in Ecology and the Environment, 20*(2), 1540-9295. https://doi.org/10.1002/fee.2435
Koirala, K. P. (2021). Multicultural classroom teaching in Nepal: perspectives and practices of a secondary level science teacher. *Cultural Studies of Science Education, 16*, 1163–1182. https://doi.org/10.1007/s11422-020-10012-w
Koirala, K. P. (2023). Science teaching in culturally diverse classrooms: Application of sociocultural knowledge at a school system in Nepal. *Journal of Science Teacher Education, 34*(5), 544–562. https://doi.org/10.1080/1046560X.2023.2206693
Ladson-Billings, G. (2014). Culturally relevant pedagogy 2.0: a.k.a. the remix. *Harvard Educational Review, 84*(1), 74–84. https://doi.org/10.17763/haer.84.1.p2rj131485484751
McCarty, T. L., & Lee, T. S. (2014). Critical culturally sustaining/revitalizing pedagogy and Indigenous education sovereignty. *Harvard Educational Review, 84*(1), 101–136. https://doi.org/10.17763/haer.84.1.q83746nl5pj34216
Meylan, A. (2014). Epistemic emotions: A natural kind? *Philosophical Inquiries, 2*(1), 173–190. https://doi.org/10.5451/unibas-ep44372
Miles, M. B., Huberman, M. A., & Saldana, J. (2014). *Qualitative data analysis: A methods sourcebook* (2nd ed.). Sage.
National Statistics Office [NSO]. (2023). *National Population and Housing Census 2021: National Report on Caste/ethnicity, Language & Religion*. Retrieved September 23, 2023, from https://censusnepal.cbs.gov.np/results/files/result-folder/Caste%20Ethnicity_report_NPHC_2021.pdf
O'Neill, T., Finau-Faumuina, B. M., & Ford, T. U. L. (2023). Toward decolonizing STEM: Centering place and sense of place for community-based problem-solving. *Journal of Research in Science Teaching, 60*(8), 1755–1785. https://doi.org/10.1002/tea.21858
Regmi, D. R. (2007). *Later medieval Nepal* (1st ed.). Rupa.
Saldaña, J. (2012). *The coding manual for qualitative researchers* (3rd ed.). Sage.
Tolbert, S. (2015). "Because they want to teach you about their culture": Analyzing effective mentoring conversations between culturally responsible mentors and secondary science teachers of indigenous students in mainstream schools. *Journal of Research in Science Teaching, 52*(10), 1325–1361. https://doi.org/10.1002/tea.21240
UNESCO. (2022). *Basics texts: Convention for the safeguarding of the intangible cultural heritage (2022 ed.)*. Retrieved January 10, 2023, from https://ich.unesco.org/doc/src/2003_Convention_Basic_Texts-_2022_version-EN.pdf.
Upadhyay, B. (2009). Teaching science for empowerment in an urban classroom: Using Hmong students' funds of knowledge. *Equity and Excellence in Education, 42*(4), 217–232. https://doi.org/10.1080/10665680902779366
Upadhyay, B., Atwood, E., & Tharu, B. (2020). Actions for sociopolitical consciousness in a high school science class: A case study of ninth grade class with predominantly indigenous students. *Journal of Research in Science Teaching, 57*(7), 1119–1147. https://doi.org/10.1002/tea.21626
Upadhyay, B., Atwood, E., & Tharu, B. (2021). Antiracist pedagogy in a high school science class: A case of high school science teacher in an Indigenous high school. *Journal of Science Teacher Education, 32*, 518–536. https://doi.org/10.1080/1046560X.2020.1869886

Upadhyay, B., Barton, A. C., & Zahur, R. (2005). Teaching science in a poor urban school in Pakistan: Tensions in the life history of a female elementary teacher. *Science Education, 89*(5), 725–743. https://doi.org/10.1002/sce.20067

Upadhyay, B., Chaudhary, M. T., Gautam, D., & Tharu, B. (2018). Maghi: A case study of Indigenous Tharu cultural heritage for democratic practice and STEM education in Nepal. In J. Liljeblad & B. Verschuuren (Eds.), *Indigenous perspectives on sacred natural sites: Culture, governance and conservation* (pp. 136–151). Routledge.

Upadhyay, B., Regmi, J., & Sharma, U. (2010). Seeking for empowerment and social justice through science education in Nepal. In Y.-J. Lee (Ed.), *The world of science education: Handbook of research in Asia* (pp. 393–414). Sense.

Upadhyay, B., & Sadykova, S. (2024). Building science teacher leaders for indigenous schools: Lessons from a science professional development workshop in Nepal. *Education Sciences, 14*(9), 964–990. https://doi.org/10.3390/educsci14090964

Valenzuela, A. (1999). *Subtractive schooling: US-Mexican youth and the politics of caring*. SUNY Press.

Wehi, P. M. (2009). Indigenous Ancestral Sayings Contribute to Modern Conservation Partnerships: Examples Using Phormium Tenax. *Ecological Applications, 19*(1), 267–275. http://www.jstor.org/stable/27645964

Yosso, T. (2005). Whose culture has capital? A critical race theory discussion of community cultural wealth. *Race Ethnicity and Education, 8*(1), 69–91. https://doi.org/10.1080/1361332052000341006

Bhaskar Upadhyay is a professor of Comparative and International Development Education and STEM Education. His research explores issues of inclusion, equity, and culture in STEM education in the Global South contexts. He also examines STEM education in K-12 and higher education spaces through a globalization and social justice perspective.

Cuc Kim Vu is a Ph.D. candidate at the University of Minnesota. Her primary research interests are inclusive biology education and broadening participation in STEM fields.

Stephanie S. Erickson is a Ph.D. candidate in curriculum and Instruction in STEM education at the University of Minnesota. Her research interests include designing secondary science curriculum around culturally relevant problems and phenomena, particularly in the areas of Earth and Space science.

Chapter 21
What Does Educational Equity Mean? Views of the Harvard Teacher Fellows Program Members

Tarso B. Ferrari, Amadeu M. Bego, and Iñigo Rodríguez-Arteche

21.1 Introduction

In 2015, the United Nations launched the "2030 Agenda for Sustainable Development", which includes 17 Sustainable Development Goals (SDG) and 169 targets to be achieved by the 193 member countries by 2030 (United Nations, 2015). This agenda is both urgent and ambitious, marked by its wide scope and universal aspirations. Among its many principles and commitments, the 2030 Agenda advocates for a world where everyone has equitable and universal access to quality education, grounded in respect for human rights and human dignity.

Achieving goals of such scale and ambition demands sustained efforts and integrated actions over the medium and long term, particularly from developing countries. As such, implementing this new socioeconomic agenda involves mandatory actions in the field of education. Without education that fosters global solidarity, it is impossible to uphold the universal and equitable principles that form the foundation of social justice (Freire, 1995). Similarly, without education for sustainability, we cannot restore peaceful coexistence with the planet for future generations.

Particularly, SDG 4 seeks to ensure inclusive and equitable quality education (EQE) while promoting lifelong learning opportunities for all. In the realm of educational policies, EQE has been a prominent focus in both official documents and academic discourse. However, advancing science teacher education that prioritizes EQE remains a major global challenge (Burgess & Patterson Williams, 2022),

T. B. Ferrari · A. M. Bego (✉)
Institute of Chemistry, Sao Paulo State University, Sao Paulo, Brazil
e-mail: tarso.ferrari@unesp.br; amadeu.bego@unesp.br

I. Rodríguez-Arteche
Department of Education, Universidad de Almería, Almería, Spain
e-mail: inigorar@ual.es

G. Cakmakci, M. F. Tasar (eds.), *Connecting Science Education with Cultural Heritage*, Contributions from Science Education Research 15,
https://doi.org/10.1007/978-3-031-90944-3_21

particularly in countries like Brazil (Bego, Ferrari, et al., 2021; Gatti, 2017) and Spain (García-Herrero et al., 2024; Porlán, 2018).

While science is often promoted as universally accessible, inequities in engagement and achievement persist. Research shows that intersecting social identities—such as gender, age, socioeconomic background, and self-perceived learner roles—can influence young learners' attitudes and participation towards science (Archer et al., 2012, 2013; Barron, 2006; Papanastasiou & Papanastasiou, 2004). Studies also highlight how science education reproduces social inequalities, with success in the subject acting as a gatekeeper to higher education and elite positions. As such, current pedagogical practices often perpetuate inequities, limiting opportunities for disadvantaged students to succeed in science (Barton & Osborne, 2001; Jordan, 2010).

On this basis, other studies have been highlighting how teachers often perceive science as a rigid, objective, and Eurocentric endeavor, perpetuated by a narrow view of the scientific method and who is considered capable of scientific work (Cachapuz et al., 2005; Elgar, 2004). This view, further reinforced by the prevalence of male, white, European scientists in textbooks, neglects diverse student identities, especially in underdeveloped countries. Thus science education must move from a lecture-based, "naive" approach to an innovative pedagogical practice aimed at empowering students to challenge prevailing norms and assume science as a diverse, accessible endeavor for all (Ladson-Billings, 2022; Moreira & Candau, 2003).

For that reason, advancing science teacher education for educational equity requires, on the one hand, overcoming crystallized traditional pedagogical practices and, on the other, fostering research on how to prepare science teachers to ensure EQE (Burgess & Patterson Williams, 2022). Nevertheless, there is no consensual meaning for EQE in the specialized literature, as the term varies in meaning depending on context and ideological perspectives, i.e., EQE is highly polysemic (Bego, Ferrari, et al., 2021; Rodriguez & Morrison, 2019).

This complexity raises a question: *How can science teachers be effectively prepared to ensure inclusive and equitable quality education for all students?* One possible approach is to study successful programs explicitly committed to preparing teachers for EQE in order to identify their theoretical, practical, and organizational principles.

In this regard, we have been investigating a unique teacher education program called Harvard Teacher Fellows (HTF)[1] which was formally and officially designed to promote EQE. What makes the HTF a unique teacher education program? Beyond several noteworthy features, the HTF is a teacher residency program with a mission explicitly centered on preparing teachers to promote EQE. Previously, we investigated the concept and role of practice within the HTF program and its implication

[1] Due to the success of the HTF program and as part of the redesign of the master's programs, the Harvard Graduate School of Education launched, for the 2022–2023 academic year, the Teaching and Teacher Leadership (TTL) program, which incorporated and expanded the characteristics of the HTF. More information: https://www.gse.harvard.edu/ideas/news/21/10/next-chapter-teacher-education.

for novice science teacher preparation (Bego, Rodríguez-Arteche, et al., 2021). In this chapter, our aim is to explore how the HTF faculties and novice science teacher fellows conceive EQE and how these perspectives relate to the program's formative cohesion, particularly regarding its focus on equity.

21.2 Methodological Procedures

In 2020, we had the opportunity to spend 8 months at Harvard Graduate School of Education, where we conducted a case study on the HTF program. Our main research question was: *What are the main theoretical and practical aspects that characterize the HTF as a program focused on promoting educational equity*?

In order to achieve this bigger goal, we carried out a qualitative study[2] that triangulated data from multiple sources: participant observations, HTF official documents, and interviews with four faculty members and five novice science teacher fellows of the program.[3] In this paper, however, we focus specifically on the interview and document data to explore *how the HTF faculties and novice science teacher fellows conceive EQE and its relationship to the program's formative cohesion, particularly regarding equity.*

For this study, we adopted the eight-category framework proposed by Zeichner and Conklin (2008) to examine key features of teacher education programs. The second category of this framework addresses how program components are conceptually connected to *beliefs about teaching, learning, and schooling*. According to the authors, this analytical category seeks to characterize how key concepts are shared among teacher educators and participants, as well as the degree of commitment to these concepts. Thus, the program's mission is a crucial aspect of this characterization, as its definition implies specific assumptions about values and political orientation.

Following the principles of triangulation (Flick, 2008), we compared the conceptions of EQE found in HTF official documents with the views expressed by four faculty members and five novice science teacher fellows, based on interview data. In this process, we relied on the following primary documents: the HTF official website; Expected Teacher Qualities from the HTF program (ETQ); Core Practices for Fellows (CP); HTF Benchmark Indicators of Effective Teaching (BI); and course syllabi.

[2] Research carried out with human beings in the context of Harvard University must follow the ethical principles established in the Belmort Report. According to the procedures defined in our investigation, it did not fall into the category that requires approval by an ethical committee. More information: https://www.ecfr.gov/current/title-45/subtitle-A/subchapter-A/part-46/subpart-A/section-46.104.

[3] The HTF was a small program with 5 faculties, and the 2019–2020 cohort consisted of 28 fellows, 5 of whom were science fellows required to complete specific Science Methods courses.

Based on the analysis of the corpus of official documents, the next analytical step involved verifying the extent to which the HTF's mission and vision were shared by faculties and science fellows, as well as the degree of their commitment to these principles. To address this, we analyzed the interviews using the Discourse of Collective Subject (DCS) technique (Lefevre & Lefevre, 2006). The DCS is grounded in the Theory of Social Representation (TSR) proposed by Moscovici (2013), which aims to produce individual statements and manifestations that are representative of a collective thought. The TSR assumes that the main conceptions of a social group can be represented through analytical categories that allow individuals to recognize themselves in a discourse. Therefore, in our case, this methodological perspective can provide insight into the degree of shared vision between faculties and novice science fellows around EQE within the HTF program.

Using the DCS approach, we adhered to specific methodological principles (Lefevre & Lefevre, 2006), which involved identifying several elements in the interview transcripts:

1. *Key-Expressions (K-E)*: These are excerpts from responses or opinions selected by the researchers that convey the essence of the speech's content. They represent segments with significant semantic value;
2. *Core Ideas (CI)*: Derived from the selected K-E, CI are succinct terms or phrases that encapsulate their meanings. The goal is to distill the essence of similar and/or complementary discourses into a single unified statement, providing a more comprehensive understanding of a particular idea. CI can represent either the K-E of a single speech or the common K-E shared across multiple speeches; in the latter case, they are also referred to as categories.

The CI are quantitatively organized according to two parameters: intensity or strength (% of participants who share a given CI) and amplitude (total number of K-E associated with a given CI). Based on these parameters, the CI can be classified into four types: *high intensity and high amplitude* (IA), when the social representation (SR) is widely shared among members and is frequently expressed in each person's speech; *high intensity and low amplitude* (Ia), when the SR is weakly emphasized in the members' discourse but is broadly distributed across most individuals; *low intensity and high amplitude* (iA), when the SR is strongly held by some members but is concentrated in a smaller subset of individuals; *low intensity and low amplitude* (ia), which indicates isolated perspectives or representations held by only a few members.

A more holistic analysis of the interviews, using the lenses of K-E and CI and focusing on how these elements intertwine in the subjects' speech, enables the formulation of a series of collective discourses. These DCS are written in the first person, integrating information from various individual extracts to create "the effect of a collective opinion, voiced through the 'mouth' of a single subject of discourse" (Lefevre & Lefevre, 2006, p. 517). The number of generated DCS depends on predefined criteria related to qualitative attributes (K-E and CI) and its quantitative parameters (intensity and amplitude).

In our case, to guide our analysis and facilitate the construction of the DCS, we conducted a preliminary reading of the data. K-E relevant to the research problem were identified. These K-E were highlighted in bold and synthesized into a summary table to inform the subsequent designation of CI. This process revealed an interconnection between the K-E, the HTF BI, and program principles, including the relationships between EQE and social justice, the mitigation of inequities, and access to qualified teacher preparation for equity. This interconnection served as the basis for CI designation. To ensure methodological rigor and trustworthiness of the analytical codes, the Brazilian researchers engaged in a continuous process of review. Additionally, the Spanish researcher independently assessed the relevance of the codes and verified the quantitative counts of the K-E. Any discrepancies were resolved through team discussion and review until a consensus was reached.

With the summary table of K-E and corresponding CI completed, we calculated the quantitative parameters (intensity and amplitude). Based on these parameters, we constructed the DCS by transcribing the K-E verbatim to preserve fidelity to the original discourse, while incorporating appropriate connectives to ensure textual cohesion.

As such, based on the question *"What is your idea about education and equity, and how does it relate to the role of educational agents?"*, we generated six DCS, three for fellows and three for faculties, which represent discourses classified as *IA*, *Ia*, and *ia*, respectively.

21.3 Results and Discussion

Primary documents reveal that the mission of HTF is "to prepare teachers to promote educational equity through skillful teaching that prioritizes academic rigor, relationship building, and continuous learning". The concept of skillful teaching is further explained in the ETQ document, which states that the program aims to develop nine key qualities in its fellows: *(1) Act for equity; (2) Engage all learners; (3) Relationship-building; (4) Inquiry-based practice; (5) Deep content knowledge; (6) Integrate technology into practice; (7) Incorporate feedback and assessment into practice; (8) Continuous growth and reflective practice; (9) Leadership and professionalism.*

To achieve these goals, a list of core practices (CP) is defined to be mastered by fellows: *modeling; group work; facilitating a discussion; analyzing student work and providing feedback; creating and leading learning environments; designing and adapting a lesson or segment of instruction; engaging in reflective practice.* The program is committed to supporting fellows in learning CP through BI. These benchmark indicators are integrated throughout the coursework and fieldwork syllabi, and the HTF faculties use them for assessment purposes. Table 21.1 provides three examples of BI directly related to EQE.

The HTF BI are used to align the intended learning results of courses, foster collaboration among faculties, inform the program's evaluation, and establish a set of

Table 21.1 Some HTF benchmark indicators of effective teaching

Indicator	Components
Meeting diverse needs	MDN1. Articulate a specific, assets-based view of students and families MDN2. Use an understanding of assets, experiences, and background to engage students in rigorous learning MDN3. Apply Universal Design for Learning (UDL) principles to guide instruction
Safe learning environment	SLE1. Demonstrate care for students' academic and non-academic needs, redress harm, and promote healing SLE2. Use efficient and effective management and organizational routines SLE3. Use and promote inclusive language and language that explicitly names injustice
High expectations	HE1. Work with students to define, develop, and maintain high expectations for the classroom community HE2. Develop and implement learning experiences that are cognitively demanding for all students HE3. Support students to identify and meet their own ambitious goals

Table 21.2 Intensity and amplitude for each core idea (CI)

	Fellows (N = 5)		Faculties (N = 4)	
Core Idea	Intensity	Amplitude	Intensity	Amplitude
1—Meet diverse needs	100%	8x	75%	7x
2—High expectations	60%	3x	50%	3x
3—Educational equity as social justice	80%	5x	75%	3x
4—Overcoming inequalities	80%	4x	25%	1x
5—Access to equitable teachers	40%	3x	50%	2x

parameters regarding teaching practices. These indicators articulate a powerful set of benchmarks that teacher educators use to assess the performance of fellows throughout the program. Based on the document analysis, we identified that the BI are integrated into both coursework and fieldwork, enabling faculties to evaluate the development of fellows' performance. This demonstrates a high level of coherence within the program's curriculum in relation to its mission and principles.

Zeichner and Conklin (2008) emphasize the importance of assessing how widely the program's mission is shared among teacher educators and other agents. Therefore, we analyzed interview responses to search for coherence in the participants' views of equity. From the speeches of both science fellows and faculties, we identified five core ideas (CI) (Table 21.2) related to professional practice (CI-1&2), educational goals (CI-3&4), and educational opportunities (CI-5). Additionally, Table 21.3 shows the different DCS units assigned to both fellows and faculties.

When comparing the DCS generated for both science fellows and faculties, we find that the speeches with *IA* share only the CI-1. The most prevalent and widespread SR is concerned with the idea that teachers should address the diverse needs of their students. In turn, the speeches with *Ia* share the CI-2 and CI-3, indicating a connection between EQE and social justice, as well as emphasizing the importance

Table 21.3 Six collective discourses of HTF science fellows and faculties

DCS 1—Fellows - IA (CI-1):
"I would define this as making education accessible to all and ensuring that the level and quality of education is as equitable as possible. It is this matter of trying to make sure that every student has the same opportunities to learn, and all students have access to learning the same material in a way that meets their needs. This does not mean that students have to have exactly the same results, but they have opportunities to learn about it"
DCS 2—Fellows - Ia (CI-2; CI-3; CI-4):
"We need to look at educational equity and what structures are in place that, maybe, are harming our students, because of the system they were built within, which is a racist system and is not equitable for every, like, socioeconomic status or racial identity. Of course, within a school and across many schools, it is also important that all students have the same opportunities to strive for academic excellence, making sure that every student has access to equal opportunities, but also has access to getting to the same success point or same performance point. I think it is kind of social justice work to be trying to improve equity in schools"
DCS 3—Fellows - ia (CI-5):
"From my perspective, students must have equal access to teachers who care a lot about their subjects, and they have the help that they need to achieve in those classes too. I think that what you provide for each student might be different, but you must ensure that you're doing everything you can to allow all the students in a class to get a success point that is equitable"
DCS 4—Faculties - IA (CI-1):
"When I think about educational equity, I think about learning about the desires and aspirations of the learner and ensuring that those, as they pertain to educational environments, that those aspirations and goals have opportunities to flourish to be met. Then, it's not only in response to what people want, but it's also we must assume that education systems embodied in schools and teachers also have a vision for what people need that is not done independently of the participants, it's not an imposition, it's not just purely a top-down thing"
DCS 5—Faculties - Ia (CI-2; CI-3; CI-5):
"Firstly, I do think of education and educators as having the ability to help students create a different world rather than just a kind of plan for the world we have now, but those are always in tension. Besides, when we talk about education equity we can focus on outcomes, which is to say that everyone leaves an education setting having learned something that they care about, everyone leaves an education setting feeling like they have value as a community member, everyone leaves thinking that they now have something more to contribute to the world. However, educational equity also means giving to each student what they must achieve academic excellence and fulfillment in their lives. So, I would define equity as having a similar goal set for different people in different groups across contexts and providing those groups different resources to meet the same goal"
DCS 6—Faculties - ia (CI-4):
"I think that where ideas of social justice intersect with ideas of equity isn't thinking about how we disrupt some of those structures that have been oppressive—that has not served all students and, in fact, have privileged some—but either how some students are completely oppressed and that marginalization that occurs inside of schools gets compounded as young people become adults. So, both schools reflect society and society also, if we think of schools as a part of the solution, is replicating schools"

of maintaining high learning expectations for all students. However, the DCS with *ia* for science fellows and faculties do not match. The SR of science fellows connects to access to equitable teachers (CI-5), while the SR of faculties is linked to overcoming inequalities (CI-4), which are considered less relevant for EQE.

As noted, the highly shared DCS among teachers and science fellows primarily revolve around the understanding of EQE as a means of addressing the diverse needs of all students while also upholding high expectations for their learning. Notably, this identification of educational needs is not exclusively approached as a top-down process but rather as a negotiation with learners, considering their aspirations and identities (DCS 4). This approach aligns with recommendations from experts in EQE within science education (Rodriguez & Morrison, 2019). However, ideas such as EQE being linked to overcoming inequalities and access to equitable teachers are merely isolated SR. This suggests that political and social themes are not so strongly emphasized in the HTF program, according to the views of its main stakeholders. In addition, these themes are not so explicitly reflected in the curriculum and key concepts of the program. These aspects, which warrant further exploration in future research, seem to indicate that associating EQE with the development of skillful teaching through practice-based teacher education may lead to a focus on *Practical Rationality* (Contreras, 2012).

Aspects related to the development of a critical sociopolitical consciousness, if not addressed by teacher educators, risk being relegated to a secondary level. Critical reflection involves evaluating concrete situations beyond individual intentions and personal actions—moving beyond reflections of "practice for practice's sake". In this way, the various dimensions that influence science teaching practice should be understood within their social, historical, and economic contexts (Archer et al., 2012; Barton & Osborne, 2001). Criticism that emphasizes the need for transformative perspectives, whether regarding institutional working conditions or broader social dynamics, is essential for fostering competent, non-alienated professional practice (Contreras, 2012; Porlán, 2018).

Despite this criticism, when considering the second category of Zeichner and Conklin's (2008) framework, we find that HTF faculties and science fellows share convergent views on their conceptions of EQE and education in general. This alignment allows us to conclude that the program's central ideas about its mission and principles are widely shared among its members. In other words, the HTF can be regarded as a well-structured teacher preparation program that remains consistent with its mission.

21.4 Conclusions

The HTF program is organized around professional practice as its core focus. The concept of EQE is related to preparing novice teachers to become transformative educators by adopting standards of skillful teaching that they are expected to enact. As a mission and a transversal theme, EQE entails a shared perspective (*shared lenses*) among program participants regarding the essential principles and concepts that should shape the community's practices, in addition to guiding the assessment parameters and teaching practice performance. Thus, the value of EQE underpins the principles of skillful teaching and helps define the core teaching qualities and practices essential to the preparation of fellows.

The explicit mission of preparing teachers to become transformative educators within their communities, combined with a framework that clearly defines the elements of EQE, seems to be essential for designing impactful formative experiences. This framework consolidates a shared conceptual foundation and language, serving as the basis for curricular design and faculty practices. As a corollary, the use of performance indicators (BI) serves a dual purpose: evaluating and monitoring the progress of fellows while also enabling adjustments of their performance and the program structure itself.

Extending this focus, the DCS highly shared among both faculties and science fellows are closely related to the conception of EQE as addressing the diverse needs of all students while maintaining high expectations for their learning. Overall, faculties and science fellows of the HTF program show confluent discourses about what they conceive as EQE. This finding led us to conclude that the main ideas surrounding EQE are widely shared among program members and align with the program's mission and pedagogical principles. Therefore, we concluded that the HTF can be regarded as a well-structured and coherent teacher education program.

Despite these strengths, ideas linking EQE to overcoming inequalities and ensuring access to equitable teachers remain isolated SR, indicating that political and social topics—inseparable from EQE—have not been a central focus of the HTF program as well. Our investigation identified that discussions about teacher education for EQE, based on performance pedagogy under the practice-based teacher education approach, may be limited to the commitments of Practical Rationality. While the HTF approach represents significant progress in science teacher education, incorporating critical perspectives could further enrich the understanding of social and institutional constraints. This deeper comprehension would enable teachers to identify and advocate for transformative actions that extend beyond the classroom and individual practices—actions aimed at improving educational quality and promoting greater equity. The awareness of teachers to critically reflect on the conditions influencing educational action should be dialectically integrated with transformative social actions. These actions, combined with skillful teaching practices, could significantly enhance professional praxis (Freire, 1995). Aligning these aspects more closely with an EQE perspective could result in a science teacher education genuinely committed to advancing social justice.

At last, while our investigation offers insights into the HTF program's approach to EQE, some limitations warrant consideration. First, the emphasis on a practice-based teacher education framework, while valuable, may constrict the conceptualization of EQE by potentially overlooking broader sociopolitical dimensions, particularly those related to systemic inequities and the imperative for transformative social action. Second, the study's reliance on self-reported data from a limited sample of fellows and faculties within a single program inherently constrains generalizability—a common constraint of case study methodology. Lastly, while the DCS method offers a nuanced approach to qualitative analysis, the researcher's interpretation of K-E and CI introduces an element of subjectivity.

Future research should emphasize the integration of diverse data sources and explore the complex interplay between individual pedagogical practices and broader systemic factors in fostering equitable science learning for all students, especially those from marginalized backgrounds, resonating with Barton and Osborne's (2001) concerns regarding marginalized discourses. Additionally, we underscore the importance of future studies that examine the dimensions of EQE and strategies for preparing more equitable science teachers, thereby contributing to achieve the SDG by nations.

Acknowledgments Grant 2022/06924-9, São Paulo Research Foundation (FAPESP). Project PID2023-150682NA-I00, Spanish Ministry of Science, Innovation, and Universities, MCIU/AEI/10.13039/501100011033/ FEDER, UE.

References

Archer, L., DeWitt, J., Osborne, J., Dillon, J., Willis, B., & Wong, B. (2012). Science aspirations, capital, and family habitus. *American Educational Research Journal, 49*(5), 881–908. https://doi.org/10.3102/0002831211433290

Archer, L., DeWitt, J., Osborne, J., Dillon, J., Willis, B., & Wong, B. (2013). 'Not girly, not sexy, not glamorous': Primary school girls' and parents' constructions of science aspirations. *Pedagogy, Culture & Society, 21*(1), 171–194. https://doi.org/10.1080/14681366.2012.748676

Barron, B. (2006). Interest and self-sustained learning as catalysts of development: A learning ecology perspective. *Human Development, 49*(4), 193–224. https://doi.org/10.1159/000094368

Barton, A. C., & Osborne, M. D. (2001). Marginalized discourses and pedagogies: Constructively confronting science for all in classroom practice. In A. C. Barton & M. D. Osborne (Eds.), *Teaching science in diverse settings: Marginalized discourses and classroom practice* (pp. 7–32). Peter Lang.

Bego, A. M., Ferrari, T. B., & Pereira, V. M., Jr. (2021). Formação de professores de química com foco na equidade educacional: conceitos, cenários e perspectivas [Chemistry teacher education for educational equity: concepts, sceneries, and perspectives]. *Debates em Educação, 13*(2), 22–48. https://doi.org/10.28998/2175-6600.2021v13nEsp2p22-48

Bego, A. M., Rodríguez-Arteche, I., & Pereira Jr., V. M. (2021). *Concepto y papel de la práctica docente en el programa Harvard Teacher Fellows [The concept and role of teaching practice in the Harvard Teacher Fellows program]*. Paper presented at the 9th Congreso Internacional sobre Investigación en Didáctica de las Ciencias, Lisboa, Portugal. Retrieved April 2025, from https://congresoenseciencias.org/wp-content/uploads/2021/09/Actas-Electronicas-del-XI-Congreso_compressed.pdf.

Burgess, T., & Patterson Williams, A. (2022). Utilizing theory to elucidate the work of creating equity for transformation within the science classroom. *Science Education, 106*(5), 1071–1083. https://doi.org/10.1002/sce.21721

Cachapuz, A., Gil-Perez, D., Carvalho, A. M. P., Praia, J., & Vilches, A. (2005). *A necessária renovação do ensino das ciências [The needed reform of science education]*. Cortez Editora.

Contreras, J. (2012). *A autonomia de professores [Teacher autonomy]* (2nd ed.). Cortez Editora.

Elgar, A. G. (2004). Science textbooks for lower secondary schools in Brunei: Issues of gender equity. *International Journal of Science Education, 26*(7), 875–894. https://doi.org/10.1080/0950069032000138888

Flick, U. (2008). *Designing qualitative research*. Sage.
Freire, P. (1995). *Pedagogy of hope: Reliving pedagogy of the oppressed*. Bloomsbury.
García-Herrero, M., Rodríguez-Conde, M. J., & Martínez-Abad, F. (2024). Factores de calidad docente asociados a la equidad educativa: formación del profesorado y estrategias docentes [Teacher quality factors associated with equity in education: Teacher development and teaching strategies]. *Revista Electrónica Interuniversitaria de Formación del Profesorado, 27*(1), 75–88. https://doi.org/10.6018/reifop.595181
Gatti, B. A. (2017). Formação de professores, complexidade e trabalho docente [Teacher education, complexity and teaching work]. *Revista Diálogo Educacional, 17*, 721–737. https://doi.org/10.7213/1981-416x.17.053.ao01
Jordan, W. J. (2010). Defining equity: Multiple perspectives to analyzing the performance of diverse learners. *Review of Research in Education, 34*(1), 142–178. https://doi.org/10.3102/0091732X09352898
Ladson-Billings, G. (2022). *The dreamkeepers: successful teachers of African American children* (3rd ed.). Jossey-Bass.
Lefevre, F., & Lefevre, A. M. C. (2006). O sujeito coletivo que fala [The collective subject that speaks]. *Interface: Comunicação. Saúde, Educação, 10*, 517–524. https://doi.org/10.1590/S1414-32832006000200017
Moreira, A. F. B., & Candau, V. M. (2003). Educação escolar e cultura(s): construindo caminhos [School education and culture(s): building up some paths]. *Revista Brasileira de Educação, 23*, 156–168. https://doi.org/10.1590/S1413-24782003000200012
Moscovici, S. (2013). *Representações sociais: investigação em psicologia social [Social representation: investigation on social psychology]*. Vozes.
Papanastasiou, C., & Papanastasiou, E. C. (2004). Major influences on attitudes towards science. *Educational Research and Evaluation, 10*(3), 239–257. https://doi.org/10.1076/edre.10.3.239.30267
Porlán, R. (2018). Didáctica de las ciencias con conciencia [Conscious Science Education]. *Enseñanza de las Ciencias, 36*(3), 5–22. https://doi.org/10.5565/rev/ensciencias.2795
Rodriguez, A. J., & Morrison, D. (2019). Expanding and enacting transformative meanings of equity, diversity and social justice in science education. *Cultural Studies of Science Education, 14*(2), 265–281. https://doi.org/10.1007/s11422-019-09938-7
United Nations. (2015). *Transforming our world: 2030 agenda for sustainable development*. Retrieved March 2024, from https://sdgs.un.org/2030agenda.
Zeichner, K., & Conklin, H. G. (2008). Teacher education programs as sites for teacher preparation. In M. Cochran-Smith et al. (Eds.), *Handbook of research on teacher education: Enduring questions in changing contexts* (pp. 269–289). Routledge.

Tarso Bortolucci Ferrari holds a PhD in Chemistry, with an emphasis on Chemistry Teaching, from the Sao Paulo State University. He also holds a Master's degree in Science Education and a Bsc degree in Chemistry, both from the Sao Paulo State University. His research interests focus on teacher education, particularly instructional planning and educational equity. He is currently a Chemistry teacher in elementary and high school and a member of the Brazilian research group "Rede de Inovação e Pesquisa em Ensino de Química" (RIPEQ).

Amadeu Moura Bego is an associate professor at the Sao Paulo State University, Brazil. His research focuses on science teacher education, educational innovation, and instructional planning. He is currently leading the Brazilian research group "Rede de Inovação e Pesquisa em Ensino de Química" (RIPEQ).

Iñigo Rodríguez-Arteche is an associate professor at the University of Almería, Spain. His research focuses on science teacher education, model-based inquiry, emotions in science teaching and learning, and non-formal science education. He is currently the principal investigator of a project funded by the Spanish Ministry of Science, Innovation, and Universities (PROCUSTO, PID2023-150682NA-I00) on teaching core practices for science teacher training.

Part IV
Societal and Global Challenges in Science Education

Chapter 22
Exploring Narratives of Change for Imagining the Future and Fostering Agency Toward Sustainability

Giulia Tasquier, Erik Knain, Hanna Røkenes, and Alfredo Jornet

22.1 Introduction

We live in times of unprecedented change, with major economic, environmental, geopolitical, societal and technological challenges that coincide and amplify each other. As the Icelandic writer Andri Snaer Magnason wrote: *"we are faced with changes much greater than what our mind is used to, more challenging than any of our previous experiences"* (Magnason, 2019)—changes that strongly encourage us to rethink our roles as agentic humans in this world. Again, as Magnason warns: *"time is running and even if Nature has left geological speed of change, our reactions seem to be still on a geological scale"* (ESERA conference 2019). This is also combined with the perspective raised by the South Korean-born Swiss-German philosopher Byung-Chul Han, who supports the idea that acceleration of social rhythms has produced, in everyday life, a widespread cult of urgency capable of eroding the possibility of control by individuals, who are forced to measure themselves against an epochal trait of uncertainty and ungovernability of the future (Han, 2019, 2021). However, as Bruno Latour (2021) stresses in the post-face of the *Atlas of Anthropocene*, the great advantage of being faced with the Anthropocene is that we no longer have to deal with a natural problem, in front of which we would be without strength and resources, but we are faced with social decisions which we can

G. Tasquier (✉)
Department of Physics and Astronomy A. Righi, ALMA MATER STUDIORUM - University of Bologna, Bologna, Italy
e-mail: giulia.tasquier2@unibo.it

E. Knain · H. Røkenes
Department of Teacher Education and School Research, University of Oslo, Oslo, Norway

A. Jornet
Department of Psychology, University of Gerona, Gerona, Spain

G. Cakmakci, M. F. Tasar (eds.), *Connecting Science Education with Cultural Heritage*, Contributions from Science Education Research 15,
https://doi.org/10.1007/978-3-031-90944-3_22

"easily" oppose. We have the responsibility and potential to build our futures and this requires unprecedented actions (IPCC, 2021).

We face changes of epochal significance and, despite growing attention from many areas as well as the demands from young people who take to the streets in vociferous protest, calling on adults to respect their responsibility and duties, classroom science still seems unable to assist students trying to make sense of these pressing socio-scientific questions. The fact that almost half of national curricula around the world do not refer explicitly to climate change issues (UNESCO, 2020) is a prime example of the delay in action within formal education regarding societal challenges. This clashes with recent reports (e.g., Bianchi et al., 2022; OECD, 2018) suggesting the need for schools to promote students' awareness of present and future issues, preparing them to deal with the complexity of real-life challenges.

In the context of complex problems such as today's sustainability and climate change, narratives become a primary and crucial means for integrating scientific knowledge, authentic inquiry, the critical thinking skills needed to pursue solutions, and the goal-oriented, affective and active engagement in action (Tasquier et al., 2022). Why do narratives matter? As Schwab and Malleret (2021) argue, as human beings *"we are storytelling creatures"*, and the stories we tell are our fundamental tool of communication and transmission, which serves as a way of understanding others (Avraamidou & Osborne, 2009; D'Orto & Tasquier, 2025). Narratives are how we make sense of life. Most importantly, compelling narratives have the power to inspire us to act (Schwab & Malleret, 2021). So, people make sense of the world and give meaning to life as reflected in the everyday *micro-narratives* (or small stories) that people tell in their social interactions (Van der Merwe et al., 2019). But why are micro-narratives so important? Schwab and Malleret (2021) support the idea that the constellation of important interrelated small stories can coalesce around a larger central story. Micro-narratives contribute more to participatory modes of sensemaking than big stories do. Furthermore, these fragments collectively disclose identities, motivations, and attitudes. They address a broad spectrum of issues aiming to shed light on what is coming and to offer some clarity on our options in terms of possible collective responses (Schwab & Malleret, 2021). Thanks to micro-narratives, young people have the opportunity to figure out who they are and envision plausible futures for themselves and their surroundings by making use of cultural resources in social relationships with the people involved in these worlds. This includes aspects such as personal and political agency, the extent to which they experience despair versus hope or detachment versus empowerment regarding the present and future, and their narratives related to climate change and transformations to sustainability. This latter aspect is the hypothesis on which we based the empirical research and the activities described in this paper, which attempt to answer the following question: *To what extent can narratives provide insights into the current attitudes of young people about their sense of agency in relation to climate change and sustainability challenges?*

22.2 Theoretical Framework

22.2.1 Background

This study is framed within the Horizon 2020 project entitled *Science Education for Action and Engagement toward Sustainability* (SEAS), which is aimed at promoting new forms of scientific literacy and skills to empower students to become agents of change. Most centrally, SEAS aims at incorporating a *transformative and agentic dimension* into scientific literacy and, especially, into teaching and learning climate change and sustainability, which is often lacking in current conceptualizations of scientific literacy (Tasquier et al., 2022).

A focus on scientific literacies has during the last decades expanded the ways in which science is conceptualized and taught in (and across) formal and informal education, not least through approaches that engage learners in addressing so-called *socioscientific issues* (SSI). By inquiring into SSIs, opportunities emerge to integrate scientific knowledge into social, political and moral dilemmas of societal and personal relevance (Sadler et al., 2016). However, with learning through SSIs there come new challenges with regards to applicability of knowledge, where science as traditionally taught in schools has no self-evident role outside of schools but needs to be negotiated along with social concerns and other sources of knowledge. Formal education seems still to struggle with confronting the crises and uncertainties of the Anthropocene (UNESCO, 2021). Extensive case studies in Norway focusing on interdisciplinary teaching and societal challenges, several including science, showed that teachers found it difficult to establish connections between disciplinary knowledge and students' real-life experiences, that students found it difficult to handle the complexity involved, and rarely evaluated content of information sources. Schools were also cautious about the sometimes-controversial aspects of societal challenges, leading to important contradictions and tensions to be downplayed, and solutions to challenges tended to emphasize individualized solutions at the expense of political and structural solutions (Furberg et al., 2023, 2024). If students and teachers are not supported in their inquiries into complex societal challenges, complex problems tend to be replaced by simpler ones (Byhring & Knain, 2016). A main model in SEAS to guide and analyze learning and transformation is the heuristic *model of the three spheres of transformation* (O'Brien & Sygna, 2013), which defines transformation as a process taking place across three interacting spheres: (1) a *practical sphere* which includes both technical and behavioural changes that contribute to the solution of climate change and sustainable issues; (2) a *political sphere* which highlights the systems and structures that facilitate or hinder transformations in the practical sphere; (3) a *personal sphere*, which highlights the importance of individual and collective worldviews, values, beliefs and paradigms that are at stake and which drive people's motives for practical and political action. In the SEAS project's approach, embedding a transformative and agentic dimension into scientific literacy means that school science has to be re-conceptualised and re-configured to support the activation of all three spheres, with the overall goal that both individuals and their institutional contexts change as participants gain agency both over their learning processes and over the institutional and political conditions that contextualise those processes. This is recognisable as an individual/collective movement that puts into a dynamic relationship the agentic role both of individuals towards the system and the system towards individuals (Tasquier et al., 2022; Tasquier & Jornet, 2023).

When changes are secured at these three levels, opportunities emerge for what in research on learning and education has been identified as *expansive learning,* which refers to the in-depth learning that takes place when individuals engage in actively and consciously addressing, transforming and re-conceptualising the goals of their collective practice and activity systems (Engeström & Sannino, 2010). From this perspective, incorporating a *transformative and agentic dimension* into scientific literacy can be considered as generating learning opportunities, for it is in the negotiating of establishing meaningful goals that conceptual, pragmatic-political and affective-motivational aspects of learning emerge. Indeed, from an expansive learning perspective, the goal is that both individuals and their institutional contexts change as participants gain agency over their learning processes and the institutional and political conditions that contextualise those processes. The SEAS' core ideas, tools and methodology are designed to support the development of literacy, engagement and empowerment for driving transformation in the three spheres. One of the approaches for exploring students' narratives is the concept of *narratives of change* which was translated into the use of a specific tool named SenseMaker® (https://cynefincentre.com; Snowden & Boone, 2007). In this context, micro-narratives serve as pedagogical means for fostering agency, making scientific concepts relatable, and linking them to sustainability challenges.

The SEAS project's use of the SenseMaker tool exemplifies, engaging students in the collection and analysis of (micro-)narratives to track emerging narratives of change and foster imagination and scenario building (Barelli et al., 2022; Levrini et al., 2021) throughout a sensemaking process (Snowden & Boone, 2007).

22.2.2 Frameworks and Core Ideas

Sensemaking

Sensemaking, as conceptualised by Snowden and Boone (2007), involves the process of interpreting and understanding complex and ambiguous situations. In educational contexts, sensemaking enables students to navigate the uncertainties and complexities of scientific and socio-scientific issues. When combined with narratives, sensemaking can be a powerful tool for fostering critical thinking and problem-solving skills (Bengtsson et al., 2020). Among the core ideas of the SEAS project, there is the assumption that learning must be approached and conceptualised in terms of "meaning-making and sense-making" (Bengtsson et al., 2020). Meaning-making is primarily understood as engagement in the construction of meaning, using symbolic and logic resources to do so. Sense-making is primarily a certain emotive response of correspondence between the experience of the world and efforts to give meaning to it (Bengtsson et al., 2020). Thus, it relates to the experience of making sense, whether perceptions of the world and attempts of meaning-making are congruent or not. Inquiry can be seen as necessary when there is a discrepancy, where such discrepancy highlights how experience does not make sense. Also, for learners to engage in inquiry, those choosing to engage in inquiry need to ascribe a certain degree of significance to the need to make sense. That is to say, while there might be a lot of non-sense experienced, not all non-sense is deemed significant or worthy of a process of inquiry. Within the project, an inquiry is seen to engage in meaning-making and sense-making. Accordingly, the process of inquiry is not pre-determined in the sense that it does not simply re-articulate the meaning of sustainability issues and challenges, according to pre-existent scientific knowledge. A core assumption is that this scientific knowledge relevant to sustainability issues and challenges is a resource in meaning-making but not a priori determinant in meaning-making and sense-making. In-situ experience and practice attempts at giving meaning to a sustainability issue/challenge fail to do so in terms of scientific knowledge exclusively. There is at play in this practical and experiential engagement with these issues/challenges an encounter of something more, and critical thinking and reflexivity are ways in which learners give meaning to, and make sense of, this surplus. Existing scientific knowledge is not a simple given in meaning-making and sense-making but may become powerful in change-making, as scientific knowledge and its associated technical ways of giving meaning are re-shaped in the light of sustainability issues/challenges. Meaning-making and sense-making in scientific inquiry try to keep ambivalence, complexity and uncertainty at bay by aiming to keeping them as entry points for engaging in sustainability challenges without treating them as external from the process of learning.

The intersection of sense-making and narratives in science education allows students to create coherent stories that link their personal experiences with broader scientific concepts. This approach not only enhances their understanding of scientific principles but also helps them to see the relevance of science in addressing real-world problems (Snowden & Boone, 2007). Narratives and sense-making are particularly effective in this regard, as they enable students to construct and communicate complex ideas in a more accessible and engaging manner (Han, 2021). By incorporating these approaches, the SEAS project aims to enhance students' capacity to understand and address the complexities of sustainability and climate change challenges.

Narratives

Narratives play a central role in the SEAS project, serving as a means to operationalise and make visible the changes and progress occurring throughout the open schooling sustainability challenges—challenges in which the school collaborates with out-of-school partners to address issues of relevance to their local communities. Narratives are well-suited to capture the individual and collective processes targeted by the project. They are more fundamental and closely tied to an individual's lifeworld than the more specialised discourses. For humans, narratives are essential for meaning-making; they are cultural tools for representing the past and projecting the future (Van der Merwe et al., 2019). Telling stories allows people to make sense of the chaotic interplay of actions, structures, gaps, and ruptures characteristic of human life (Gottschall, 2018).

In addressing complex problems, like today's sustainability and climate change challenges, narratives become crucial for integrating scientific factual knowledge, authentic inquiry, and critical thinking skills needed to pursue science-based solutions (Tasquier & Pongiglione, 2017). Narratives also foster goal-oriented, affective, and active engagement in actual actions. People figure out who they are and envision plausible futures for themselves and their surroundings by using cultural resources within social relationships with those who share their world (Laherto & Rasa, 2022; Levrini et al., 2021). Focusing on narratives also acknowledges the importance of discursive and argumentative aspects of critical thinking in the contemporary information society (Erduran, 2007). Such critical thinking is essential as young people today have access to an almost infinite amount of scientific information, including contradictory information on climate change, where evidence-based scientific consensus is presented alongside competing populist narratives (Røkenes & Jornet, 2025).

Narratives are useful for connecting people across different backgrounds. Specialised practices, such as those in scientific disciplines, build upon narrative forms, just as everyday discourse practices do. These worlds are semiotically construed and mediated by sign systems employed in talk, writing, visual representations, gestures, bodily postures, and architecture (Jornet & Roth, 2015). These systems normalise particular beliefs, values, and ways of acting and talking, and are therefore pragmatically tied to action (Tasquier et al., 2024). They are means for communication among people and are also tied to the affective and motivational aspects that underlie the development of understanding when learners encounter scientific knowledge and representations.

As learners from different contexts collaborate to address shared sustainability goals, drawing from scientific discourse and resources, they create narratives that include argumentative and rhetorical forms of both every day and scientific domains (Erduran, 2007). These narratives intertwine practical, political, and personal aspects represented in the three spheres of learning (O'Brien & Sygna, 2013). By linking the present to the future through participation in a local sustainability challenge, these cultural tools empower citizens to become agents of change using science (Tasquier et al., 2022).

Narratives of Change

In the SEAS project, narratives are operationalised according to the notion of *narratives of change*. These narratives become resources and indicators for learners to continue collaborating and building increasingly rich scientific literacies, and for consortium partners to assess progress in terms of practical accomplishments, scientific literacy, and interest in science. More generally, narratives of change refer to the stories and accounts that articulate how and why transformations occur in individuals, communities, and societies. These narratives encompass a wide range of elements, including personal experiences, historical events, cultural beliefs, and future aspirations. They provide a lens through which people can interpret their surroundings, understand the dynamics of change, and envision potential futures (Polkinghorne, 1988).

The concept of narratives of change has roots in various disciplines, including psychology, sociology, anthropology, and education, each contributing to its development and application.

In psychology, narratives of change are often linked to personal identity and development. Individuals construct and reconstruct their life stories to make sense of their experiences and guide their actions (McAdams, 1993). In sociology and anthropology, narratives of change are used to examine how communities and societies evolve, highlighting the role of culture, power, and agency in shaping social transformations (Riessman, 2008). These disciplines underline how narratives shape and are shaped by the broader social and cultural context, revealing the intricate relationship between individual and collective transformation.

In educational contexts, narratives of change are employed to enhance teaching and learning processes by making them more relevant and engaging. They serve as pedagogical tools that enable students to connect theoretical knowledge with real-world issues, thereby fostering a deeper understanding and appreciation of the subject matter. Narratives of change can be particularly effective in disciplines such as history, literature, and social studies, where they help students to explore the causes and consequences of historical events, the development of cultural identities, and the dynamics of social change.

In science and sustainability education, narratives of change facilitate a holistic understanding of complex socio-scientific issues. They help students to grasp the interconnections between scientific concepts, environmental challenges, and human behaviours. By framing scientific knowledge within the context of personal and societal narratives, educators can make abstract concepts more tangible and relevant to students' lives. This approach not only enhances cognitive understanding but also promotes emotional engagement and motivation to learn (Bruner, 1991).

In the European context, narratives of change are integral to various educational initiatives and policy frameworks aimed at fostering sustainability and social cohesion. The European Union (EU) has recognised the importance of narratives in shaping public attitudes and behaviours towards sustainability. For instance, the European Green Deal emphasises the need for a new growth strategy that transforms the EU into a modern, resource-efficient, and competitive economy. This strategy is underpinned by narratives of change that envision a sustainable future and motivate collective action towards achieving this.

The notion of narratives of change used in the project addresses the individual and collective processes of learning (Bengtsson et al., 2020). Narratives connect through a process of inquiry into sustainability, issues as experienced by a subject existing in and engaging with the world (lifeworld) also interacting with more specialised discourses of public life and disciplinary knowledge. Narratives provide means of "meaning- and sense-making", where learners draw on specialised discourses but also use common language to create cohesion in the chaotic interplay of actions, structures, gaps and ruptures as experienced in their lifeworld. Narratives opens for integrating scientific knowledge, critical and reflexive thinking and empathy into inquiry and practice that has particular goals and local circumstances in mind when addressing sustainability issues. Narratives also involve "anticipatory thinking", envisioning alternate futures drawing on diverse technical and common cultural resources by relating to people and objects making up their environment. In this way, narratives connect different individual stories (people) as well as different plots (e.g. competing scientific explanations).

To sum up, the concept of narratives of change offers several key benefits in educational contexts. In the SEAS project, the use of narratives of change is integral to its transformative and agentic approach to scientific literacy and climate change education. By engaging students in the construction and analysis of narratives, SEAS aims to enhance their understanding of sustainability issues, foster critical thinking, and empower them to take action towards a more sustainable future. Specifically, narratives of change are considered as a means of operationalising and revealing change and progress towards social transformation occurring throughout open schooling. Emerging narratives of change become visible and shareable as textual objects (narratives from students), which mediate both individual and collective learning in learning trajectories across places, time, and encounters with experiences of phenomena, people and texts.

22.3 Design of the Study and Method

The worlds that narratives create are semiotically construed and mediated by sign systems employed in the talk, writing, visual representations, gestures, bodily postures, and architecture of the learners. As learners from different contexts collaborate to address shared sustainability goals, drawing from scientific discourse and resources, they create narratives that include argumentative and rhetoric forms of both every day and scientific domains, and which together intertwine personal and collective forms of meaning-making and sense-making. When they meet in narratives, individual stories and alternative plots open up for different learning trajectories, multiplying notions of past, present, and future. In the SEAS project, narratives of change are operationalised to render visible (indicators of) change, both as a form of transformation of the learner (transformative learning) and as project output.

In this study, we analyse data collected through the use of a specific digital tool, SenseMaker® (https://cynefincentre.com) developed by Prof. Dave Snowden on the Cynefin sensemaking framework (Snowden & Boone, 2007), which aims to explore and understand the ambiguous dynamics of complex social systems by capturing and making sense of people's attitudes, perceptions, and experiences. It is used by researchers and decision-makers for monitoring and evaluating; mapping ideas, mind-sets and attitudes; and detecting trends and weak signals (Caspari, 2020; CognitiveEdge, 2021). SenseMaker was re-adapted within the SEAS project with the aims of: engaging young people in sustainability challenges, keeping track of emerging narratives of change during scientific learning (Levrini et al., 2021; Bengtsson et al., 2022), as well as fostering imagination and scenario building through the use of micro-narratives. SenseMaker links micro-narratives with human sensemaking. In this way, decision makers are enabled to listen more deeply and capture the dreams, fears, concerns, ideas, trends, and early warning signs of changes within a group of people. They can identify patterns and themes in the stories and identify collectively-held beliefs and contradictions. Methodologically, SenseMaker collects narratives from respondents and combines these with quantitative data and pattern visualisation. This quantitative data is created by the respondents themselves who self-define the narratives they have provided by *signifying* the responses they gave. Through this *self-signification* (Caspari, 2019), SenseMaker is able to combine the qualitative data with the respondent's own self-generated quantitative data. Importantly the power of interpretation sits with respondents, not with third parties e.g., researchers, who interpret the qualitative data on behalf of the respondents.

More specifically, research questions are built into the SenseMaker instrument as *signifiers* i.e., concepts that anchor the plotting of respondents' micro-narratives in space. Building on these signifiers, a *signification framework* is created. This consists of predetermined questions that guide the inquiry, solicit micro-narratives, and clarify respondents' interpretations of the narratives they provide.

Signification takes place when respondents answer questions about the micro-narrative they shared. Numerical coordinates are associated with the signifiers in the backend. This means qualitative and quantitative data can be assessed in parallel.

When the set of signifiers from all of the respondents are plotted, they display mathematical patterns that reveal a map of the social landscape. This becomes the basis for exploring the relevance of patterns, underlying relationships, norms, and dynamics in the social system being investigated (Van der Merwe, 2019).

This set of data that emerges provides a rich snapshot of reality and helps analysers/leaders to make sense of complex, ambiguous, and rapidly changing environments in order to form insights into meaning making, to find effective opportunities to intervene in a system, and to monitor changes (https://sensemaker.cognitive-edge.com/).

The SEAS SenseMaker capture was co-designed with a group of researchers form SEAS together with the SenseMaker team[1] (Bengtsson et al., 2022).

The construction of the SEAS SenseMaker questionnaire is designed to:

- Keep track of emerging ***narratives of change*** during learning and development processes
- Examine the development of ***narratives of change*** and how different aspects of scientific literacy trigger transformational action/change

[1]The first initial pilot SenseMaker capture was developed in close collaboration with cCHANGE, University of Oslo partner, Anne Caspari at EZC Partners (https://ezc.partners/), and Beth Smith at The Cynefin Centre.

The structure of the questionnaire followed the common structure of a typical SenseMaker item with a (1) beginning with micro-narratives and (2) questions for giving *self-signification* of the narratives.

The goal of the initial micro-narratives (1) was to seek stories that provide insights into the current perspectives of young people about their sense of agency in relation to climate change and sustainability challenges. This includes, for instance, personal agency and political agency, and the extent to which they experience hope versus despair, and detachment versus empowerment, regarding the present and future and their narratives related to climate change and transformations to sustainability.

The *self-signification* questions (2) were organised in *multiple choice questions*, aimed at clarifying aspects of the micro-narrative or capturing the participants' demographics; *dyads*, sliders used to indicate the relative strength of a belief, or outcome on a continuum between two opposite extremes; *triads*, triangles with labelled corners used to convey the relative importance of three concepts. Participants are invited to place a dot in a triangle to indicate the relative strength or influence of the concepts in the corners relative to their micro-narrative.

The questionnaire was translated into all the languages involved in the project—English, Italian, Norwegian, Swedish, German—and the participants could choose which language version to use. More details about the questionnaire and types of questions are available in Bengtsson et al. (2022). The SenseMaker tool was specifically adapted to emphasize scientific literacy and the interplay between scientific knowledge/discourses and discourses emphasizing agency. Guided by the SenseMaker prompts, participants reflected on how scientific concepts and practices informed their narratives. For example, the initial prompts in the SenseMaker questionnaire were designed to elicit stories that connected scientific observations with personal and collective actions.

So, what can be observed from analysis of the questionnaire? This paper focuses only on the answers emerging from Italy and Norway. This is because this paper is the pipeline of a group of project papers (Tasquier et al., 2022; Tasquier & Jornet, 2023) which aimed to compare those two open schooling networks from many perspectives.

Specifically, this paper presents some findings related to students' micro-narratives collected through the SenseMaker questionnaire, which reveals individual/collective dynamics activated by the activities and the impact on the students' ability to imagine future scenarios and take agency toward climate change and sustainability. The data were analysed by a group of five researchers through qualitative analysis for cross-checking, corroborating evidence and evaluating the extent to which all evidence converges (Anfara et al., 2002). Given the sample and research issues, we employ the approach of *reflexive thematic analysis* where a mixed inductive/deductive operative method of looking at the data and highlighted themes is used (Braun & Clarke, 2006). The sample consists of approximately 200 questionnaires, with about one-third from Italy and two-thirds from Norway.

22.4 Results

This chapter presents the thematic analysis of micro-narratives from Italian and Norwegian students collected via the SenseMaker questionnaire. The analysis aims to understand how individual and collective dynamics are activated by climate change-related activities, and how these impact students' ability to imagine future scenarios and take agency towards climate change and sustainability. The study employs a reflexive thematic analysis approach (Braun & Clarke, 2006) using both inductive and deductive methods to explore the data, with cross-checking and corroboration by five researchers to ensure reliability and convergence of evidence (Anfara et al., 2002). To ensure reliability and consistency, the analysis involved multiple iterative cycles, where the team of five researchers cross-checked findings and collaboratively resolved discrepancies. This process ensured that the themes identified were grounded in the data and reflected a shared understanding across the research team. However, inter-rater reliability metrics were not calculated, as the reflexive thematic analysis prioritizes the iterative and interpretive nature of theme generation over rigid quantification. While the SenseMaker questionnaire facilitates the collection of both qualitative and quantitative data, this paper focuses exclusively on the qualitative dimension. The decision was guided by the research aim, which sought to delve deeply into the nuanced, subjective narratives of students to understand their sense-making processes and perceptions of agency. Quantitative findings, although available, were beyond the scope of this analysis.

Overall, the narratives reveal complex interactions between individual and collective agency in the students' responses to climate change. Both Italian and Norwegian students demonstrated awareness of climate issues, though their responses varied in focus and intensity according to their cultural and environmental contexts. These differences shed light on the ways in which young people relate to sustainability and climate change, both in terms of daily practices and larger systemic concerns.

The analysis is structured in two sections, as follows:

- **Themes emerging from the students' micro-narratives (Sect. 22.4.1)**: This section presents the themes that emerged from the two different samples, focusing on the unique perspectives and experiences within each group. It also includes a comparative analysis to highlight similarities, differences, and the interplay between their approaches to sustainability and climate change
- **Overall considerations and further findings (Sect. 22.4.2)**: This section synthesizes the reflections from the thematic and comparative analysis, integrating the perspective of the three spheres model (personal, practical, political) to provide a deeper understanding of students' perceptions, challenges, and agency in the context of sustainability and climate change.

22.4.1 Themes Emerging from Students' Micro-narratives

This analysis presents the themes that emerged from the two different types of samples, focusing on the unique perspectives and experiences within each group, followed by a comparative analysis to highlight similarities, differences, and the interplay between their approaches to sustainability and climate change.

The analysis of the Italian students' micro-narratives identifies three foci of attention (I.1, I.2, I.3), which are presented below:

I.1. Personal Responsibility and Daily Practices
Italian students frequently described efforts to integrate sustainable practices into their everyday lives. Actions like recycling, reducing water consumption, and buying second-hand items were recurring themes, reflecting a growing individual awareness of their environmental impact. However, these practices were often framed within the context of small, personal actions rather than collective or community-wide efforts. The narratives suggest that for many students, their sense of responsibility towards sustainability was primarily focused on the individual level.

For example, one student described the shift in their behaviour after learning about the environmental impact of long showers: "I've learned to think about the environment, not just my personal well-being". Similarly, another student expressed pride in purchasing second-hand clothing, initially feeling embarrassed but ultimately embracing it as an eco-conscious decision. These examples underscore how individual actions can shift from personal discomfort to pride, contributing to a growing sense of agency.

I.2. Frustration with Societal and Family Inaction
A significant theme in the Italian responses was frustration with the broader society and, in particular, family members who did not participate in sustainable practices. Several students noted intergenerational conflicts regarding climate action, with younger students feeling that older family members were dismissive of their environmental concerns. One student highlighted their frustrations with family members who refused to recycle, noting: "They say that recycling won't change anything. This made me reflect on how much we need everyone's cooperation." This reflects the collective frustration with the slow pace of change at the societal level, despite individual efforts.

I.3. Observations of Environmental Degradation
Italian students frequently described personal encounters with environmental degradation, including reduced snowfall, polluted natural spaces, and extreme weather events. For many, these experiences were moments of heightened awareness, leading to deeper reflection on the urgency of climate action. These direct observations of climate change in their local environments served as a catalyst for reflection and concern, although students often expressed feelings of helplessness in the face of such large-scale challenges.

The analysis of the Norwegian students' micro-narratives identifies three foci of attention (N.1, N.2, N.3), which are presented below.

N.1. Connection to Nature and Environmental Change
Norwegian students exhibited a strong connection to nature, with many narratives centred around their personal experiences in outdoor environments. Students frequently described observing seasonal changes, such as warmer winters or diminished snowfall, which they directly linked to climate change. This emphasis on nature reflects Norway's cultural and geographical context, where the outdoors plays a significant role in daily life.

One student described the emotional impact of noticing changes in the winter season: "The winters have become warmer, and it doesn't snow as much. It feels strange because it used to snow every year." These narratives suggest that for Norwegian students, the visible changes in their local environments served as powerful reminders of the global climate crisis.

N.2. Emphasis on Collective Action
Unlike their Italian counterparts, Norwegian students placed greater emphasis on community-oriented efforts to combat climate change. Several students discussed participating in local environmental campaigns or cleanup efforts, indicating a sense of collective responsibility towards sustainability. For these students, community action was a key theme, with individual efforts viewed as part of a larger societal commitment to addressing climate change.

For instance, one student mentioned their involvement in a community cleanup effort, noting: "We organised a cleanup of our local park, and it felt good to contribute. " This contrasts with the Italian students' focus on individual responsibility and highlights the importance of collective action in Norwegian students' approaches to sustainability.

N.3. Economic and Political Awareness
Norwegian students also expressed concerns about the economic and political dimensions of climate change, frequently pointing to the role of government and global corporations in driving large-scale environmental degradation. Some students noted that, while they were committed to personal action, they felt that systemic change was necessary for any real impact to occur. One student emphasised the need for global cooperation, stating: "Even if I do my part, it won't matter unless the bigger countries and corporations change too. "

This awareness of the political and economic factors driving climate change suggests that Norwegian students view sustainability not just as a personal responsibility, but as a global issue that requires cooperation from all levels of society.

The second step after having identified themes for the two different samples, was to make a comparative analysis in order to explore how their experiences, attitudes, and actions towards climate change and sustainability differ. The analysis reveals key insights into individual and collective dynamics, shaped by cultural, environmental, and societal factors in each country. By comparing the thematic findings, we aim to understand how these young people perceive their role in addressing climate change, how they imagine future scenarios, and the extent to which they feel empowered to take action. Both groups demonstrate a strong awareness of environmental issues, but their approaches to personal and collective responsibility differ significantly, influenced by their unique contexts. The comparative analysis identifies three foci of attention: (C.1) Focus on Individual vs. Collective Action; (C.2) Connection to Nature; (C.3) Frustration with Systemic Inaction.

C.1. Focus on Individual vs. Collective Action

A key distinction between the Italian and Norwegian students' narratives lies in their approaches to individual versus collective action in tackling climate change. Italian students primarily focus on personal responsibility, with a strong emphasis on individual behaviours such as recycling, reducing water consumption, and making sustainable consumer choices (e.g., buying second-hand clothes). These actions reflect an internalisation of environmental responsibility, where change starts at the individual level. However, Italian students often expressed frustration with the lack of collective engagement, especially from family members, which can create a sense of isolation in their efforts. This frustration suggests that, for many Italian students, the challenge lies not in their own willingness to act but in the difficulty of mobilising others around them. In contrast, Norwegian students demonstrate a greater focus on community-driven efforts and the importance of collective action. Many Norwegian narratives describe participation in local environmental initiatives, such as cleanup campaigns or efforts to protect local ecosystems. This reflects a cultural emphasis on community and societal responsibility, where individual actions are seen as part of a larger, collective movement towards sustainability. Norwegian students are more likely to view their actions within the context of broader societal cooperation, indicating a stronger alignment between individual efforts and collective goals.

The difference in focus between individual and collective action likely stems from the cultural and societal contexts of each country. Italy, with its dense urban environments, may foster a more individualistic approach to sustainability, where students see personal action as their primary tool for change. In Norway, with its strong tradition of environmental stewardship and outdoor engagement, collective action and community-based initiatives are more deeply embedded in social consciousness. This distinction highlights the importance of cultural frameworks in shaping how young people perceive their roles in addressing global issues like climate change.

C.2. Connection to Nature

The connection to nature is another key difference between the Italian and Norwegian students. Norwegian students' narratives often reflect a deep awareness of seasonal and environmental changes in their immediate surroundings. For instance, students frequently mentioned awareness of changes in snowfall patterns, shorter winters, or disruptions in natural cycles that they attributed to climate change. This strong connection to the environment suggests that direct experiences of nature are a significant factor in shaping their understanding of climate change. For many Norwegian students, the natural world serves not only as a backdrop for their lives but as an essential part of their identity and a powerful motivator for climate action.

Italian students, by contrast, focus more on urban or day-to-day environmental issues, such as waste management, pollution, or water consumption. While some students mentioned observing climate-related changes in nature (e.g., reduced snowfall in the mountains), their narratives were less likely to describe direct, personal experiences with nature as a key factor in their environmental awareness. Instead, Italian students are more concerned with the tangible, immediate impacts of unsustainable practices in their urban or suburban environments. The urban context seems to influence their perception of climate change as a challenge that primarily manifests through pollution and resource use rather than as a disruption of natural cycles.

This difference in connection to nature may stem from the geographical and cultural contexts in which these students live. Norway's vast natural landscapes and outdoor culture likely foster a more direct relationship with nature, making the effects of climate change more visible and immediate to students. In contrast, Italy's more urbanised and densely populated regions may focus students' attention on the challenges of managing environmental resources in human-dominated settings. These differences suggest that place-based experiences play a critical role in shaping how young people relate to environmental issues and the kinds of solutions they prioritise.

C.3. Frustration with Systemic Inaction

Both Italian and Norwegian students expressed frustration with the slow pace of systemic change in addressing climate change. However, the focus of this frustration differed between the two groups. Italian students tended to express dissatisfaction with local and familial inaction, often describing conflicts with family members who did not share their commitment to sustainability. This frustration highlights the interpersonal challenges Italian students face in mobilising their immediate social circles to take action on climate change. Their narratives suggest a tension between their desire to act and the apathy or resistance they encounter from those around them. They also show a greater presence of the closer family as the students' immediate collective surrounding.

Norwegian students, on the other hand, directed their frustration more towards global and political inaction. Their narratives often emphasised the need for international cooperation and criticised the lack of decisive action from governments and corporations. This reflects a broader awareness of the global scale of the climate crisis and a belief that while individual and community actions are important, they are not sufficient without significant changes at the systemic level. Norwegian students appeared to place greater trust in collective efforts but were critical of the larger structures that they feel must change to complement individual and local initiatives.

This difference in focus reflects the broader societal and political discourse in each country. In Norway, a country where the attention on environmental policies is more widespread, students may feel that government and corporate actions are the next frontier for meaningful change. In Italy, where environmental activism may be more fragmented or unevenly adopted across different communities, the frustration with local inaction reflects the challenge of building a cohesive response to climate change at the grassroots level. These differing frustrations highlight how social and political contexts shape young people's perceptions of where the most urgent action is needed.

22.4.2 Overall Considerations and Further Findings

The comparative analysis of the Italian and Norwegian students' micro-narratives offers valuable insights into how young people perceive and relate to climate change and sustainability. Italian students often emphasise individual responsibility, focusing on personal actions such as recycling and resource conservation. Their narratives reflect a sense of personal agency, but also frustration when collective or familial support for sustainable practices is lacking. In contrast, Norwegian students prioritise community engagement and collective efforts, revealing a cultural disposition towards shared responsibility and societal involvement in addressing environmental issues. Despite their different approaches, both groups express a deep concern for the global nature of climate change, recognising that individual or local actions are insufficient without broader systemic change. This points to the necessity of fostering both individual and collective agency to effectively tackle the climate crisis.

A more fine-grained thematic analysis of the narratives reveals additional layers of complexity, particularly in the ways students construct their understanding of climate change. Many of the stories start with local and personal experiences, often from the past or present, which help students frame the problem in tangible terms. These micro-narratives highlight a range of actors involved in the climate change issue, from family members and peers to larger societal entities like governments and corporations. In recounting these stories, students create a space of agency where they can imagine themselves as part of the solution. However, there is also a recurring theme of detachment or otherness, where students describe the problem as somewhat external to their personal lives, reflecting a gap between recognition of the issue and personal involvement.

As the narratives progress from personal experiences to imagined future scenarios, several features become apparent. There is a plurality of future visions, touching on different aspects of climate change, ranging from personal to systemic levels. Interestingly, these futures tend to fall into two distinct categories: polarised and non-polarised visions. Some students present highly polarised futures, either dystopian or utopian, where climate change leads to extreme outcomes, often emphasising either total collapse or perfect environmental harmony. These starkly contrasting futures reflect the uncertainty and anxiety that many young people feel about the global climate crisis.

In contrast, the non-polarised futures share a common feature: they emphasise the importance of a change in mindset and the need for balance among the various actors involved in addressing climate change. Whether the focus is on individual actions or systemic reforms, these narratives highlight the importance of alliances between personal choices and political or collective reactions. Students argue for the necessity of cooperation between individuals, communities, and governments, suggesting that only through such an equilibrium can meaningful progress be made. This sense of interdependence between individual and collective efforts suggests a growing awareness that the climate crisis cannot be addressed in isolation but requires a holistic approach that integrates personal responsibility with institutional and societal change.

In conclusion, this comparative analysis underscores the influence of cultural, geographical, and societal factors in shaping students' approaches to sustainability and climate action. While Italian students lean towards individualism, often grappling with frustration over a lack of collective support, Norwegian students exhibit a more community-oriented approach, emphasising the role of collective agency. Despite these differences, both groups recognise the necessity of broader systemic change to complement their actions. Moreover, as students move from narrating past experiences to imagining future scenarios, a common theme emerges: the recognition that solving climate change requires a fundamental shift in mindset, one that balances individual actions with collective and political responses. This analysis reveals that fostering agency in young people involves not only encouraging personal responsibility but also promoting a vision of interconnectedness, where individual choices and systemic changes work together to create a sustainable future.

Looking at the themes with the lens of the model of the three spheres of transformation, the analysis reveals how the personal, practical, and political spheres interact to shape their perceptions of climate change and sustainability.

From the perspective of the *Personal Sphere*, both groups of students articulated individual experiences and responsibilities as central to their engagement with sustainability. Italian students emphasized personal actions like recycling and water conservation, often expressing frustration when these efforts were not mirrored by others. For instance, one student noted, "I've started turning off taps to save water, but my family doesn't seem to care about these small changes." These narratives highlight the need for fostering individual agency while addressing the relational dynamics that can either support or hinder sustainable practices.

Concerning the *Practical Sphere*, Norwegian students displayed a stronger focus on collective and community-based actions. They frequently mentioned participating in environmental clean-ups and local initiatives. One student reflected, "Working together on a park cleanup made me feel like part of something bigger". This contrast with Italian students' individualistic focus underscores the importance of cultural and societal contexts in shaping practical engagement.

Finally, concerning the *Political Sphere*, many Norwegian students expressed awareness of systemic and policy-level barriers to climate action, identifying global corporations and governments as key actors. One student remarked, "Even if we do our part, it won't matter unless governments take real action." Italian students, meanwhile, showed more frustration with local and familial inaction, indicating a perceived gap between personal efforts and broader systemic support. These findings suggest the necessity of educational strategies that connect individual actions with systemic change, empowering students to see themselves as both contributors and advocates.

As an additional consideration, this analysis was also able to highlight an accent on the topic of futures across the spheres. Indeed, students' narratives often transitioned from past and present experiences to imagined future scenarios. These ranged from dystopian fears to hopeful visions, highlighting the interplay between personal responsibility, community action, and systemic reform. For example, one Norwegian student imagined, "If governments, communities, and individuals all work together, we can create a future where nature thrives". In contrast, an Italian student expressed concern: "What's the point of trying if the bigger powers don't change?". These reflections emphasize the interconnectedness of the three spheres and the importance of fostering a holistic mindset in climate education.

22.5 Discussion

The central hypothesis of this research—that micro-narratives offer young people an opportunity to understand their identity, envision plausible futures, and engage with cultural and social resources around climate change and sustainability—has been explored through the analysis of student stories. The analysis reveals that micro-narratives are indeed a powerful tool for capturing young people's reflections on personal and political agency, as well as their emotional responses—ranging from hope to despair and from empowerment to detachment—regarding the present and future in the context of climate change. These personal and interpersonal dimensions of the students' sensemaking intersect with their scientific knowledge about the complex issue of achieving more sustainable futures.

One of the clearest findings is that, despite the opportunity for young people to reflect on their roles and futures, many narratives are permeated by a sense of pessimism and emotional distance. Students frequently express anxiety, fear, and mistrust, which are dominant emotions when thinking about the future. This aligns with the hypothesis that micro-narratives can reveal not only students' potential to engage with sustainability challenges but also the psychological and emotional barriers that prevent them from doing so. The students' responses indicate a widespread loss of hope, where their narratives reflect more on external factors and other people's actions than on their own potential to make change. This sense of detachment complicates the hypothesis that narratives help students build a sense of personal agency, as many of them appear to perceive the future as something shaped by others, not by their own actions.

However, the polarised responses regarding what is needed to solve climate change also provide valuable insights. Some students emphasise the importance of individual actions, such as small everyday efforts that collectively contribute to positive outcomes. Others focus on systemic solutions, calling for large-scale political and economic interventions as the only means of achieving sustainability. This polarisation, though restrictive in some respects, still provides evidence that young people are grappling with the complexity of the climate crisis. These stories reflect how they perceive themselves within the broader landscape of climate action: some see their actions as part of a larger collective effort, while others feel overwhelmed by the scale of systemic challenges. Thus, while many students express detachment, others indicate an awareness of their potential to contribute to change through individual or collective efforts.

Moreover, the findings suggest that narratives can provide deep insights into students' current perspectives towards agency, but they also expose significant gaps in how students relate to climate change and sustainability challenges. While the micro-narratives help students articulate their experiences, they sometimes reveal a lack of personal involvement and an inability to connect their daily actions to broader systemic changes. This highlights a critical educational challenge: how to encourage students to see their own actions as interconnected with the larger political and societal changes necessary for climate resilience, while also connecting these with scientific discourses and perspectives. This gap may have relevant connections to the findings of the Norwegian case studies referred to in the theoretical framing above, suggesting a challenge in connecting everyday experiences and scientific discourses, if 'everyday experiences' include students' thoughts about cause and solutions to sustainability challenges. If students are more aware of political and structural aspects of the current sustainability crises than what emerges in classroom discourses, it means both that there is a gap between students' lifeworlds and school, but also that there is a potential for bringing these aspects into teaching. Maybe students are better prepared for this complexity than current teaching practices open for, but of course this notion must be explored in further research.

Interestingly, the few non-polarised narratives offer a more nuanced understanding of how students might bridge the gap between personal and political agency. These stories emphasise the importance of changing mindsets and fostering collaboration between different stakeholders—individuals, governments, corporations—in order to achieve long-term sustainability. While these stories were in the minority, they demonstrate that some students are capable of seeing climate change as a multi-dimensional problem that requires integrated solutions, not just one-shot actions. This reinforces the hypothesis that micro-narratives can nurture a sense of complexity, where students begin to grasp the interconnectedness of individual choices and systemic structures.

Thus, the findings suggest that while micro-narratives provide a rich opportunity for students to reflect on their roles in the sustainability landscape, they also expose the challenges young people face in feeling empowered to take action. The stories reveal a spectrum of engagement, from deep emotional detachment to hopeful agency, with many students navigating the tension between feeling personally responsible and relying on others to lead the way. These insights underscore the importance of narrative-based learning, which can help students bridge the gap between despair and empowerment, detachment and agency, and between individual actions and systemic transformations.

22.6 Conclusions

This research set out to explore the extent to which micro-narratives provide insights into young people's current attitudes toward agency in relation to climate change and sustainability challenges. The findings support the hypothesis that narratives offer a powerful lens for examining how students make sense of their personal and political roles in a world shaped by climate change. Through their stories, students revealed their emotions, beliefs, and ideas about the future, demonstrating the complex interplay between personal responsibility and societal forces.

However, the analysis also reveals significant challenges in fostering a sense of agency among young people. While many students are aware of the gravity of climate change, their stories often reflect a sense of helplessness or detachment, with responsibility placed on external actors like governments or corporations. This suggests that while narratives can provide a space for reflecting on futures and identities, they do not always lead to a strong sense of personal involvement in driving change. Micro-narrative analysis suggests that students' emotional responses, such as fear and anxiety, may influence their engagement with sustainability issues by shaping their perception of agency and limiting their ability to see themselves as active participants in the solution.

At the same time, the polarisation in students' views of the future—whether focused on individual actions or systemic solutions—highlights the complexity of fostering climate literacy. Many students struggled to integrate these two approaches, reflecting a broader educational challenge in helping young people understand that sustainability requires both individual and collective efforts. The few narratives that called for a change of mindset and collaboration among all actors show that some students are beginning to grasp the importance of holistic solutions that bring together personal choices and political reforms.

In conclusion, micro-narratives offer a valuable means for young people to reflect on their place in the world, but they also expose significant gaps in their ability to see themselves as agents of change. Educators must continue to develop strategies that help students move from a position of detachment to one of engagement, fostering a sense of complexity and interconnectedness in their approach to climate change. Through narrative-based learning, students can begin to understand the interplay between individual and collective action, developing the critical skills and emotional resilience needed to envision plausible futures and take meaningful action toward sustainability. The SenseMaker questionnaire has proven to be a useful tool in this process, offering both educators and researchers a window into the emotional and cognitive landscapes of young people as they confront one of the greatest challenges of our time.

Acknowledgement, Data Statement and Authors' Contribution The data discussed in this manuscript were collected through a SenseMaker questionnaire, which was developed as part of the Horizon 2020 Research and Innovation Programme under grant agreement No 824522 (SEAS project). The data have been analysed and processed by the authors of this manuscript. We acknowledge all participants in the project for their collective contributions to the realisation of this research.

The SenseMaker questionnaire was designed as part of the Horizon 2020 Research and Innovation Programme under grant agreement No 824522 (SEAS project). The raw data have been stored in a private folder and are not available as open access. GT, EK, and AJ conceptualised the work; GT, EK, and AJ discussed the methodology; GT carried out the analysis, which was triangulated by the other authors. GT led the writing of the original draft, with EK and AJ collaborating in the co-writing, reviewing, and editing of the paper. HR triangulated the analysis, read, and approved the paper. All authors have read and agreed to the published version of the manuscript.

References

Anfara, V. A., Brown, K. M., & Mangione, T. L. (2002). Qualitative analysis on stage: Making the research process more public. *Educational Researcher, 31*(7), 28–38. https://doi.org/10.3102/0013189x031007028

Avraamidou, L., & Osborne, J. (2009). The role of narrative in communicating science. *International Journal of Science Education, 31*(12), 1683–1707.

Barelli, E., Tasquier, G., Caramaschi, M., Satanassi, S., Fantini, P., Branchetti, L., & Levrini, O. (2022). Making sense of youth futures narratives: Recognition of emerging tensions in students' imagination of the future. *Frontiers in Education, 7*, 911052. https://doi.org/10.3389/feduc.2022.911052

Bengtsson, S., Jornet Gil, A., Van Poeck, K., & Knain, E. (2020). "D2.2 – Definitions of shared pool of concepts," in Document delivered do the EC about SEAS project, (SEAS).

Bengtsson, S., Östman, L., Van Poeck, K., Goodwin, L., Sygna, L., Arctander, H., O'Brien, K., & Knain, E. (2022). D4.2—*Updated guideline manual describing the use of each of the SEAS digital tools*. Report delivered do the EC about SEAS project (SEAS). University of Oslo.

Bianchi, G., Pisiotis, U., & Cabrera Giraldez, M. (2022). *GreenComp—The European sustainability competence framework*. EN, Publications Office of the European Union.

Braun, V., & Clarke, V. (2006). Using thematic analysis in psychology. *Qualitative Research in Psychology, 3*, 77–101. https://doi.org/10.1191/1478088706qp063oa

Bruner, J. (1991). The narrative construction of reality. *Critical Inquiry, 18*, 1–21. https://doi.org/10.1086/448619

Byhring, A. K., & Knain, E. (2016). Intertextuality for Handling Complex Environmental Issues. *Research in Science Education, 46*, 1–19. https://doi.org/10.1007/s11165-014-9454-6

Caspari, A. (2019). *Driving development vs scaling change.* Blog, www.mindshift-Integral.com, or at www.ezc.partners/blog/ "Scaling Microchanges".

Caspari, A. (2020). *Sensemaking/sensemaker collecting micronarratives in cities*. Integral urban series on thriveable cities. Volume 21: Coming of age in a post-COVID19 world: Dare to dream.

CognitiveEdge. (2021). *SenseMaker—Why SenseMaker?* SenseMaker website. Retrieved March 17, 2021, from https://sensemaker.cognitive-edge.com/why-sensemaker/

D'Orto, E., & Tasquier, G. (2025). Narrativity and climate change education: Design of an operative approach. *Sustainability, 17*(4), 1587. https://doi.org/10.3390/su17041587

Engeström, Y., & Sannino, A. (2010). Studies of expansive learning: Foundations, findings and future challenges. *Educational Research Review, 5*(1), 1–24.

Erduran, S. (2007). Methodological foundations in the study of argumentation in science classrooms. In S. Erduran & M. P. Jiménez-Aleixandre (Eds.), *Argumentation in science education* (Science & technology education library) (Vol. 35). Springer.

Furberg, A., Rødnes, K. A., Silseth, K., Arnseth, H. C., Kvamme, O. A., Lund, A., Sæther, E., Vasbø, K., Ernstsen, S., Fundingsrud, M. W., Øistad, J. H., & Slungård, K. (2023). *Fagfornyelsen i møte med klasseromspraksiser: Læreres planlegging og gjennomføring av undervisning om tverrfaglige tema [The curriculum Fagfornyelsen in the face of classroom practices: Teachers' planning and implementation of teaching on interdisciplinary topics]*. Report 6, University of Oslo.

Furberg, A., Silseth, K., Rødnes, K. A. , Arnseth, H. C., Kvamme, O. A., Rasmussen, I., Sæther, E., Vasbø, K. B., Pettersen, K., Øistad, J. H., & Lund, A. (2024). *Fagfornyelsen i møte med klasseromspraksiser: Elevers læring i møte med de tverrfaglige temaene [The curriculum Fagfornyelsen in the face of classroom practices: Students' learning in relation to the interdisciplinary themes]*. Report 7, University of Oslo.

Gottschall, J. (2018). *L'istinto di narrare. Come le storie ci hanno reso umani*. Bollati Boringhieri.

Han, B. C. (2019). *La scomparsa dei riti: una topologia del presente*. Edizioni nottetempo.

Han, B. C. (2021). *The palliative society: Pain today*. Polity Press.

IPCC. (2021). *Climate change 2021: The physical science basis*. Cambridge University Press.

Jornet, A., & Roth, W.-M. (2015). The joint work of connecting multiple (re)presentations in science classrooms. *Science Education, 99*(2), 378–403.

Laherto, A., & Rasa, T. (2022). Facilitating transformative science education through futures thinking. *On the Horizon, 30*(2), 96–103.

Latour, B. (2021). Postface. In F. Gemenne & A. Rankovic (Eds.), *Atlante dell'Antropocene*. Mimesis Edizioni.

Levrini, O., Tasquier, G., Barelli, E., Branchetti, L., Laherto, A., Palgrem, E., Wilson, C., & Branchetti, L. (2021). Recognition and operationalization of Future-Scaffolding Skills: Results from an empirical study of a teaching-learning module on climate change and futures thinking. *Science Education, 105*(2), 281–308.

Magnason, A. S. (2019). *On time and water*. Iperborea Edizioni.

McAdams, D. P. (1993). *The stories we live by: Personal myths and the making of the self*. William Morrow & Co..

O'Brien, K., & Sygna, L. (2013). Responding to climate change: The three spheres of transformation. In *Proceedings of transformation in a changing climate, 19–21 June 2013, Oslo, Norway*. University of Oslo.

OECD. (2018). The future of education and skills—education 2030.

Polkinghorne, D. E. (1988). *Narrative knowing and the human sciences*. State University of New York Press.

Riessman, C. K. (2008). *Narrative methods for the human sciences*. Sage.

Røkenes, H., & Jornet, A. (2025). Affective contradictions and narratives of change: A critical, cultural-historical approach to hope and action in Education for Sustainability. *Science Education*. https://doi.org/10.1002/sce.21965

Sadler, T. D., Romine, W. L., & Topçu, M. S. (2016). Learning science content through socio-scientific issues-based instruction: a multi-level assessment study. *International Journal of Science Education, 38*(10), 1622–1635.

Schwab, K., & Malleret, T. (2021). *The great narrative—For a better future*. Forum.

Snowden, D., & Boone, M. E. (2007). A leader's framework for decision making. *Harvard Business Review*.

Tasquier, G., & Jornet, A. (2023). A conversation on the pragmatics of open schooling: Reflections after two European projects. *Education in Science, 293*, 22–23.

Tasquier, G., Knain, E., & Jornet, A. (2022). Scientific literacies for change making: Equipping the young to tackle current societal challenges. *Frontiers in Education, 7*, 689329. https://doi.org/10.3389/feduc.2022.689329

Tasquier, G., & Pongiglione, F. (2017). The influence of causal knowledge on the willingness to change attitude toward climate change: Results from an empirical study. *International Journal of Science Education, 39*(13), 1846–1868.

Tasquier, G., Pongiglione, F., & Ricci, E. (2024). Rethinking agency in the context of climate change: a three-lens approach. *Science & Education*, under review.

UNESCO. (2020). *Education for sustainable development: A roadmap.* UNESCO report.

UNESCO. (2021). *Getting every school climate-ready: How countries are integrating climate change issues in education.*

Van der Merwe, S. E., Biggs, R., & Preiser, R. (2019). Making sense of complexity: Using SenseMaker as a research tool. *Systems, 7*(2). https://doi.org/10.3390/systems702002

Giulia Tasquier is associate professor in Physics Education at the Department of Physics and Astronomy of the University of. She serves as member of the ESERA executive board (2022-2026) and conference manager of ESERA conference 2019 held in Bologna. Her research interests include design and implementation of teaching materials on climate change; agency in climate change; open schooling.

Erik Knain is full professor in Science Education at the University of Oslo, at the Department of Teacher Education and School Research. His research interests include inquiry-based science teaching, socio-scientific issues, sustainability education and open schooling.

Hanna Røkenes is a psychologist from the University of Helsinki and holds a master degree in Education from the University of Oslo. Her present research and teaching have been focusing on collective agency in local communities and teacher education, particularly in Open schooling sustainability education. Currently, she is in the process of finishing her doctoral degree in pedagogy at the University of Oslo.

Alfredo Jornet is a Ramón y Cajal Researcher at the Department of Psychology at the University of Girona. His current research focuses on knowledge co-creation across university, school, and society towards social and climate justice in times of crisis.

Chapter 23
The Voice of Policymakers, Professional Development Leaders, and Teachers Who Engage in Climate Change Education: A Cross-Country Comparison

Daphne Goldman, Orit Ben Zvi Assaraf, Vaille Dawson, Efrat Eilam, Tuba Gokpinar, Nofar Naugauker, Gusti Agung Paramitha Eka Putri, Agung Wijaya Subiantoro, Sakari Tolppanen, Peta White, and Helen Widdop Quinton

D. Goldman (✉)
Faculty of Education, Beit Berl College, Kfar Saba, Israel
e-mail: dafnag@beitberl.ac.il

O. Ben Zvi Assaraf · N. Naugauker
School of Education, Ben Gurion University of the Negev, Beer-Sheva, Israel
e-mail: ntorit@bgu.ac.il

V. Dawson
Graduate School of Education, The University of Western Australia, Perth, WA, Australia
e-mail: vaille.dawson@uwa.edu.au

E. Eilam · H. Widdop Quinton
The College of Arts and Education, Victoria University, Melbourne, VIC, Australia
e-mail: Efrat.eilam@vu.edu.au; helen.widdopquinton@vu.edu.au

T. Gokpinar
Institute of Education, University College London, London, UK
e-mail: tgokpinar@essex.ac.uk

Gusti Agung Paramitha Eka Putri
Nossal Institute for Global Health, University of Melbourne, Melbourne, VIC, Australia
e-mail: mitha.eka@unimelb.edu.au

A. W. Subiantoro
Department of Biology Education, Universitas Negeri Yogyakarta, Yogyakarta, Indonesia
e-mail: agung_wijaya@uny.ac.id

S. Tolppanen
School of Applied Educational Sciences and Teacher Education, University of Eastern Finland, Joensuu, Finland
e-mail: sakari.tolppanen@uef.fi

P. White
Faculty of Arts and Education, Deakin University, Burwood, VIC, Australia
e-mail: peta.white@deakin.edu.au

G. Cakmakci, M. F. Tasar (eds.), *Connecting Science Education with Cultural Heritage*, Contributions from Science Education Research 15,
https://doi.org/10.1007/978-3-031-90944-3_23

23.1 Introduction

Anthropogenic global climate change (CC) is currently approaching a tipping point, where natural forces in the climate system react to cause further warming. Every person, across the globe, will be impacted in some way by CC. The understanding that climate change education (CCE) is crucial in the face of today's climate emergency is now widely established and accepted among experts, activists, educationalists as well as citizens (Reid, 2019). It is widely agreed that young people in today's world need to be equipped with the appropriate knowledge and skills, attitudes and values, which, together, will both enable and motivate them to address and cope with the challenges CC poses in informed and responsible ways. Despite this, a study of how countries are incorporating CC within their national curriculum indicates much room for improvement: in a study of 100 countries, nearly one half made no reference to CC in their curriculum, and in the rest, while CC was mentioned, it was not addressed in depth (UNESCO, 2021a). Furthermore, a UNESCO survey of educators found that over a third (36%) of these educators reported that they had not received either pre-service or in-service training on CC and sustainable lifestyles. While ~40% felt confident in teaching the cognitive aspects of CC, much less felt confident about addressing the behavioural domain, and identified knowledge of suitable pedagogies as a challenge to their competencies. Education systems need to respond to these gaps by incorporating CCE as a core curriculum component (UNESCO, 2021a). Teachers need the competencies, confidence and agency to incorporate CC within their teaching.

The study presented here builds on a previous study (Dawson et al., 2022) that compared the middle-school curricula in science and geography of six countries—Australia, Israel, Finland, Indonesia, Canada, and England—in relation to inclusion of CC. The study found that the middle-school curricula of all six countries contained subject matter pertaining to CC, but that the curricula varied widely in the extent to which that content was mandated and structured. Although all curricula contained implicit opportunities for teachers to choose to teach CC, the content was not always explicitly related to climate science. For example, CC was mentioned in the Australian curriculum, in the context of geography, as an example of a challenge to food security. Such findings raised several questions, for example: What is CCE? What are effective approaches for its implementation? What are the challenges associated with implementing CC curricula? What types of resources are necessary for implementation?

Therefore, the current study continues our multi-country comparison of CCE and expands it by focusing on the perspectives of three key educational roles involved in developing and implementing CCE: policymakers, providers of professional development (PD) and teachers who teach CCE.

Climate Change Education: Challenges to Implementation

The critical role of schools in educating young people about CC and preparing them to be informed, responsible citizens is widely recognized. However, research suggests that education systems in many countries are falling short in addressing this crucial task (e.g., Dawson et al., 2022; UNESCO, 2021a). This gap reflects complex, interconnected challenges across multiple levels of the education system. The following section briefly highlights some of the factors contributing to this gap. This is not intended as a comprehensive literature review, but rather to contextualize the focus of this cross-country study which explores the perspectives of policymakers, teachers and PD providers involved in developing and implementing CCE.

At the policy level, despite governmental commitments to promote CCE, there remains a lack of global comparative data to effectively monitor and evaluate progress in CCE implementation. Existing data reveals that CCE provision is still insufficient. Mckenzie (2021) reviewed CCE and communication included in the reports of 194 countries to the UNFCCC (UN Framework Convention on Climate Change) to examine how CCE is addressed in these submissions. While 95% of the countries reported some focus on CCE, the emphasis was primarily on public awareness, with significantly fewer references to the formal education system. Regardless of education level, the focus was mainly on cognitive learning rather than social-emotional or action-oriented learning processes or outcomes. A UNESCO survey on how countries are incorporating CC into their national curricula (UNESCO, 2021a) found that nearly one half of the 100 countries reviewed made no reference to CC in their curriculum. For those countries that did mention it, the topic was not addressed in depth.[1] These findings suggest that in most countries, there is considerable room for improvement in the development and implementation of CCE curricula.

Another broad survey found that the subjects most likely to include environment-related content were biology, science and geography with limited cross-curriculum implementation, and the depth of inclusion was shallow (UNESCO, 2021b). An in-depth cross-country comparison of CC in the compulsory middle school science and geography curricula of six countries found that while CC appeared in the formal curriculum of all these countries, and all the curricula contain implicit opportunities for teachers to elect to teach CC, the content was not always explicitly related to climate science. CC was often presented as a context or an example of other science concepts rather than as a distinct topic (Dawson et al., 2022). This study also found that the emphasis was primarily on the cognitive aspect of knowledge of causes with less attention given to the socio-emotional dimension or action-oriented and behavioural competences, which are essential for fostering responsible citizenship. The predominance of cognitive learning is also evident in other studies (e.g., Enke & Budke, 2023; UNESCO, 2021a)

The challenges to implementation extend into teacher preparedness and professional capacity. Studies indicate a gap between teachers' perceptions of the importance of CCE and its actual implementation in practice (Ben Zvi Assaraf et al., 2024). A global survey of over 58,000 teachers on their readiness to teach education for sustainable development and global citizenship revealed that while 70% of the teachers considered teaching CC important, only 42% felt fully prepared to do so, while 32% felt moderately ready and 26% felt not ready. Additionally, these

[1] For a comprehensive review of CC in national curricula, readers are referred to Dawson et al. (2022).

teachers felt more confident about teaching the cognitive dimensions of CC, but less confident regarding teaching the socio-emotional perspectives and even less about addressing the behavioural aspects (UNESCO, 2021c). Lack of content knowledge and expertise have been identified central in teachers' reluctance to address CC in their classrooms (e.g., Eilam, 2022; Plutzer et al., 2016). Eilam (2022) argues that one of the main problems contributing to insufficient representation of CC in school curricula is its poor epistemological conceptualization by educators. Compounding the issue of curricular ambiguity, educators are often hesitant to introduce complex controversial, and value-laden issues that have a socio-political context into their classrooms (e.g., Ho & Seow, 2015; Nation & Feldman, 2022). One reason for this reluctance is a limited sense of competence in their argumentation skills, which hampers their ability to facilitate critical discussions and foster their students' argumentation about climate science in a balanced manner (Liu & Roehrig, 2019).

High quality professional development can enhance teachers' competencies, particularly in understanding the scientific consensus on CC, addressing misconceptions around CC and fostering students' critical thinking skills about CC as a controversial socio-environmental and socio-scientific issue (Ennes et al., 2021). Clark et al. (2020) explored the effects of such a program, with three science teachers who participated in a 3-year PD initiative aimed at helping teachers adapt instruction to the Next Generation Science Standards. Despite the key role of preparing teachers (both pre-service and in-service) to teach CC, a broad survey of educators found that about half (55%) reported receiving training on CC during either pre-service or in-service teacher training programmes (UNESCO, 2021b).

The challenges confronting CCE implementation span all levels, from national policy to teacher preparedness, and are interconnected. National policy regarding CC in school curricula shapes the delivery of CCE and influences what teachers teach and how they teach it. It also affects teachers' opportunities for pre-service training and ongoing professional development. Given that the challenges exist at multiple levels, this study focuses on the perspectives of stakeholders at various levels within the education system.

Key Objectives

The objective of this cross-country comparative study is to analyze the perceptions of policymakers, PD leaders, and science teachers involved in implementing CCE. In doing so, it also seeks to identify opportunities and challenges for effective CCE implementation. So far, limited research has given attention to the perceptions of key CCE actors—policy makers and educators—involved in the development and implementation of CCE. This study (Ben Zvi Assaraf et al., 2024) offers a response to this gap.

23.2 Research Design and Method

Research Approach

This study employed a qualitative phenomenographic research approach, which is designed for "mapping the qualitatively different ways in which people experience, conceptualize, perceive, and understand various aspects of, and phenomena in, the world around them" (Marton, 2005, p. 143). The phenomenographic approach was selected for this cross-country comparative study as it allows for an in-depth and nuanced inquiry into the different ways in which the stakeholders in CCE conceptualize CCE and perceive its learning objectives, approaches to implementing it and challenges faced in practice.

Participants

Using a purposive sampling method (Creswell & Plano Clark, 2011), 33 participants were recruited across the five countries, based on their engagement with CCE. The participants were divided into three groups: educational policy makers or influencers, educators involved in providing teacher PD, and science teachers who teach CC in their classes.[2] Noteworthy, across the five countries, representative of the three groups varied. For example, while in Israel the participating policymakers were high-ranking Ministry of Education officials who develop and oversee CC curriculum in the national curriculum, in Australia, participants in this group had been involved in policy-making in the past but no longer hold decisions-making positions in relation to CCE, and in Finland, these participants were highly regarded figures with whom schools regularly consult, since there are no official policymakers in their system.

Data Collection

Data was gathered using in-depth semi-structured interviews, employing a protocol jointly developed by the research team. The interviews explored: How the educational practitioners conceptualize CCE in the Year 7–10 curriculum in relation to (a) the primary learning goals; (b) appropriate approaches for implementation and (c) the enablers and inhibitors associated with implementing CC curricula. The interviews extended 60–75 min, were conducted in the respective language of the participants, and were recorded and fully transcribed back to English.

Data Analysis[3]

The transcripts were analysed thematically and inductively, creating themes and sub-themes, in keeping with a qualitative phenomenographic approach (Marton, 2005), in a three-stage procedure that included: (1) initial coding by the authors for their respective country, and sharing this coding with the research team to ensure consistency in role classifications and coding; (2) after insuring response consistency within the three roles, emergent themes were identified in each country for cross-country examination. Three multi-national teams then conducted a comparative cross-country analysis of a different set of themes; (3) All authors reviewed other teams' analysis.

[2] For additional data on the participants in each of the countries and their coding, see Ben Zvi Assaraf et al. (2024).

[3] For detailed data analysis procedure, see Ben Zvi Assaraf et al. (2024).

23.3 Findings

The iterative analysis procedure elicited five themes with several sub-themes, concerning CCE conceptualization, as shown in Fig. 23.1. These are: (1) Perceptions of CC, with two sub-themes (epistemology—conceptualizing CC as a body of knowledge, ideal representation of CC in the curriculum); (2) Perceptions regarding actual representation/inclusion in the school curriculum; (3) Perceptions concerning CC contents and pedagogies, with two sub-themes (CC content knowledge, CC pedagogies); (4) Support for teaching—teaching resources and PD; and (5) inter-sectorial collaborations. This paper does not address the theme of inter-sectorial collaborations. Elaboration on this theme in one of the countries can be found in Naugauker et al. (2023).

The themes are briefly summarized as follows: Regarding the theme 'Perceptions of CC knowledge', there was consensus regarding the importance of CCE; however, concerning epistemology, there was terminology vagueness about the nature of CC knowledge and some distinctions among countries came up. For example, a Finnish PD provider distinguished between climate education and climate change as follows: "I know that in geography there is a lot of teaching about climate change. However, this is different from climate education, which includes value discussion, emotional aspects, societal aspects, etc." (FI PD2). The Indonesian participants differentiated between global warming and climate change in relation to Grade 7 science (the subject that addresses this area), and an Israeli participant raised the issue of shifting terminology from the greenhouse effect to global warming and then climate change, this inconsistency creating confusion. However, in the sub-theme 'Ideal representation of CC in the curriculum', there was a high level of agreement across all five countries that CC is a topic, a theme, or a big idea that potentially spans multiple subjects and should, thus, be taught in a multidisciplinary way. As stated by an Australian PD provider, "…it's more than just science or just math or just geography, or just art. I think it's an integrated area. So multidisciplinary…" (AU D3). Additionally, many participants emphasized that CC can only be fully

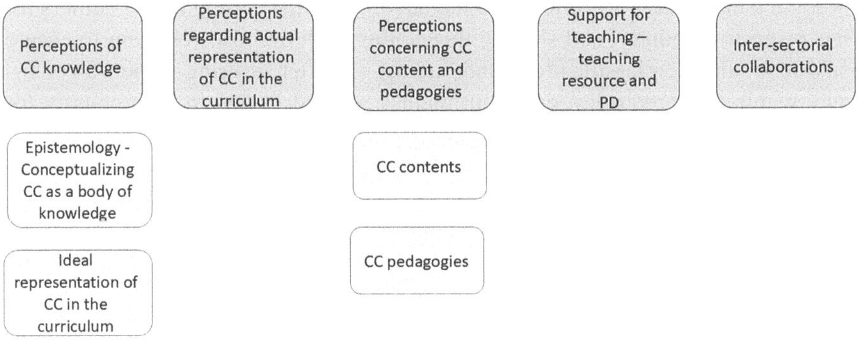

Fig. 23.1 Themes and sub-themes from interviews with 33 key actors (policymakers, PD developers, and teachers) involved in CCE in five countries

understood through a human perspective and holistic examination of not just the science but also the economics, politics, values and ethics. The challenge that this potentially poses for teachers was acknowledged by the participants, with a call for support of their competencies. As critiqued by a Finish PD provider, "It is not natural for teachers to implement multidisciplinary education, so it would be important to support them in this" (FI PD2). The participants also addressed the inclusion of CC in science, geography or as a standalone subject, as well as CC's relationship with sustainability education and the sustainable development goals.

With regards to the 'Actual representation of CC in the curriculum', CCE seems to be in a transitional phase, as it is present in the curricula of only some countries and its status is variable. The distinction between what "can" be taught and what "must" be taught was addressed by multiple participants across the countries. For example, an Australia PD provider noted that "You can completely avoid sustainability or climate change, or any of those issues ... so I could manage a whole curriculum, tick off everything I needed to do, and never mention sustainability or social justice, and really never mention climate impact" (AU D3). In Indonesia, for example, CC appears as a core component in Year 7 Science, as an Indonesian teacher explained: "It has become a policy in our national education. Students must be introduced to the impacts of climate change so that ... students make efforts to prevent the adverse effects of climate change" (IN T2). The participants in all five countries called for the curriculum to make more explicit connections to CC.

Concerning 'CC contents and pedagogies', the participants specified different contents (or scientific principles) that are important for the students to learn and made only a general reference to relevant skills. However, there was an agreement about the need to teach the scientific foundation. Participants in all countries and groups produced a rich repertoire of pedagogies for teaching CC, emphasizing the importance of implementing CCE via constructivist, learner-centered pedagogies that engage students in active learning, experiencing phenomena in their authentic settings, connecting CC to local issues, and cultivating critical thinking and activism, in the aim that CC, as science education, is meaningful for the learners. Participants from four out of five of the countries and all three participant groups made references to the emotional aspects of CC and importance of social emotional learning in CCE. As claimed by a Finish teacher "It is not only about knowledge; it's also about emotions. The emotional side is not easy to address in schools" (FI T2), thus raising concerns regarding the teachers' efficacy in teaching this topic. An English policymaker noted the importance of classroom culture, "Being able to discuss and knowing that they're in a relatively safe environment where if they want to argue that they didn't believe in CC ... The teacher is going to keep control of the class and make sure they're not insulted by the other fellow students ... so you do need a good climate and culture" (EN P1).

Regarding the theme 'Support for teaching', the participants generally agreed that there are abundant resources for teaching CCE, but these are not always easy to access, expressing the need for guidance in selecting appropriate resources and implementing them in the classroom. They also addressed issues concerning the provision of effective PD and explained that PD is available for teachers, but they do not take advantage of it. Participants in Australia and Indonesia complained about the lack of PD and support for teaching, emphasizing that "it's got to be professional learning" (AU P1), and that "so far there have been no supporting factors to teach this material" (IN D1). Furthermore, the findings suggest that while there is a "buffet" of online resources, it is often difficult for teachers to find time to familiarize themselves with the materials and train in using them. In Israel, the findings suggest a consensus across the three role types, by which the Ministry of Education must show its commitment to supporting teachers in teaching CC. A policy maker stated: "It is clear to us that this is our commitment, I also see this as a commitment by the teachers" (IS P4). Similarly, another Israeli policymaker stated: "We must, we must train teachers to teach climate change" (IS P1).

23.4 Discussion and Conclusions

The aim of this study was to conduct a cross-country comparison of the perceptions held by key role-players involved in CCE. Consistent with previous research (UNESCO, 2021b), the findings confirm that the educators view CC as a critical component of the curriculum. However, its current position in the curriculum does not fully correspond with their perception of CCE as a complex multidisciplinary and multi-domain topic that requires implementation in a multidisciplinary cross-curricular approach. Eilam (2022) examined the problems associated with the cross-curriculum approach, arguing that an epistemological weakness—namely educators' limited conceptualization of CC—poses a significant obstacle to developing and implementing effective CC curricula. She advocates for recognizing CC as a discrete body of knowledge, that should not be subsumed and fragmented under other disciplines but, instead, included in the curriculum as a coherent independent disciplinary subject. However, integrating CC as an independent subject is challenging, given the already overloaded national curricula. The positioning of CC within the curriculum remains a critical issue, with significant implications for teacher preparation and professional development. Participants of this study also demonstrated epistemological vagueness regarding the nature of CC, which, as Eilam (2022) has argued, may hinder the effective development and implementation of CC curricula.

While the participants acknowledged the importance of the scientific basis of CC, they also emphasized its multidisciplinary and multi-domain nature, extending beyond science to encompass socio-economic aspects, ethical and emotional dimensions. Accordingly, CCE requires a comprehensive holistic approach that bridges science and the social sciences, addressing cognitive, affective, and behavioural dimensions. This presents challenges, as many science educators believe their role is limited to conveying factual climate science information. Monroe et al., (2019), reviewed CCE strategies and found very few educational programs that intentionally approached CC from both the social and science dimensions. Additionally, their review of CCE interventions revealed that none focused on both social- and environmental justice-oriented outcomes. In line with this, Meira Cartea (2020) argues that CCE must extend beyond Climate Literacy, which is inherently limited as it centers on the scientific understanding of climate and its human-induced alterations, framing the learning primarily within the realm of natural and physical sciences. Instead, the educational response must transcend this focus to address the concept of "change", emphasizing education for action and social transformation. This approach necessitates the inclusion of the social sciences and humanities. Along this vein, CC can be characterised as a socio-scientific issue (SSI). The SSI approach offers a framework for CCE that explicitly considers, in addition to the scientific foundation, the socio-cultural dimensions of complex environmental issues, such as socioeconomic status, gender, cultural norms, emotions, values and ethics (Zeidler & Newton, 2017). SSI aims to develop socio-scientific reasoning, which is essential for tackling CC. Indeed, teaching CC as an SSI is not a new approach (e.g., Sadler et al., 2011).

The rich repertoire of constructivist and student-centered pedagogies for teaching CC reported by the participants across the countries that participated in this study are essentially non-CC domain specific (Ben Zvi Assaraf et al., 2024) and are well established in the context of science education and environmental and sustainability education (e.g., Hollweg et al., 2011; Monroe et al., 2019; NAAEE, 2019). Student-centered teaching methods that engage students in active learning, as well as making CC personally relevant to students, have also emerged as strategies for effective learning about CC (Monroe et al., 2019).

The findings of this study also highlight the importance of the emotional aspects specifically associated with CC that teachers need to respond to through social-emotional learning (SEL) strategies. The role of the affective domain in teaching CC is not new. The SSI framework provides a lens to consider how emotions, beliefs and values influence the individual's motivation to act on CC (Zeidler & Newton, 2017). Effective CCE recognizes the need to address the range of emotions (such as fear, hopelessness, helplessness) that can surface specifically around CC and which are collectively termed 'climate anxiety' (e.g., NAAEE, 2024). The "bicycle model" proposed for CCE positions emotions as the lamp—lighting the way forward (Cantell et al., 2019). In their study examining how CCE practitioners perceive the usability of this model in presenting the dimensions of CCE, emotions emerged as one of the most important CCE elements. Participants in the present study emphasized the need for teacher support in managing the emotional dimension in CCE, further underscoring the role of the affective domain in CCE.

Finally, the combined findings of this study underscore the need for strong administrative leadership to conceptualize and position CCE effectively within national curricula. This includes providing PD to guide teachers in implementation strategies, and direct and instruct them in allocating and utilizing the various resources for teaching CCE.

This study offers some contributions. It addresses the educational hierarchy—from the classroom teacher up to the policymakers—a scope largely missing in prior research. Research on policymakers' perceptions is particularly scarce. A limitation of the study is its small sample size for representing the different CCE role players in each country. Additionally, while country specific attributes may influence the findings,[4] the consistency of several of the findings, within the role players in each country and among countries, highlights areas for improvement in CCE. There is need for a comprehensive CC curriculum that addresses cognitive, affective and behavioural goals integrating the scientific, social and emotional dimensions. There is need to provide PD that takes a comprehensive approach, addressing not only the science basis but also the social emotional aspects of CC and guides resource allocation and its use. Effective CCE requires strong administrative leadership and policy direction to establish it as a critical educational priority. This need extends from centralized educational systems to de-centralized education systems, where local stakeholders (including government and teachers) have more autonomy and responsibility in decision making concerning the aims, content and how education is organised and assessed, and, thus, play a key role in aligning CCE with local traditions and curricula. It appears that these calls for CCE leadership are timely. Since this study, several participating countries, including Israel, Australia, Finland and the UK, have taken steps to strengthen their CCE curricula and resources. For example, Israel has introduced a mandated CC curriculum (Keshet & Eilam, 2022), while the other countries have enhanced their curricular directives and resourcing for CCE.

References

Ben Zvi Assaraf, O., Dawson, V., Dillon, J., Eilam, E., Gokpinar, T., Goldman, D., Naugauker, N., Putri, G. A. P. E., Subiantoro, A. W., Tolppanen, S., White, P., & Widdop Quinton, H. (2024). Climate change education implementation: The voices of policymakers, professional development providers, and teachers in five countries. *International Journal of Science Education*. https://doi.org/10.1080/09500693.2024.2314572

Cantell, H., Tolppanen, S., Aarnio-Linnanvuori, E., & Lehtonen, A. (2019). Bicycle model on climate change education: Presenting and evaluating a model. *Environmental Education Research, 25*(5), 717–731. https://doi.org/10.1080/13504622.2019.1570487

Clark, H. F., Sandoval, W. A., & Kawasaki, J. N. (2020). Teachers' uptake of problematic assumptions of climate change in the NGSS. *Environmental Education Research, 26*(8), 1177–1192. https://doi.org/10.1080/13504622.2020.1748175

Creswell, J. W., & Plano Clark, V. L. (2011). *Designing and conducting mixed methods research*. Sage.

Dawson, V., Eilam, E., Tolppanen, S., Ben Zvi Assaraf, O., Gokpinar, T., Goldman, D., Putri, G. A. P. E., Subiantoro, A. W., White, P., & Widdop Quinton, H. (2022). A cross-country comparison of climate change in middle school science and geography curricula. *International Journal of Science Education, 44*(9), 1379–1398. https://doi.org/10.1080/09500693.2022.2078011

Eilam, E. (2022). Climate change education: The problem with walking away from disciplines. *Studies in Science Education, 58*(2), 231–264. https://doi.org/10.1080/03057267.2021.2011589

Enke, K. A., & Budke, A. (2023). Preparing students for a changing world: How geography curricula in Europe are tackling climate change. *Frontiers in Education, 8*, 1216790. https://doi.org/10.3389/feduc.2023.1216780

[4] Ben Zvi Assaraf et al. (2024) elaborated on the influence of country-specific attributes.

Ennes, M., Lawson, D. F., Stevenson, K. T., Peterson, M. N., & Jones, M. G. (2021). It's about time: Perceived barriers to in-service teacher climate change professional development. *Environmental Education Research, 27*(5), 762–778. https://doi.org/10.1080/13504622.2021.1909708

Ho, L. C., & Seow, T. (2015). Teaching controversial issues in Geography: Climate change education in Singaporean schools. *Theory & Research in Social Education, 43*(3), 314–344. https://doi.org/10.1080/00933104.2015.1064842

Hollweg, K. S., Taylor, J. R., Bybee, R. W., Marcinkowski, T. J., McBeth, W. C., & Zoido, P. (2011). *Developing a framework for assessing environmental literacy*. North American Association for Environmental Education. Available at: http://www.naaee.net

Keshet, G., & Eilam, E. (2022). The Ministry of Education curriculum for educating climate change. *Ecology and Environment, 13*(2) [in Hebrew]. Available at: https://magazine.isees.org.il/pdf/51709.pdf?1734104789

Liu, S., & Roehrig, G. (2019). Exploring science teachers' argumentation and personal epistemology about global climate change. *Research in Science Education, 49*(1), 173–189. https://doi.org/10.1007/s11165-017-9617-3

Marton, F. (2005). Phenomenography: A research approach to investigating different understandings of reality. In R. R. Sherman & W. B. Rodman (Eds.), *Qualitative research in education: Focus and methods* (pp. 140–160). Routledge.

McKenzie, M. (2021). Climate change education and communication in global review: Tracking progress through national submissions to the UNFCCC Secretariat. *Environmental Education Research, 27*(5), 631–651. https://doi.org/10.1080/13504622.2021.1903838

Meira Cartea, P. A. (2020). Climate change and education. In W. L. Filho, A. M. Azul, L. Brandli, P. G. Özuyar, & T. Wall (Eds.), *Climate action, encyclopaedia of the UN sustainable development goals* (pp. 109–120). Springer.

Monroe, M. C., Plate, R. R., Oxarart, A., Bowers, A., & Chaves, W. A. (2019). Identifying effective climate change education strategies: A systematic review of the research. *Environmental Education Research, 25*(6), 791–812. https://doi.org/10.1080/13504622.2017.1360842

Nation, M. T., & Feldman, A. (2022). Climate change and political controversy in the science classroom. *Science & Education, 31*, 1567–1583. https://doi.org/10.1007/s11191-022-00330-6

Naugauker, N., Ben-Zvi-Assaraf, O., Goldman, D., & Eilam, E. (2023). Implementing climate change education: The role of inter-sectorial collaborations. In K. Korfiatis, M. Grace, & M. Hamman (Eds.), *Shaping the future of biological education research—Selected papers from the ERIDOB 2022 Conference* (pp. 205–217). Springer Nature.

North American Association for Environmental Education (NAAEE). (2019). *Guidelines for excellence—K-12 environmental education*. NAAEE. Available at: https://eepro.naaee.org/sites/default/files/eepro-post-files/k-12_ee_guidelines_for_excellence_2019_2.pdf

North American Association for Environmental Education (NAAEE). (2024). *Guidelines for excellence—Educating for climate action and justice*. NAAEE. Available at: https://eepro.naaee.org/sites/default/files/2024-09/Guidelines-%237-Educating-for-Climate-Action-and-Justice-Accessible_0.pdf

Plutzer, E., McCaffrey, M., Hannah, A. L., Rosenau, J., Berbeco, M., & Reid, A. H. (2016). Climate confusion among US teachers. *Science, 351*, 664–665. https://doi.org/10.1126/science.aab3907

Reid, A. (2019). Climate change education and research: Possibilities and potentials versus problems and perils? *Environmental Education Research, 25*(6), 767–790. https://doi.org/10.1080/13504622.2019.1664075

Sadler, T. D., Klosterman, M. L., & Topcu, M. S. (2011). Learning science content and socio-scientific reasoning through classroom explorations of global climate change. In T. D. Sadler (Ed.), *Socio-scientific issues in the classroom: Teaching, learning and research* (Contemporary trends and issues in science education) (Vol. 39, pp. 45–77). Springer.

UNESCO. (2021a). *Getting every school climate-ready. How countries are integrating climate change issues in education*. https://doi.org/10.54675/NBHC8523

UNESCO. (2021b). *Learn for our planet—A global review of how environmental issues are integrated into education*. https://unesdoc.unesco.org/ark:/48223/pf0000377362

UNESCO. (2021c). *Teachers have their say—Motivation, skills and opportunities to teach education for sustainable development and global citizenship*. Available at: https://doi.org/10.54675/YXRW9784

Zeidler, D. L., & Newton, M. H. (2017). Using socioscientific issues framework for climate change education: An ecojustice approach. In D. P. Shepardson, A. Roychoudhury, & A. S. Hirsch (Eds.), *Teaching and learning about climate change—A framework for educators* (pp. 56–66). Taylor & Francis.

Daphne Goldman, Associate Professor in the Department of Environmental Science and Agriculture, the Faculty of Education, Beit Berl College, Israel. Her research looks at theoretical and applied aspect of environmental and sustainablity education (ESE) including environmental education policy in Israel. Much of her research in this area focuses on the influence of educational interventions on Environmental Literacy and Citizenship across various educational settings and populations. She is co-founder of the National Conference on Environmental Education and has been involved in several key initiatives including the accreditation of Beit Berl College as a "Green Campus".

Orit Ben Zvi Assaraf is Full Professor in the School of Education, and member of the School of Sustainability and Climate Change, Ben-Gurion University of the Negev, Beer-Sheva Israel.

Vaille Dawson is a Professor of Science Education in the Graduate School of Education at the University of Western Australia in Australia where she teaches preservice secondary science education, supervises doctoral students and conducts classroom based education research. Her primary research expertise spans climate change education, genetics education, socioscientific issues and argumentation, scientific literacy, teacher education, and teaching in disadvantaged schools in Australia, India, and Indonesia. She has conducted research at the primary, secondary and tertiary levels.

Efrat Eilam, Associate Professor of Science Education at Victoria University, Australia. Her primary research focuses on climate change education.

Tuba Gokpinar is a Social Impact Researcher at the University of Essex. Her current research interests include the development of impact toolkits to measure social value for adults with learning disabilities. Her PhD from the Institute of Education, University College London focused on science aspirations, attitudes, and capabilities among minority students in the UK. Dr. Gokpinar is also an Honorary Research Fellow at UCL, contributing to projects on science attitudes & aspirations and climate change education.

Nofar Naugauker has been a master's student at Ben-Gurion University of the Negev, Israel. Her thesis focused on the Challenges in Climate Change Education from the Perspectives of Educationists and Policymakers Involved in Implementation of CCE in the Israeli Education System.

Gusti Agung Paramitha Eka Putri is based at Nossal Institute for Global Health, University of Melbourne. She holds a PhD in Education from Victoria University Australia. Her primary research interests are creative pedagogies, learning for sustainability, intercultural research, and collaborative curriculum design.

Agung Wijaya Subiantoro, Doctor of Science Education and Research Associate in the Department of Biology Education, Universitas Negeri Yogyakarta, Yogyakarta, Indonesia.

Sakari Tolppanen is senior researcher in the School of Applied Educational Sciences and Teacher Education, Philosophical Faculty, University of Eastern Finland, Joensuu. His core research interests are in education for sustainable development and climate change education.

Peta J. White is an Associate Professor in Science and Environmental Education at Deakin University. Her current research follows three narratives: science and biology education; sustainability, environmental, and climate change education; and collaborative/activist methodology and research.

Helen Widdop Quinton is a Lecturer and Research Fellow with the Institute of Sustainable Industries and Liveable Cities (ISILC) at Victoria University, Melbourne, and a Research Associate with the Sustainability, Environment and the Arts in Education (SEAE) Research Centre at Southern Cross University. Her research interests are primarily focused at the intersections of socioecological learning, situated knowledges, youth participatory research, and place-based pedagogies.

The manufacturer's authorised representative in the EU is Springer Nature Customer Service Centre GmbH, Europaplatz 3, 69115 Heidelberg, Germany. If you have any concerns regarding our products, please contact ProductSafety@springernature.com

Printed and bound by CPI Group (UK) Ltd, Croydon, CR0 4YY
30/09/2026
02225592-0016